ELECTRICAL SENSORS AND TRANSDUCERS

James R. Carstens, P.E.

Assistant Professor
School of Technology
Michigan Technological University

Regents/Prentice Hall, Englewood Cliffs, N.J. 07632

Carstens, James R.
    Electrical sensors and transducers / James R. Carstens.
        p.   cm.
    Includes bibliographical references and index.
    ISBN 0-13-249632-1
    1. Transducers.   2. Detectors.   I. Title.
    TK7872.T6C38   1992
    681'.2--dc20   91-43325
                                                                            CIP

Editorial/production supervision and
   interior design: Eileen M. O'Sullivan
Cover design: Ray Lundgren Graphics, LTD.
Prepress buyer: Ilene Levy
Manufacturing buyer: Ed O'Dougherty

To Marge and Dorwin

 © 1993 by REGENTS/PRENTICE HALL
A Division of Simon & Schuster
Englewood Cliffs, New Jersey 07632

Printed in the United States of America
10 9 8 7 6 5 4 3 2 1

ISBN 0-13-249632-1

PRENTICE-HALL INTERNATIONAL (UK) LIMITED, *London*
PRENTICE-HALL OF AUSTRALIA PTY. LIMITED, *Sydney*
PRENTICE-HALL CANADA INC., *Toronto*
PRENTICE-HALL HISPANOAMERICANA, S.A., *Mexico*
PRENTICE-HALL OF INDIA PRIVATE LIMITED, *New Delhi*
PRENTICE-HALL OF JAPAN, INC., *Tokyo*
SIMON & SCHUSTER ASIA PTE. LTD., *Singapore*
EDITORA PRENTICE-HALL DO BRASIL, LTDA., *Rio de Janeiro*

CONTENTS

Contents

PREFACE

It is very difficult to find a text that treats transducers in a concentrated manner. That is, to find information on transducers, you usually have to refer to a related text dealing with automatic controls or process controls. There you may find a chapter or two devoted to the subject of sensing. This makes life especially difficult for those engineers, technicians, or people in general who are curious—who wish to find out something about how transducers work but cannot get their hands on solid information.

This book is an effort to comfort those people in their frustrating efforts. This book is meant to serve as a refresher course in the subject in addition to being a textbook. It is not meant to be a complete, highly technical dissertation on the subject. Using a book of this kind for that particular purpose can only serve "to confuse and cloud Man's mind" (to quote a childhood radio idol of mine, Lamont Kranston, "The Shadow"). Students in particular have difficulty at times sifting through the classroom rhetoric, not having the advantage that the instructor has had during years of accrued experience. Therefore, I hope that this book will be used not only for the purpose of gaining knowledge but also for refreshing old memories. The subject matter is suitable for a beginning course text for two-year and possibly for four-year students.

Much of the information given in this book is based on my experience in industry. Its style is based on my several years in the classroom transferring to my students what I know on the subject of sensors and transducers. Hopefully, I have succeeded in carrying on that effort in this book.

Naturally, a project like this is not the product of one person but is also the product of others who were present with their expertise when help was needed. These people gave unselfishly of their effort. My thanks to the School of Technology secretaries, Phylis and Pam, for handling all the necessary paperwork and documents associated with book writing; thanks also to my students, Kris Baldrica, Greg Powell, and Shawn Smith for producing most of this book's illustrations on AutoCad—a tremendous undertaking to say the least. The most thanks go to my wife, Sandy, for putting up with me during the pursuit of this project. I can only hope that she feels that it was worth it.

J.R.C.
Houghton, Michigan

1

INTRODUCTION TO SENSORS AND TRANSDUCERS

CHAPTER OBJECTIVES

1. To define the sensor and transducer.
2. To review physics needed to understand sensor principles.
3. To define the measurands sensed by the transducer.
4. To discuss methods used to describe sensor performance.

1-1 BEFORE WE GET STARTED: SOME COMMENTS ABOUT UNITS

Confusion reigns as to which systems of units should be used in an engineering discussion or report. Despite the good intentions of many, the United States has done a questionable job in converting to the obviously superior metric system. What with the three most popular systems all coexisting in this country—the English (sometimes referred to as the British system), the CGS, and the MKS systems, not to mention a fourth, the SI or "System International" (from a translation from French)—many people have been led down the path of insanity in very short order. (If you are interested in obtaining more information on the SI system of measurements, contact the National Bureau of Standards, Washington D.C., and ask for NBS Special Publication 330, their latest edition.)

In an attempt to cope with this rather serious problem of unit indecisiveness, we will state both the English units and the SI units whenever both references are necessary or when space allows. Table 1-1 is a list of the four systems and some of

TABLE 1-1   COMPARISON OF THE ENGLISH AND METRIC SYSTEMS

	English	Metric		
		MKS	CGS	SI
Length	yard (0.914 m)	meter (39.37 in.)	centimeter (2.54 cm = 1 in.)	meter
Mass	slug (14.6 kg)	kilogram	gram	kilogram
Force	pound (4.45 N)	newton (100,000 dyn)	dyne	newton
Power	horsepower (746 W)	watt	watt	watt
Energy	ft-lb (1.336 J)	joule	erg	joule
Temperature	°F (1.8°C + 32)	°C	°C	kelvin (273.2 + °C)
Time	second	second	second	second
Volume	gallon (3.785 L)	liter	cubic centimeter (1000 cm^3 = 1 L)	liter

the more commonly used conversions needed to leap from one system to another. In addition, when the English units are stated, we will use the most appropriate units when more than one is possible. For instance, if a machining process is being discussed, rather than using feet for measuring the length of a small precision part, inches will be stated instead. Abbreviations for units used in this book will be stated as shown in Table 1-2. Other, more specialized units not listed in Table 1-2 will be pointed out and explained as they are encountered.

1-2 DEFINITIONS OF SENSOR AND TRANSDUCER

People who work with sensors and transducers every day tend to use these two words somewhat interchangeably in their speech. There is no hard and fast rule as to what distinguishes an electrical transducer from an electrical sensor. Ask any number of engineers or scientists for their definitions of these two terms and you will probably get that many different answers. For our purposes in this book I will make the following distinctions: An *electrical sensor* is a device that converts a quantity or an energy form into an electrical output signal. The form of that output signal, whether it is an ac or a dc signal, whether the output is a change in resistance or a change in capacitance, whether the signal is digital or analog, to name just a few, is determined by the sensor's electromechanical and/or chemical makeup. On the other hand, an *electrical transducer* is a device comprised of a sensor whose output signal is modified or conditioned to suit a particular application or need by its user. Furthermore, that transducer may have an onboard power source for its proper

TABLE 1-2 UNIT ABBREVIATIONS USED IN THIS BOOK

English units		SI units	
inch	in.	gram	g
foot	ft	centigram	cg
pound	lb	kilogram	kg
slug	slug	millimeter	mm
mile	mi	nanometer	nm
horsepower	hp	centimeter	cm
foot pound	ft-lb	kilometer	km
degrees Fahrenheit	°F	joule	J
second	s	second	s
gallon	gal	liter	L
		degrees Celsius	°C
		kelvin	K
		cubic centimeter	cm^3
		newton	N
		watt	W
		dyne	dyn
		mole	mol

operation. Figure 1-1 will help to illustrate this major difference. The transducer depicted uses a sensor as one of its many possible operating elements, whereas a sensor has no internal separate operating elements, just the sensing element itself. A transducer depends on the sensor to produce the quantity or energy conversion. It is also capable of modifying the sensor's output signal to convert that signal into a more desirable form. In Figure 1-1 the sensing element that is shown is what is referred to as a photo-resistive sensor whose electrical resistance varies with the light intensity that is striking this element. The sensor's output is in the form of a changing

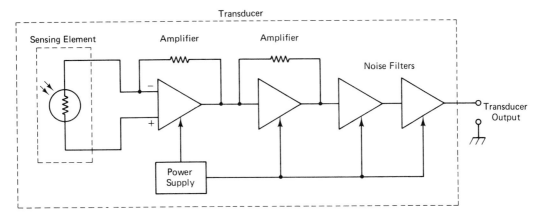

Figure 1-1 Typical transducer configuration. (From James R. Carstens, *Automatic Control Systems and Components,* copyright 1990, p. 86. Reprinted by permission of Prentice Hall, Englewood Cliffs, NJ.)

dc resistance, as already mentioned. However, because it is generally more desirable to work with changing voltages and currents instead of changing resistance, this particular transducer uses amplifiers to convert the resistance to these quantities. In addition, these amplifiers have the task of modifying the signal's amplitude to more desirable signal levels. The noise filters that are shown have the job of "stripping off" or diminishing any electrical noise that might be present on the modified signal coming from the sensor. Not all transducers have this elaborate provision or capability.

Now one might argue that a gasoline engine could also be classified as a transducer. After all, it does convert one form of energy, chemical energy, in the form of gasoline in this case, into another form, kinetic energy or motion. Therefore, we have to be a little more precise in our transducer definition. We have to qualify our statement by saying that a transducer has a very predictable output based on the energy or quantity input that it is converting. This output can be measured and quantified, usually in very precise amounts. In other words, the transducer can be calibrated. An engine is not considered that precise an instrument in terms of ready calibration. Also, many transducers can easily be transported and installed in systems or circuits. Although admittedly that in itself is a somewhat imprecise criterion, it is a simply applied criterion that may be useful in identifying transducers.

In summary, then, transducers are used to measure or sense certain quantities or energy forms. They can be precisely calibrated if desired and, for the most part, are easily transportable. In the material that follows, we discuss the types of outputs that are available from sensors and transducers.

1-3 FORMS OF TRANSDUCER OUTPUTS

Looking at the output terminals of a transducer, let's now see what varieties of electrical outputs can be obtained there. There is probably a transducer available from a manufacturer that can produce the signal that you may have in mind for your application. In the discussion that follows, we look at the three major categories of data signal forms that presently exist: (1) analog, (2) digital or pulse, and (3) carrier.

1-3.1 Analog Signal

Very simply stated, an analog signal is a continuous (not interrupted) data signal comprised of a flow of current or a voltage level whose amplitude, frequency, or phase relationship with some reference signal contains data information. This information represents a proportional relationship or analog of the input measurand that is controlling the signal. In Figure 1-2 we see an aneroid barometer that has a variable potentiometer attached. The variable-voltage output of this potentiometer is a direct representation of the varying air pressure being sensed by the aneroid. This output is an analog equivalent of the varying air pressure. As a matter of fact, the resultant voltage curve can easily be transformed into a pressure reading merely by noting how

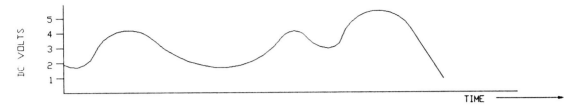

VOLTAGE OUTPUT AS MEASURED BETWEEN TERMINALS 'A' AND 'B' OVER A PERIOD OF TIME

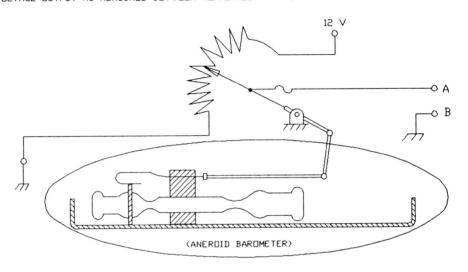

(ANEROID BAROMETER)

Figure 1-2   Aneroid barometer with analog output signal.

much voltage change results for a given change in air pressure. Stated more easily, simply divide the output voltage change of the transducer by its input pressure change, which created the change in output voltage. The ratio that results is often called the instrument's *sensitivity*. This concept is discussed in detail in Section 1-9.2.

1-3.2 Digital Signal

A digital signal is comprised of a series of interrupted flows, or pulses, of current or voltage levels. Each pulse or series of pulses contains the encoded information corresponding to the input data. A decoding process must then be used to decipher the desired information. The transducer shown in Figure 1-2 can be modified to produce a digital output signal.

1-3.3 Carrier Signal

A carrier is an electromagnetic wave of constant amplitude that acts as an electromagnetic vehicle for transporting data. Data being transmitted by a carrier may be either analog or digital, depending on the methods used to modulate the carrier. We discuss this topic further in Section 11-2.

1-4 DEFINING THE MEASURAND

The *measurand* for a transducer is the input quantity being measured by that transducer. For instance, if a transducer is being used to measure the reflected light of a distant planet or the radiated light from a star, as is sometimes done onboard space satellites, the incoming light to the sensor portion of that transducer would be the measurand. Light is, of course, just one of several different broad categories of measurands that are typically sensed by sensors. A list of primary measurand categories is given below. These are called *primary measurands* because they are considered to be at the most elementary level of existence. Other measurands certainly exist but are comprised of a combination of these primary measurands. For example, to measure the quantity, acceleration, a sensor designed to be sensitive to the measurand, force, would probably be used in combination with a known mass to indicate acceleration. This is because of the $F = ma$ relationship that exists among force, mass, and acceleration (Newton's law). Consequently, the transducer would, in reality, be sensing force, the primary measurand in the transducer's design. However, the desired measurand, acceleration, is dependent on the force-sensing aspect of the transducer and, as a result, becomes the *secondary measurand*. The primary measurands sensed by electrical transducers are:

1. Position
2. Motion (velocity, acceleration, jerk)
3. Sound
4. Light
5. Electromagnetic radiation (other than light)
6. Particle radiation (other than light)
7. Temperature
8. Flow
9. Force (pressure)
10. Potential difference

1-5 MEASURAND DESCRIPTIONS AND THEIR PHYSICAL BEHAVIOR (A REVIEW OF PHYSICS)

Each measurand almost requires its own unique method for detection; perhaps this is what makes sensors so interesting and challenging to study. New methods are continuously being found to detect these measurands more efficiently and accurately. However, before we discuss these methods and the ones presently being used, we should define the measurands listed above so that we will have a better understanding of the methods used for detection.

1-5.1 Position

The position or location of an object must be defined by its *distance* and *direction* relative to a fixed reference point; usually, but not always, this is the sensor itself. For instance, we can say that a ship at sea has been detected 12.78 miles (20.56 km) from the detection site onshore at a compass bearing of 48.67° (see Figure 1-3). Or, as another example, we can say that a measuring probe on a CNC (computer numerically controlled) machine has located the center of a hole to be drilled on a machined part that is located 125.81 mm in the x-direction and 27.89 mm in the y-direction as measured from the origin on the machined piece.

Position can also be specified as an angular quantity (a quantity often designated by the Greek lowercase letter, θ). For example, we can say that a shaft has been rotated 228.3° in a counterclockwise direction. Convention dictates that angular measurements originate from a horizontal line extending through the shaft's center and begin at the right side of the shaft (Figure 1-4). *Positive* angles are rotated *counterclockwise* from this particular point, and *negative* angles are rotated *clockwise* from this point.

In some instances, the direction of measurement is omitted when the distance or location is stated since the direction is obvious for the observer to determine. For instance, during a baseball game when the sports announcer mentions the speed of the pitched ball as measured with a speed gun, the pitcher's pitching speed is usually given without stating the direction of the pitch. The listener assumes that the ball

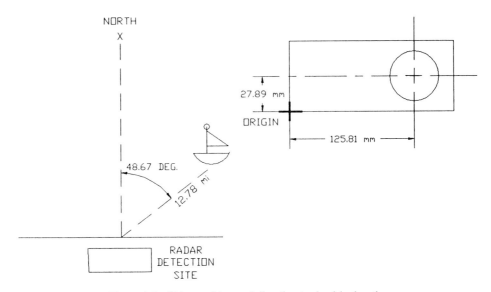

Figure 1-3   Using position and direction to descirbe location.

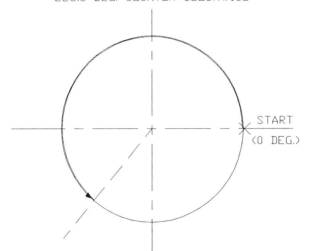

Figure 1-4 Using angular measurement for describing location.

is being thrown toward home plate without having to be told. Similar assumptions are often made in industry, too. If the direction being measured with a sensor is obvious, that portion of the description is frequently omitted. All of this may sound rather simplistic and mundane at this point, but as we will find out later, direction-indicating capabilities for a sensor can become a technical nightmare in design.

A typical measurement of position may be made using a wide assortment of position-measuring transducers. One method is found in industry in determining the location or amount of extension of a piston rod in a hydraulic cylinder. The amount of extension may be determined by an attached linear potentiometer whose output resistance is an analog of the extension amount.

1-5.2 Motion

Linear velocity is defined as displacement per unit time, or

$$v = \frac{s}{t} \tag{1-1}$$

where v = velocity (ft/s, m/s)

s = displacement (ft, m)

t = time (s)

The following example demonstrates the foregoing principle.

Example 1-1

An orbiting earth research satellite can complete one orbit in approximately 91 min at an altitude of 178 mi. What is its orbiting linear velocity in miles per hour? Assume that the satellite's orbit is circular in shape.

Solution: Since the earth's radius is 4000 mi and the satellite is 178 mi above the earth, the total radius of the satellite's orbit is 4000 + 178, or 4178 mi. We can now find the circumference of its orbit since this would be the total distance traveled in one complete orbit:

$$\text{circumference} = \pi \times d, \qquad \text{where } d = 4178 \text{ mi} \times 2$$

or

$$\text{circumference} = \pi(4178 \times 2)$$
$$= 26{,}251 \text{ mi}$$

To find the satellite's velocity, we use eq. (1-1):

$$\text{velocity} = \frac{v}{t} = \frac{26{,}251}{91} \frac{\text{mi}}{\text{min}}$$
$$= 288.5 \text{ mi/min} \quad \text{or} \quad 17{,}310 \text{ mi/h}$$

Just as displacement is a vector quantity, so velocity is a vector quantity. That is, to complete its description, a direction must usually be specified together with the magnitude of the velocity. Rotational velocity, usually symbolized by the Greek lowercase letter, ω, is defined as angular displacement per unit time, or simply

$$\omega = \frac{\theta}{t} \tag{1-2}$$

where ω = angular velocity (deg/s, rad/s)

$\quad\quad\quad \theta$ = angular displacement (deg, rad, etc.)

$\quad\quad\quad t$ = time (s)

Example 1-2

In Example 1-1 find the angular velocity, in rad/s, of the satellite.

Solution: Since the circumference of the satellite's orbit is equal to 360° of travel or 2π radians, and it took the satellite 91 min to traverse this angle, using eq. (1-2) we can find its angular velocity.

$$\omega = \frac{\theta}{t} = \frac{2\pi \text{ rad}}{91 \text{ min}}$$
$$= 0.069 \text{ rad/min} \quad \text{or} \quad 0.0012 \text{ rad/s}$$

Both linear and angular velocity may be measured by using a variety of transducers similar to those used for measuring displacement. The only difference is in their application. In measuring velocity, displacement is measured over a known period of time, thus allowing for the velocity to be calculated.

Another component of motion is *acceleration*, defined as a change in velocity. *Linear acceleration* is defined as

$$a = \frac{\Delta v}{t} \qquad (1\text{-}3)$$

where a = acceleration (ft/s^2, m/s^2)

Δv = change in velocity (ft/s, m/s)

t = elapsed time over which change in velocity took place (s)

Angular acceleration, α, is defined similarly:

$$\alpha = \frac{\Delta \omega}{t} \qquad (1\text{-}4)$$

where α = angular acceleration (rad/s^2)

$\Delta \omega$ = angular velocity (rad/s)

t = elapsed time over which change in angular acceleration took place (s)

Example 1-3

A rotating shaft's angular velocity increases at a steady rate such that its angular acceleration remains constant. If the shaft's velocity increased by 40 rpm in 14.5 s, find the resultant average angular acceleration in rad/s^2.

Solution: The shaft's change in angular velocity is 40 rpm or $40 \times 2\pi = 80\pi$ rad/min or $80\pi/60$ rad/s (1.333π rad/s). Then, according to eq. (1-4),

$$\alpha = \frac{\Delta \omega}{t} = \frac{1.333 \text{ rad/s}}{14.5 \text{ s}}$$

$$= 0.092 \text{ rad/s}^2$$

Acceleration may be measured by a unique transducer specifically designed for the detection and measurement of acceleration. This type of transducer contains a mass whose displacement amount is a function of the applied acceleration. The displacement amount may be detected by a displacement-type sensor.

The third component of motion is *jerk*, defined as a change in acceleration according to the following equation:

$$J = \frac{\Delta a}{t} \qquad (1\text{-}5)$$

where J = jerk (ft/s^3, m/s^3)

Δa = change in linear acceleration (ft/s^2, m/s^2)

t = elapsed time over which change in acceleration took place (s)

Jerk is somewhat rare and is not often mentioned in engineering literature. It is mentioned here merely to complete our discussion of motion. Angular jerk is even rarer in discussions, although it is a bona fide quantity. We also exclude it from our discussion.

Again, as is often the case with displacement measurements, motion specifications sometimes omit the direction of motion that is taking place. This is often done in those cases where the direction is obvious or perhaps the motion is limited to one direction. For a motion sensor to have the capability to indicate direction of motion in addition to its magnitude indication may complicate the sensor's design considerably.

1-5.3 Sound

Sound may be defined as a series of pressure waves emanating from a vibrating source and whose propagation is supported by a fluid, typically air, or by a solid material. The speed of propagation can easily be determined for the material being used. The following equations may be used:

For solids:

$$c = \sqrt{\frac{Y}{\rho}} \qquad (1\text{-}6)$$

where c = velocity of sound (ft/s, km/s)

Y = Young's modulus, a constant (lb/in^2, N/km^2)

ρ = mass density (slugs/in^3, kg/km^3)

For liquids:

$$c = \sqrt{\frac{B}{\rho}} \qquad (1\text{-}7)$$

where c = velocity of sound (ft/s, km/s)

B = bulk modulus, a constant

ρ = mass density (slugs/in^3, kg/km^3)

For gases:

$$c = \sqrt{\frac{\Gamma R T}{M}} \qquad (1\text{-}8)$$

where c = velocity of sound (ft/s, m/s)

Γ = ratio of heat capacity at constant pressure to heat capacity at constant volume (a constant, therefore dimensionless)

R = universal gas constant (J/mol-K)

T = temperature of fluid (K)

M = molecular mass (km/mol)

Table 1-3 lists the various constants mentioned in eqs. (1-6) to (1-8) for several common materials.

TABLE 1-3 CONSTANTS FOR SOME COMMON MATERIALS (AT 20°C)

	γ	B	M	R	ρ	Y
			English units			
	$\dfrac{c_p}{c_v}$	$\dfrac{\text{lb}}{\text{in}^2}$	$\dfrac{\text{slugs}}{\text{mol}}$	$\dfrac{\text{lb}}{\text{mol-R}}$	$\dfrac{\text{slugs}}{\text{ft}^3}$	$\dfrac{\text{lb}}{\text{in}^2}$
Air	1.4				2.33×10^{-3}	
Water	1.01	3.19×10^5			1.937	
Iron					15.093	2.6×10^6
Lead					34.938	2.3×10^6
Aluminum					5.13	10.0×10^6
			SI units			
	$\dfrac{\text{N}}{\text{m}^2}$	$\dfrac{\text{kg}}{\text{mol}}$	$\dfrac{\text{J}}{\text{mol-K}}$	$\dfrac{\text{kg}}{\text{m}^3}$	$\dfrac{\text{N}}{\text{m}^2}$
Air		28.8×10^{-3}	8.3149		
Water	2.2×10^9	18.02×10^{-3}		998.3	
Iron				7,079	1.9×10^{11}
Lead				11,370	1.6×10^{12}
Aluminum				2,643	7.3×10^{10}

Example 1-4

Calculate the velocity of sound in water using the constants given in Table 1-3. Do the same for air. Assume all conditions at 68°F (20°C).

Solution: First, for water [using eq. (1-7) and referring to Table 1-3],

$$c = \sqrt{\frac{B}{\rho}} = \sqrt{\frac{2200 \times 10^6 \times 10^6 \text{ N/km}^2}{998.3 \times 10^6 \text{ N-km/kg}}}$$

$$= \sqrt{2.204 \times 10^6 \text{ N-km/kg}}$$

Note: The newton (N) is equivalent to the *kg-km/s²* (since $F = ma$). Therefore,

$$c = \sqrt{2.204 \times 10^6 \text{ kg-km}^2/\text{kg-s}^2}$$

$$= 1.485 \times 10^3 \text{ km/s}$$

Now, for air [using eq. (1-8) and Table 1-3],

$$c = \sqrt{\frac{\Gamma R T}{M}} = \sqrt{\frac{(1.4)(8.31 \text{ J/mol-K})(293 \text{ K})}{28.8 \times 10^{-3} \text{ kg/mol}}}$$

$$= \sqrt{118.36 \times 10^3 \text{ J/kg}}$$

Note: The joule is equivalent to the *meter-newton*. Therefore,

$$c = \sqrt{11.836 \times 10^4 \text{ m-N/kg}}$$

$$= \sqrt{11.836 \times 10^4 \text{ m-kg-m/s}^2\text{-kg}}$$

$$= 3.440 \times 10^2 \text{ m/s} \quad \text{or} \quad 344 \text{ m/s}$$

We must also reckon with sound intensity and how it is measured. Sound intensity is not a linear phenomenon; that is, doubling its magnitude or amount does *not* double its intensity or loudness as perceived by the human ear. Instead, sound is logarithmic. The unit of sound intensity is the *decibel* (abbreviated dB). The relative comparison of the power intensities of two sound levels is given by

$$dB = 10 \log \frac{P_1}{P_2} \qquad (1-9)$$

where dB = unit of loudness (dB)

P_1 = power content of first sound source (hp, watts)

P_2 = power content of second sound source (hp, watts)

Example 1-5

Calculate the decrease in decibel output for an engine whose power output has been decreased from 178 hp to 145 hp.

Solution: From eq. (1-9),

$$dB = 10 \log \frac{P_1}{P_2} = 10 \log \frac{145}{178}$$

$$= 10 \log 0.8146$$

$$= 10 \times -0.0891$$

$$= -0.891 \text{ dB}$$

Here is an interesting example of the logarithmic response of the human ear. Even though the engine's power in Example 1-5 was reduced by almost 19%, the noise reduction amounted to less than 1 dB, a change hardly detectable by the human ear. As a matter of fact, the engine's power could have been reduced by 50% and the ear would only begin to detect the reduction in sound.

As a further example, if a sound measurement were being conducted on a noise source, say a motor, inside a room where a microphone was being used for the sound pickup device and an ac power meter was hooked up to the microphone's output for an indicating device, the true noise output of the motor would equal the recorded meter indication "minus" the *ambient noise* level of the room itself. The ambient noise condition is the noise condition that would exist if no noise test were being conducted; it is the background noise. This noise may be the result of other noise sources, such as people talking, outside motor traffic, or radios playing.) The intuitive thing to do would be merely to record the other extraneous noise sources and subtract their values from the total reading obtained with the meter. However, this would be an incorrect assumption because of the logarithmic nature of sound. Instead, you would have to "subtract" the background noise logarithmically. Figure 1-5 is a chart that allows this to be done. Notice that the curve is an exponential function allowing the direct conversion of the difference between two noise sources to the number of decibels that can actually be subtracted from the total sound source plus background noise.

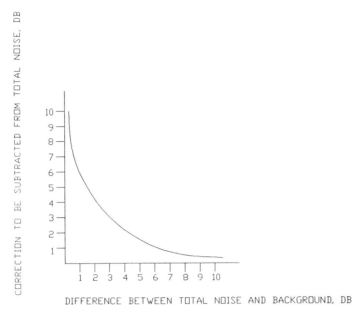

Figure 1-5   Describing the logarithmic nature of sound.

It is interesting to note that in electronics this logarithmic relationship also exists among current, voltage, and power comparisons; these are comparisons that are made frequently in circuit design. Equation (1-5) is used for relative power comparisons, whereas a similar equation, with the constant 10 in eq. (1-9) replaced by a constant value of 20, is used for making current or voltage comparisons. The reason for this lies in the fact that power varies with the *square* of voltage or current, resulting in a factor of 2 when taking the log of this squared factor.

Sound may be treated as if it were a wave type of phenomenon. That is, sound emissions from a sound source travel in concentric ever-expanding shells or pressure fronts (Figure 1-6). However, these wavefronts arrive at our ears at a particular frequency or combination of frequencies. Since the fronts travel at a particular velocity determined by the equations given earlier, we can associate a wavelength with each frequency knowing the speed of the sound [eq. (1-6)]:

$$\lambda = \frac{c}{f} \tag{1-10}$$

where λ = wavelength of sound (ft, m)

c = speed of sound (ft/s, m/s)

f = frequency of sound (Hz)

Example 1-6

The frequency of a certain sound was found to be 535 Hz. Assuming the velocity of sound to be 1090 ft/s, find the equivalent wavelength of the sound (i.e., find the distance between the crests of each wavefront).

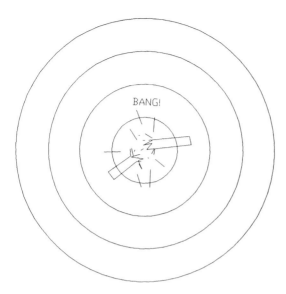

Figure 1-6 Wavefronts surrounding a sudden noise such as a firecracker.

Solution: Assuming from our work in Example 1-4 that the sonic velocity for air is 344 m/s, we can then use eq. (1-10) to find that

$$\lambda = \frac{c}{f} = \frac{344 \text{ m/s}}{535 \text{ Hz}}$$

$$= 0.643 \text{ m}$$

Doppler shift.   An interesting effect takes place when dealing with a moving sound source or with a moving listening instrument. The relative velocity existing between the sound source and an observer can either add or subtract from the velocity of sound, depending on the direction of travel relative to the observer. This velocity alteration of the sound source's frequency, called the *Doppler effect*, can be observed and measured precisely.

It is easy to observe the results of the Doppler effect on sound. As you travel through a town while a church bell is chiming or a local nearby college campus is playing a tune on its outdoor carillon, listen to the notes being played. The sounds emanating from the carillon are all out of tune. They may be distorted to the point where you cannot recognize the tune; furthermore, the effect is worsened by an increase in the car's speed. Any relative motion between you and the sound source changes the timing between the individual wavefronts. Any movement of yours between wavefronts that lengthens or shortens the time it takes for the front to reach you will certainly change the frequency of the front. The following equations allow you to calculate the observed frequency knowing the source's frequency and the velocity direction of the observer relative to the source.

If there is relative movement causing the observer and source to increase their separation, then

$$f_o = f \frac{v_{\text{air}}}{v_{\text{air}} + v_s} \tag{1-11}$$

If there is relative movement causing the observer and source to decrease their separation, then

$$f_o = f \frac{v_\text{air}}{v_\text{air} - v_s} \qquad (1\text{-}12)$$

where f_o = observed frequency (Hz)

$\qquad f$ = source frequency (Hz)

$\qquad v_\text{air}$ = velocity of sound in air (m/s, ft/s, etc.)

$\qquad v_s$ = relative velocity of source (m/s, ft/s, etc.)

Example 1-7

A siren emits a constant note whose frequency was found to be 535 Hz. What would its frequency be if measured from a moving car traveling at 55 mi/h toward the siren? Assume the velocity of sound in still air to be 1090 ft/s.

Solution: Since all units are given in the English system, let's stay within that system for our solution. Since the car is moving toward the source, we will use eq. (1-12). First, let's convert 55 mi/h to ft/s.

$$\frac{55 \text{ mi/h} \times 5280 \text{ ft/mi}}{3600 \text{ s/h}} = 80.7 \text{ ft/s}$$

Therefore,

$$f_o = f \frac{v_\text{air}}{v_\text{air} - v_s} = (535 \text{ Hz}) \frac{1090 \text{ ft/s}}{1090 \text{ ft/s} - 80.7 \text{ ft/s}}$$

$$= 535 \text{ Hz} \times 1.08$$

$$= 577.8 \text{ Hz}$$

Sound transducers may be used for measuring distances much as radar is used for the same application. Sound pulses may be transmitted from a sound emitter and reflected from distant objects back to a microphone receiver located near the emitter. The time interval is recorded for the sound travel and the distance is then calculated and displayed.

1-5.4 Light

To approach the physics of light raises some very interesting questions—probably more questions than perhaps there are answers for. I mention this only because during the course of explaining how certain sensors work in detecting light, apparent contradictions within the physical explanation of light come into the picture.

Light may be considered as either an electromagnetic occurrence in nature or as comprised of energy-carrying particles, or discrete energy packets as they are sometimes called. Unfortunately, about the time when you think it is behaving as one form, it turns out to be in the other form. But this is the nature of light. The medium it happens to be traveling through or interacting with may well have a role in determining its particular state.

It was because of this chameleon-like characteristic that it was necessary to develop the quantum theory of light. Essentially, the theory boils down to this: The propagation of light is best described by electromagnetic phenomena, whereas the absorption or emission characteristics of light interacting with material are better handled using the particle theory. The quantum theory also allows us to equate or convert light's one mode of existence to the other through a very simple equation that is discussed below (see also the discussion in Sections 11-1 and 11-2).

Electromagnetic light form.   Figure 1-7 illustrates the electromagnetic spectrum, of which visible light (to the human eye) is a very small portion. Its location within the spectrum places it between the extremely short wavelengths of microwaves and the even much shorter wavelengths of x-rays. However, notice that there is a substantial area on either side of this narrow band that is occupied by the infrared and ultraviolet spectrums. The light-sensitive sensors that we will be discussing in later chapters operate somewhere between the span of frequencies or wavelengths encompassing these two regions.

Light may be identified either by its wavelength or by its frequency. The relationship between the two quantities is

$$c = \lambda f \tag{1-13}$$

where c = speed of light, 300,000 km/s or 186,291 mi/s

λ = wavelength (km, mi)

f = frequency (Hz)

This is the same relationship as that expressed in eq. (1-10) for sound waves.

Example 1-8

Determine the frequency of orange light. Assume a wavelength of 625 nm.

Solution: The wavelength of orange light may be rewritten as 625×10^{-9}, or 625×10^{-12} km. Therefore, according to eq. (1-13),

$$c = \lambda f = 300,000 \text{ km/s} = (625 \times 10^{-12} \text{ km}) \cdot f$$

$$f = \frac{\sqrt{300 \times 10^3} \text{ km/s}}{625 \times 10^{-12} \text{ km}}$$

$$= 0.480 \times 10^{15} \text{ Hz}$$

Light-particle form

When speaking of light as being a particle, the name given to that particle is *photon*. Just as any other particle, the photon has energy and momentum as it travels. However, its mass is zero. Consequently, this is no ordinary particle. Einstein's

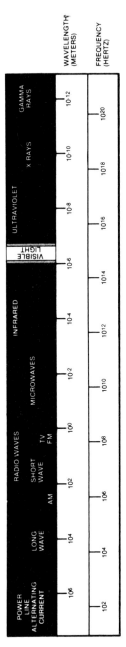

Figure 1-7 Electromagnetic spectrum. (Courtesy of the American Radio Relay League, Inc., Newington, CT.)

theory of relativity states that any particle having energy must also have momentum even though it has no mass. This momentum, p, is determined as

$$p = \frac{hf}{c} = \frac{h}{\lambda} \qquad (1\text{-}14)$$

where p = momentum (kg-m/s)

h = Planck's constant (6.62×10^{-34} J-s; see discussion below)

f = frequency (Hz)

c = speed of light (m/s)

λ = wavelength of light (m)

The energy contained by one photon is simply

$$E = hf \qquad (1\text{-}15)$$

where E = energy (J)

h = Planck's constant (J-s)

f = frequency (Hz)

Example 1-9

Calculate the energy content of each photon in orange light in Example 1-8.

Solution: We found from Example 1-7 that orange light has a frequency of 0.480×10^{15} Hz. Therefore, according to eq. (1-15), the energy content of each photon would be

$$E = hf = (6.62 \times 10^{-34} \text{ J-s})(0.480 \times 10^{15} \text{ Hz})$$

$$= 3.178 \times 10^{-19} \text{ J}$$

Equation (1-15) simply states that *all the energy contained by a photon is determined solely by its frequency*. The proportionality constant relating this energy, E, and its frequency, f, is *Planck's constant*, named in honor of Max Planck, the German physicist who conceived it.

Equations (1-14) and (1-15) have very important implications when explaining the photoemission characteristics of certain light sensors. This is discussed more extensively in Chapter 15.

The color of light is determined by its frequency or wavelength. The wavelengths of some of the common colors perceived by the human eye are as follows:

Red: from 700 to 650 nm (1 nm = 10^{-9} m)

Orange: from 650 to 600 nm

Yellow: from 600 to 550 nm

Green: from 550 to 500 nm

Blue: from 500 to 450 nm

Violet: from 450 to 400 nm

It can be seen that as the wavelength shortens, the light color changes, going from red to the violet.

The transducers used for measuring light intensity vary widely in their function and operation. Many of these light-sensitive devices are often referred to as *photocells*. Examples of usage are found in photographic and video cameras, security systems, and in certain communication applications.

1-5.5 Electromagnetic Radiation

As we use the term in this book, *electromagnetic radiation sensing* will cover all forms of radiation other than light and temperature, even though light and temperature are forms of electromagnetic radiation. The reason for this approach is that methods of detection for temperature and light are substantially different compared to the methods of detection used for other electromagnetic phenomena. Consequently, heat and light have been set aside in their own categories of treatment.

Referring again to the electromagnetic spectrum in Figure 1-7, we see a tremendously wide range of wavelengths, ranging from approximately 10,000,000 m (a frequency of 30 Hz) to 1×10^{-12} m (a frequency of 3×10^{20} Hz). One of the things that we will want to address as we discuss the functions of the various sensors being used today is: How does one limit or hold that sensor's frequency response capabilities to a specific range? Obviously, if we want to detect x-rays, we want a detector that is responsive only to x-rays and not to the radio transmissions coming from a nearby AM or FM radio station.

1-5.6 Particle Radiation

The detection of highly charged and uncharged particles, that is, atomic particles that have been excited or otherwise given energy from some outside force, is a fairly recent development in sensing technology. Examples of charged particles are alpha particles, beta particles, certain helium atomic nuclei, and protons. An example of an uncharged particle is the neutron. (Gamma radiation and x-rays are considered to be electromagnetic forms of radiation and therefore do not have charge associated with them.) There are certainly other forms of energetic particles in addition to the ones just mentioned; however, their detection can be quite involved, requiring extremely elaborate and often very large detection facilities. These facilities are not usually classified as transducers, although that is precisely what they are designed to do—to transduce or sense. For the purposes of this discussion, we will stick to "instrument-sized" or portable transducers, as we agreed to do in Section 1-2.

There are no generalized laws or mathematical expressions that govern the behavior or detection of energetic particles that we can apply here as we have done for the other measurands. Instead, when we approach this discussion in Chapter 17, we will go into much greater detail.

1-5.7 Temperature

This is the second special detection category of electromagnetic radiation. Temperature detection has to do with sensing of radiant heat from a source. This detection can be done either through direct contact with the heating source, or remotely, without direct contact with the source, using radiated energy instead. Either method involves that portion of the electromagnetic spectrum immediately to the red side of the visible spectrum. This area extends from about 700 nm (7×10^{-11} m) to about 1×10^{-3} m (1 mm).

There are three methods of heat transfer that allow heat to travel from one location to another: (1) convection, (2) conduction, and (3) radiation. The method of primary interest to sensor design is radiation. The amount of radiated heat received depends on the medium through which it must travel. This is one characteristic of an electromagnetic transmission that makes heat so similar to, say, a radio wave. There are many substances that absorb heat waves, just as there are substances that absorb radio waves. Similarly, there are substances that do not seem to affect either. As a matter of fact, the most efficient path for heat or radio transmissions is a vacuum. A good example of this fact is the heat radiated from the sun to earth. This radiation is transmitted virtually unattenuated over a distance of 1.496×10^8 km (93,000,000 mi).

The relationship that exists between the power transmitted through heating, the area at the receiving site being heated, and the temperature of that area is

$$P \propto AT^4 \tag{1-16}$$

where P = power

A = area

T = temperature (absolute)

The proportionality sign, $\propto$, can be eliminated, so that an equation may be produced by substituting the proportionality constant, σ, which has the value of 5.67×10^{-8} W/m²-K⁴. This constant is called *Stefan's constant*. In addition, the proportionality expression in eq. (1-16) must be further modified by inserting an emissivity factor, ϵ, which describes the ability of the area's surface to emit or absorb heat (surfaces that are good absorbers of heat are also good emitters of heat). The value of ϵ will be anywhere between 0 and 1, depending on the type of surface area. Our revised equation now becomes

$$P = \sigma \epsilon A T^4 \tag{1-17}$$

where P = radiated power (W)

σ = Stefan's constant, 5.67×10^{-8} W/m²-K⁴

ϵ = emissivity factor (0 to 1)

A = area (m²)

T = absolute temperature (K)

Example 1-10

Find the radiated power, in watts, contained in the heat radiated from an iron ingot coming from a furnace. The ingot's dimensions are 10 ft × 4 ft × 4 ft, and the surface temperature is 1950°F or 1066°C (1339 K). Assume an emissivity factor of 1.

Solution: From the dimensions given we can calculate the ingot's total surface area to be 192 ft^2, or 17.837 m^2. Then, according to eq. (1-17),

$$P = \sigma\epsilon AT^4 = (5.67 \times 10^{-8} \text{ W/m}^2\text{-K}^4)(1)(17.837 \text{ m}^2)(1339 \text{ K})^4$$

$$= 3.251 \times 10^6 \text{ W}$$

or 3.251 MW of power.

Note that the equation above requires the use of an absolute temperature value. This is typical of many engineering applied equations, where absolute values determined from an absolute-zero energy state reference point must be used rather than using relative values determined from some other energy-level reference point. An example similar to this one is the use of a perfect vacuum for a reference level in pressure reading referencing.

The temperature scales used in industry today are the Fahrenheit, Celsius, Rankine, and Kelvin scales. Figure 1-8 compares the four and shows how to convert from one temperature-measuring system to another.

Many heat-sensitive sensors perform temperature measurements based on their ability to respond to radiant heat energy. However, there are others that derive their operation from the fact that when heated to a sufficient temperature, a surface will emit a certain color. The color, whether red, white, yellow, or any other, is a direct indication of the energy absorbed by that surface. Since energy and temperature are related through eq. (1-16), we have a reliable means of finding the

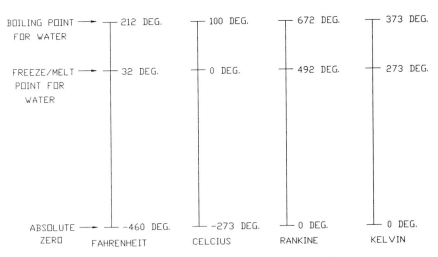

Figure 1-8 Comparison of the four temperature scales.

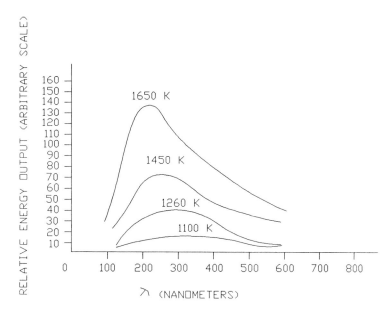

Figure 1-9   Relationship between light's wavelength, energy content, and temperature.

temperature of heated surfaces or bodies merely by observing this radiated color. Figure 1-9 shows a relationship between the energy content (i.e., energy absorbed) of a heated body, the wavelength of emitted energy (light), and some selected temperatures. Actually, the area beneath each curve represents the total energy content for the given temperatures. In practice, when observing the higher temperatures, especially 1800 K or higher, the total energy content is much higher, causing the actual radiated color to become whitish. Since *all* colors are being radiated rather equally here, they mix, which tends to create white light. Below 1800 K there tends be more of a predominate color level. At the much higher temperatures beyond 1800 K or so, it becomes necessary to use special filters, or a special device such as a spectroscope, to break the white light down into its spectrum to determine the actual predominate color, and therefore the existing temperature.

1-5.8 Flow

The measurement of flow, or flow rate, has to do with the quantity measurements of solids, liquids, or gases in motion flowing past a fixed point in a given time. As a result, typical units of measure associated with flow measurements are pounds per minute, newtons per second, and cubic feet per second. These are all weight flow rate measurements. Units of measurements such as grams per second, slugs per minute, and kilograms per hour are all *mass flow rate* measurements (assuming that we are using the gram-mass system here, not the gram-force system). Units such as gallons per minute and cubic centimeters per second are *volumetric flow rates*.

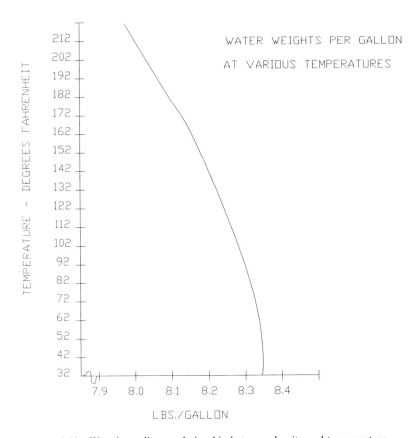

Figure 1-10 Water's nonlinear relationship between density and termperature.

Remembering the relationship $F = ma$ (or perhaps more appropriately, $W = mg$, where W = weight, m = mass, and g = 32.17 ft/s^2), this allows you to convert freely back and forth between units of mass flow rate and weight flow rate. Converting from mass or weight flow rates to volumetric flow rates can be somewhat trickier. To begin with, volumetric flow rate is temperature dependent. That is, the weight density of a liquid or gas (typical units are lb/ft^3, lb/gal, N/m^3, dyn/cm^3) varies with the fluid's temperature. More specifically, it is the volume that varies with temperature. Water is about as good an example as any fluid for illustrating this example, especially since its temperature dependency is definitely nonlinear and it is easy to observe and document. Look at Figure 1-10. At 42°F the weight density of pure water is 8.35 lb/gal. However, at 200°F the weight density becomes 8.04 lb/gal, a decrease of about 3.7%.

The volume of a solid or liquid substance subjected to a change in temperature varies according to the relationship

$$V_f = V_i(1 + \gamma \, \Delta t) \tag{1-18}$$

where V_f = final volume (m^3)

V_i = initial volume (m^3)

γ = coefficient of volumetric expansion (m^3/m^3-K, or simply K^{-1})

Δt = change in temperature to produce change in volume (K)

Example 1-11

In the iron ingot problem in Example 1-10, find the room-temperature volume of the ingot after leaving the furnace. Assume a coefficient of volumetric expansion of 36×10^{-6} K^{-1} and a room temperature of 20°C (293 K).

Solution: First, we must convert the dimensions of the ingot given in feet to dimensions in meters. The metric dimensions would then be 3.048 m $\times$ 1.219 m $\times$ 1.219 m. These are the initial dimensions upon leaving the furnace; therefore, the ingot's initial volume, V_i, is 4.529 m^3. Equation (1-18) will now give us the final volume after the ingot has cooled.

$$V_f = v_i(1 + \gamma \Delta t) = 4.529 \text{ m}^3[1 + (36 \times 10^{-6} \text{ K}^{-1})(293 \text{ K} - 1339 \text{ K})]$$

$$= 4.529 \text{ m}^3[1 + (-0.0377)]$$

$$= (4.529 \text{ m}^3)(0.962)$$

$$= 4.357 \text{ m}^3$$

Table 1-4 lists values of γ for a few common materials.

To summarize, then, the weight density of any substance, whether a solid, liquid, or gas, is defined as

$$\omega = \frac{W}{V} \tag{1-19}$$

TABLE 1-4   VALUES OF γ (COEFFICIENT OF VOLUMETRIC EXPANSION)[a]

Substance	γ(K^{-1})
Pyrex glass	9×10^{-6}
Iron	36×10^{-6}
Copper	51×10^{-6}
Aluminum	72×10^{-6}
Ice[b]	153×10^{-6}
Mercury	180×10^{-6}
Water	210×10^{-6}

[a] Average values for the temperature range 0 to 100°C.

[b] The value for ice is for the temperature range -10 to 0°C.

where ω = weight density of substance (lb/ft^3, N/m^3)

W = weight of substance (lb, N)

V = volume of substance (ft^3, m^3)

Remember: In all cases, V must be measured at the desired temperature.

Example 1-12

A substance weighing 40 N was found to have a volume of 1.27 m^3. What is its weight density?

Solution: From eq. (1-19) we see that

$$\omega = \frac{W}{V} = \frac{40 \text{ N}}{1.27 \text{ m}^3}$$

$$= 31.5 \text{ N/m}^3$$

When measuring gases we have a special problem in flow rate measurement. Not only are gas volumes affected by temperature similarly to water, as demonstrated above, they are also affected by applied pressure. The following equation, known as the *ideal gas law*, shows the interdependency of pressure, volume, and temperature for a so-called "ideal gas." An ideal gas is one that obeys the following equation (most gases do, especially at low and moderately high pressures):

$$PV = nRT \qquad (1\text{-}20)$$

where P = pressure (expressed in atm; see Example 1-13 below)

V = volume (L)

n = number of moles of gas (mol)

R = universal gas constant (see the discussion below)

T = absolute temperature (K)

The value of the universal gas constant is 0.08021 L-atm/mol-K. This is the universal gas constant explained in eq. (1-8) (but listed differently in Table 1-3). Because of the different system of units used in eq. (1-20), the value appearing here is changed from that shown in the table. The universal gas constant may be considered to be approximately the same for all gases, especially for gases at high temperatures and low pressures. To utilize the gas constant value in Table 1-3 or the SI equivalent just mentioned, the following units must be used:

P = pressure (N/m^2)

V = volume (m^3)

n = number of moles of gas (mol)

R = universal gas constant (8.3149 J/mol-K)

T = absolute temperature (K)

Example 1-13

A 6-m^3 container contains 3 mol of oxygen that has been heated to 45°C. Find the created pressure inside the container.

Solution: Use eq. (1-20) and solving for P:

$$P = \frac{nRT}{V} = \frac{(3 \text{ mol})(8.3149 \text{ N-m/mol-K})(45 + 273 \text{ K})}{6 \text{ m}^3}$$

$$= 1322.1 \text{ N/m}^2$$

The number of moles (more accurately stated as the kilogram-mole in SI units) can be calculated simply by dividing the mass of the gas, expressed in kilograms, by its molecular weight. The molecular weights for many gases may be obtained from a periodic table of elements or from a chemistry text.

Another very useful relationship to remember for ideal gases is the following:

$$\frac{P_1 V_1}{T_1} = \frac{P_2 V_2}{T_2} \qquad (1\text{-}21)$$

where P_1 = initial pressure (expressed in atmospheres or in absolute pressure; see the note below)

P_2 = final pressure (expressed in atmospheres or in absolute pressure; see the note below)

V_1 = initial volume (ft^3, m^3)

V_2 = final volume (ft^3, m^3)

T_1 = initial absolute temperature (R, K)

T_2 = final absolute temperature (R, K)

Note: An atmosphere (atm) is any pressure divided by the value 14.69 psia (pounds per square inch absolute), the pressure of the earth's atmosphere at sea level at standard conditions. *Absolute pressure is equal to the gage pressure added to the existing barometric pressure at the time of the reading.*

Example 1-14

A gas having a volume of 147 m^3, a temperature of 15°C, and a pressure of 1.150 atm is heated to 180°C and expanded to a volume of 310 m^3. Find the new pressure for the gas.

Solution: Letting 1.15 atm = P_1, 147 m^3 = V_1, and (15 + 273 K) = T_1, and letting 310 m^3 = V_2 and (180 + 273 K) = T_2, we can use eq. (1-21) to solve for P_2:

$$\frac{P_1 V_1}{T_1} = \frac{P_2 V_2}{T_2} = \frac{(1.15 \text{ atm})(147 \text{ m}^3)}{228 \text{ K}} = \frac{(P_2^2)(310 \text{ m}^3)}{453 \text{ K}}$$

Then we solve for P:

$$P = \frac{(1.15 \text{ atm})(147 \text{ m}^3)(453 \text{ K})}{(288 \text{ K})(310 \text{ m}^3)}$$

$$= 0.858 \text{ atm}$$

The fluid's flow rate, the size of conduit (pipe, duct, etc.) that it is flowing in, and the fluid's velocity are all related by the simple relationship

$$Q = Av \qquad (1\text{-}22)$$

where Q = flow rate (ft^3/s, m^3/s)

A = cross-sectional area of conduit (ft^2, m^2)

v = velocity of fluid (ft/s, m/s)

Example 1-15

A pipe of diameter 0.75 m contains water flowing at a velocity of 1.2 m/s. Find the water's flow rate.

Solution: According to eq. (1-22),

$$Q = Av = \frac{\pi r^2}{4} v = \frac{\pi (0.75 \text{ m})^2}{4} (1.2 \text{ m/s})$$

$$= 0.530 \text{ m}^3/\text{s}$$

Often, flow rates are measured indirectly by the amount of pressure drop created by a restriction within the conduit. Figure 1-11 illustrates such a device. The restriction in this case is an orifice that has a manometer attached across it for indicating pressure drop. (A manometer is a pressure-measuring device containing a liquid of known specific gravity used for determining pressure head values for

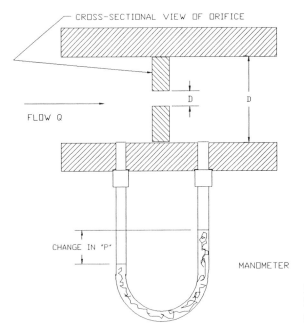

Figure 1-11 Measuring a liquid's flow rate using pressure drop.

liquids and gases. The height of the liquid's column within the manometer, the column being supported by the measured pressure, is a direct indication of the measured gas or liquid's head pressure.) The relationship between pressure drop and flow rate is described by the general equation

$$Q = AK\sqrt{2g\,\Delta p} \tag{1-23}$$

where Q = flow rate (ft³/s, m³/s)

A = cross-sectional area of orifice (ft², m²)

K = proportionality constant or discharge coefficient (no dimensions: must be given or calculated; see below)

g = gravitational constant, 32.17 ft/s² or 9.805 m/s²

Δp = pressure drop (ft of H_2O, m of H_2O)

Note: The value of K depends strongly on the method used to create the pressure drop in a flow rate monitoring system (i.e., it will depend on whether the restriction is an orifice or a venturi-type restriction). This information can be obtained from any mechanical engineering textbook on fluid flow.

Example 1-16

Determine the flow rate for a pipe containing an orifice having a K value of 0.61 and a pressure drop of 0.092 m as read with a water-filled manometer. The orifice's diameter is 0.35 m.

Solution: Applying eq. (1-23), we obtain

$$Q = AK\sqrt{2g\,\Delta p} = \frac{\pi(0.350\ \text{m})^2}{4}\,(0.61)\sqrt{(2)(9.805\ \text{m/s}^2)(0.092\ \text{m})}$$

$$= (96.2 \times 10^{-3}\ \text{m}^2)(0.61)(1.343\ \text{m/s})$$

$$= 0.0781\ \text{m}^3/\text{s}$$

Flow transducers utilizing the methods described above are found everywhere in industry. Orifice-type flow meters are used extensively in the chemical and food industries. Positive-displacement transducers are used by utilities for measuring natural gas consumption, sewage processing, and in laboratory applications where precise flow rate measurements are required.

1-5.9 Force

The unit of force in the English system is the pound (lb); the unit of mass in this system is the slug. In the SI system the unit of force is the newton (N); the unit of mass is the kilogram (kg). Mass and force are related by Newton's equation, which states that

$$F = ma \tag{1-24}$$

where F = force (lb, N)

m = mass (slugs, kg)

a = acceleration (ft/s², m/s²)

To help understand the difference between the quantity of force and mass, it is necessary to realize that the weight of a substance is determined by the gravitational constant. As an example, a substance weighing 1 pound on earth will weigh only about one-sixth as much on the moon. This is because the moon's gravitational constant at its surface is only one-sixth the value of earth's gravitational constant. On the other hand, if that substance has a mass of, say, 3 slugs here on earth, it will still have a mass of 3 slugs on the moon—or on any other planet or satellite. (Another quantity, *inertia*, must not be confused with either mass or force. Inertia is a quantity that has become synonymous with mass in everyday language. To move or accelerate a very large massive body we say that we have to "overcome that body's inertia" in order to move it. However, in technology, inertia has a very specific meaning. Inertia, more properly referred to as the *moment of inertia*, is a quantity having to do with that body's rotation about an axis, and its mass.) The following example demonstrates the relationship between mass, force, and acceleration.

Example 1-17

A mass of 16 kg is accelerated to a value of 10 m/s². Find the force needed to create this acceleration.

Solution: Use eq. (1-24):

$$F = ma = (16 \text{ kg})(10 \text{ m/s}^2)$$

$$= 160 \text{ N}$$

This equation is often found stated in the form

$$W = mg \tag{1-25}$$

where W = weight (lb, N)

m = mass (slugs, kg)

g = gravitational constant (32.17 ft/s², 9.805 m/s²)

Force and pressure are related through the expression

$$F = pA \tag{1-26}$$

where F = force (lb, N)

p = pressure (lb/in², N/m²)

A = area over which pressure is distributed (in², m²)

Example 1-18

Calculate the pressure generated by a force of 125 N when applied to an area of 15 m².

Solution: According to eq. (1-26),

$$F = pA = 125 \text{ N} = (p)(15 \text{ m}^2)$$

Solving for p yields

$$p = \frac{125 \text{ N}}{15 \text{ m}^2}$$

$$= 8.333 \text{ N/m}^2$$

Acceleration is usually measured by a specially designed transducer called an *accelerometer*. This device contains an internal mass whose deflection can be easily measured. This deflection is the result of an applied acceleration.

Pressure transducers vary widely in design. Many of these devices merely respond to the applied force, which is later converted to a pressure indication. The *strain gage* is an example that utilizes the change in electrical resistance of a wire grid, which varies inversely with the applied pressure (or force).

1-5.10 Potential Difference

Electrochemical applications

The measurement of the potential difference between two measuring elements is often associated with sensors used for chemical detection; these are sensors used for chemical identification purposes or for chemical concentration measurements. Good examples of this application are in measurement of the pH of solutions and in making oxygen-reduction measurements of fluids necessary for proper industrial process control. Both of these processes are explained in detail in later chapters. The chemistry involved is quite complex and each transducer utilizing an electrochemical process is quite specialized; therefore, the chemistry will not be covered here.

The change in electrical conductivity and hence the change in potential difference across a material is also often used as an indicating means. The basis for this type of detection evolves around Ohm's law for direct current and voltages:

$$E = IR \tag{1-27}$$

where E = potential difference (dc volts)

I = current (dc amperes)

R = resistance (Ω)

Example 1-19

A humidity sensor working on the principle that its internal resistance varies with humidity is supplied with 12.6 V dc for its operation. For a given humidity condition it was found that its internal resistance was 245 kΩ. What is the value of current flowing through the sensor?

Solution: According to Ohm's law, eq. (1-27), the current may be calculated in the following manner:

$$E = IR = 12.6 \text{ V} = (I)(245,000 \ \Omega)$$

Solving for I, we obtain

$$I = \frac{12.6 \text{ V}}{245,000 \ \Omega}$$

$$= 0.0000514 \text{ A}$$

$$= 51.4 \times 10^{-6} \text{ A}$$

$$= 51.4 \ \mu\text{A}$$

Electrostatic applications

In addition to the measuring of moving charges of electricity, the measuring or sensing of static electrical charges is also very important. Generally, the one relationship most often depended on by sensing devices for their proper operation is *Coulomb's law*. This law states the relationship that exists between the charges on two particles separated by a distance, and the attracting or repelling force generated, as a result, between these two particles. In equation form, the law states that

$$F \text{ (attraction or repulsion)} = \frac{kQ_1 Q_2}{r^2} \qquad (1\text{-}28)$$

where F = attraction or repulsion force (N)

k = proportionality constant (9.0×10^9 N-m^2/C^2)

Q_1 = charge on first particle [coulombs (C)]

Q_2 = charge on second particle [coulombs (C)]

r = distance of separation between charges (m)

Example 1-20

Two charged particles, one containing a charge of $+10$ C and the other, $+25$ C, are separated by a distance of 0.17 m. Calculate the repulsion force between them.

Solution: Apply eq. (1-28):

$$F = \frac{kQ_1 Q_2}{r^2} = \frac{(10 \text{ C})(25 \text{ C})}{(0.17 \text{ m})^2}$$

$$= 8650.5 \text{ N}$$

1-6 PRINCIPAL DETECTION METHODS OF ELECTRICAL TRANSDUCERS

Now that we have identified the various measurands that we will be detecting, we will now want to discuss the various principles of operation used in designing electrical transducers. The principal methods used for sensing in electrical transducers are:

1. Capacitive
2. Electromagnetic
3. Inductive
4. Magnetostrictive
5. Photoconductive
6. Photovoltaic
7. Piezoelectric
8. Potentiometric
9. Reluctive
10. Resistive
11. Strain gage
12. Thermoelectric

Additional methods are discussed in later chapters.

It is interesting to note that as you look at the list above, there are five methods of transduction that do not require power supplies for their operation (at least in theory). This is because these methods—the electromagnetic, magnetostrictive, photovoltaic, piezoelectric, and thermoelectric transducers—generate their own power source. To function properly, all the other methods in this list require some form of power source.

1-7 CALLING A TRANSDUCER BY ITS PROPER NAME

With the list of measurands in Section 1-4 and the list of sensor methods in Section 1-6, it is natural that some confusion could develop over the proper name of any given sensor. Should the sensor be identified by the measurand it was designed to detect, or should it be called by its principal design function? As it turns out, a layperson using a transducer usually identifies that transducer by its measurand. In other words, if the transducer is being used to sense and record air velocity, the transducer device would probably be called an air velocity sensor. What is even more likely, it would be identified by its marketed name, for instance, an Acme Breeze-O-Meter. However, technically trained persons are more inclined to identify the transducer by first stating the measurand being sensed, followed by the transduction method being used. Assuming that a thermoelectric element was the sensing element in the Acme device above, the transducer would then be identified by the statement: "This is a motion (or velocity)-detecting thermoelectric transducer."

1-8 THE TWO MAJOR CATEGORIES OF SENSORS

In general, sensors can be divided into two very broad categories: *contacting* sensors and *noncontacting* sensors. These categories refer to the sensor's necessity of having to be in direct contact with or not having to be in direct contact with the measurand it is trying to sense. There is no advantage in using sensors of the contacting type if a noncontacting type is available that meets the same required performance criteria. This category exists only because there is no other practical means of sensing available; it is a technological limitation. A contacting sensor will eventually suffer from mechanical wear due to friction. If we look at the measurand list in Section 1-4, the measurands that could cause wear due to direct contact are position, motion, flow, and force. It is very unlikely that the other measurands would create any notable wear in their sensors. There has been much technological development work done over the past several years to create noncontacting sensing methods for these particular measurands. The results of this work are discussed in subsequent chapters.

1-9 METHODS USED TO DESCRIBE SENSOR PERFORMANCE

To determine the performance capabilities of sensors, many performance criteria have been established by industry. Armed with these criteria a person can make direct comparisons between several sensors to determine which will perform the best. Unfortunately, there is a lack of standardization in the industry as to what constitutes a complete set of performance data, but for the most part, most of these data can be understood if you also come armed with a relatively good knowledge of electronics and mechanics.

1-9.1 Frequency Response and Bandwidth

The frequency response of a device refers to its ability to respond adequately to a given sinusoidal input signal. The adequacy is determined by the manufacturer's design specifications. If the device is designed to respond to a range of frequencies, as is most often the case, the *bandwidth* of that device is specified. Often, this specification also includes the *3-dB down points* of that bandwidth.

Figure 1-12 illustrates a frequency response curve that may be found for a typical transducer device. As is frequently found in these curves, there is a relatively flat portion of the curve that exists before each end of the curve tends to drop off at a fairly rapid rate. This flat-area portion is indicated at the top of the curve. Note, however, that there is also a portion at each end which is also included with the original flat portion, bringing the total indicated bandwidth to a value of 19,985 Hz in our example. Often what is done is to include that additional bandwidth enclosed by a 29.3% reduction in signal output in the device being tested. When converted to the decibel scale, the same scale discussed earlier in Section 1-5.3, this reduction amount is the equivalent of a 3-dB reduction in output signal. This is also equivalent

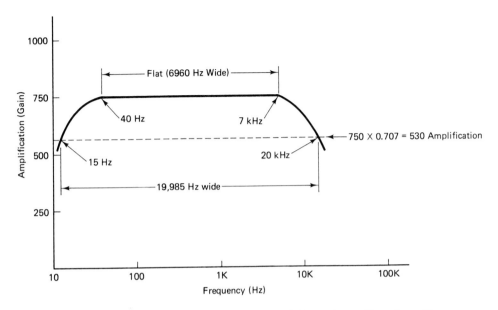

Figure 1-12 Frequency response curve for a typical transducer. (From James R. Carstens, *Automatic Control Systems and Components*, copyright 1990, p. 56. Reprinted by permission of Prentice Hall, Englewood Cliffs, NJ.)

to reducing the output power of the device by one-half. As a result, the 3-dB down points are often referred to as the *half-power points* on a frequency response curve. The original justification for the existence of these points is based on the fact that such a reduction in signal is barely detectable by the average human ear. The reason for this lack of detection is due to the ear's logarithmic response.

1-9.2 Sensitivity

The sensitivity of a transducer refers to its ability to generate an output response for a given change in the measurand. Mathematically, this is expressed as

$$\text{sensitivity} = \frac{\text{change in output indicator}}{\text{change in measurand that produced change in indicator}} \tag{1-29}$$

where a change in the output indicator is measured in output scale units such as degrees, millimeters, digits, divisions, and radians.

Example 1-21

An electronic thermometer reading changed from 55.0°F to 56.7°F, a change of 2.6 divisions on its output scale as a result of a 3.1°F actual change in the measured temperature. What is the instrument's sensitivity?

Solution: Using eq. (1-29) gives us

$$\text{sensitivity} = \frac{\text{change in output indicator}}{\text{change in measurand that produced change in indicator}}$$

$$= \frac{2.6 \text{ div}}{3.1°F}$$

$$= 0.829 \text{ div/°F}$$

1-9.3 Static and Dynamic Error

The static error of a sensor is defined simply as

$$\text{static error} = \text{measured value} - \text{actual value} \qquad (1\text{-}30)$$

The term *static error* means that the measured value of the measurand does not change with time; it is relatively constant. *Dynamic error*, on the other hand, is calculated in precisely the same manner as static error. However, the characteristic of the measured value is such that it *does* change with time. Consequently, the determined error also changes with time.

Example 1-22

If a reluctive-type rpm transducer used for measuring the revolutions per minute of a rotating shaft measured a consistent rotation rate of 1190 rpm, where in fact the rpm was actually found to be 1255 rpm, find the transducer's static error.

Solution: From eq. (1-30), the static error is

$$\text{static error} = \text{measured value} - \text{actual value}$$

$$= 1190 \text{ rpm} - 1255 \text{ rpm}$$

$$= -65 \text{ rpm}$$

1-9.4 Percent Accuracy

The percent accuracy of a sensor or transducer is defined as a percentage comparison of the static error to the actual value. In other words,

$$\% \text{ accuracy} = \frac{\text{measured value} - \text{actual value}}{\text{actual value}} \times 100 \qquad (1\text{-}31)$$

where % accuracy is expressed as a percentage.

Example 1-23

Find the percent accuracy in the device described in Example 1-22.

Solution:

$$\% \ \text{Accuracy} = \frac{\text{measured value} - \text{actual value}}{\text{actual value}} \times 100$$

$$= \frac{1190 \ \text{rpm} - 1255 \ \text{rpm}}{1255 \ \text{rpm}} \times 100$$

$$= -5.18\%$$

1-9.5 Responsiveness

The responsiveness of a transducer is the ratio resulting from dividing the change in a measured quantity needed to produce an indicated change in the transducer's output by the measured quantity itself, expressed as a percentage. Mathematically,

$$\text{responsiveness} = \frac{\text{change in measured quantity to produce indicated change}}{\text{measured quantity}}$$

(1-32)

$$\times \ 100$$

Example 1-24

A strain gage pressure transducer reads 375 psig (pounds per square inch, gage) on a pipeline. A change of 1.5 psig in the measurand is necessary to cause a change in the output reading. Find the transducer's responsiveness.

Solution: Referring to eq. (1-32) yields

$$\text{responsiveness} = \frac{\text{change in measured quantity needed to produce indicated change}}{\text{measured quantity}}$$

$$= \frac{1.5 \ \text{psig}}{375 \ \text{psig}} \times 100$$

$$= 0.4\%$$

1-9.6 Reproducibility

Reproducibility in a transducer refers to its ability to indicate identical values of the measurand at its output each time a measurement is made, assuming that all environmental conditions are the same for each measurement. Another way to state this is: the degree of agreement that a value of a measured variable has when measured at different times. All instruments, including transducers, possess a certain amount of inherent uncertainty in their ability to reproduce the same output readings time after time. Variations in output readings may be unpredictable and random. Table 1-5 shows an example of this characteristic using the readings obtained from a pressure transducer. This reproducibility may be expressed as a dynamic error.

Another way to quantify reproducibility is through graphing. A graph such as

TABLE 1-5 EXAMPLE OF OUTPUT VARIATIONS

Reading	True (psi) pressure	Gage (psi) reading
1	152.5	150.7
2	152.5	151.4
3	152.5	151.9
4	152.5	153.0
5	152.5	151.1
6	152.5	151.8
7	152.5	152.4
8	152.5	151.7
9	152.5	153.0
10	152.5	151.9

the one in Figure 1-13 may be used to show the variations of output readings for a given number of identical input measurand values.

There is no equation for determining reproducibility. Instead, it is a quality possessed by a transducer rather than a quantity.

1-9.7 Range

A transducer's range refers to the two stated values, the lower-limit value of the measurand and the upper-limit value of the measurand, between which the transducer is designed to operate.

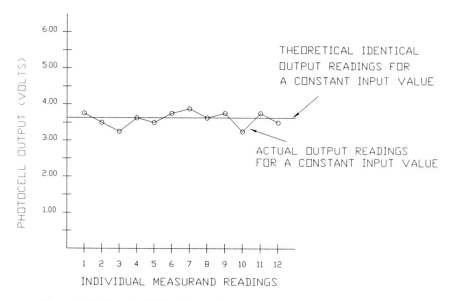

Figure 1-13 Reproducibility of a transducer's output for several different readings of the same measurand value.

Example 1-25

What is the range of a sensor that is designed to sense a temperature from -10 to $+250°F$?

Solution: The range of this sensor is the two stated limiting temperature values, $-10°F$ and part $+250°F$.

1-9.8 Span

The *span* of a transducer or sensor refers to the arithmetic difference between the two stated range values.

Example 1-26

What is the span of the sensor described in Example 1-24?

Solution: The span of this sensor is simply the arithmetic difference between the two range values: $250°F - (-10°F)$, or $260°F$.

1-9.9 Deviation from Linearity

A sensor's deviation from linearity is determined by plotting its output against a linear output and then stating the amounts of departure from the linear curve. These statements are usually given for certain spans of operation, as shown in the following example.

Example 1-27

Figure 1-14 shows the output of a flow-metering reluctance-type sensor. Express the deviation from linearity for this device based on the resultant curve shown.

Solution: The following linearity deviation information can be obtained from Figure 1-14 (all figures are maximum figures within each span):

$$0\text{–}2 \text{ L/min} = -2 \text{ mV}$$
$$2\text{–}4 \text{ L/min} = -3 \text{ mV}$$
$$4\text{–}6 \text{ L/min} = +8 \text{ mV}$$
$$6\text{–}8 \text{ L/min} = +8 \text{ mV}$$
$$8\text{–}10 \text{ L/min} = -10 \text{ mV}$$

1-9.10 Hysteresis

Stated simply, *hysteresis* refers to the characteristic that a transducer has in being unable to repeat faithfully, in the opposite direction of operation, the data that have been recorded in one direction (i.e., ascending or descending in value). The data used for ascertaining the hysteresis characteristic are derived from static data points, that is, data points that do not change in value with time. Consider Figure 1-15. A pressure gage was used to record the information given in the graph. Two sets of data

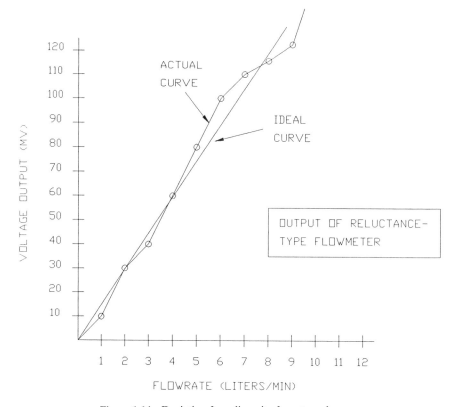

Figure 1-14   Deviation from linearity for a transducer.

were recorded. One set was recorded with the pressure increasing with time, starting at time = 0, pressure = 0. The other set of data was recorded beginning at maximum time and maximum pressure and decreased over the same time increments. Theoretically, the pressure readings should coincide within each set of data. But in reality, a different set of pressure readings resulted for the same time increments. The resultant curves form what is often referred to as the *hysteresis loop* for that particular instrument. In the case of our pressure gage, there is an apparent "stretching" or perhaps "slop" in the mechanical linkages coupling the mechanical sensor to the indicating needle. In the case of solid-state components, where there are no moving mechanical parts, this hysteresis can be caused by temporary polarization of materials within the component itself that directly affects the buildup or decrease of electrical charges inside the crystalline makeup of the solid-state materials.

1-9.11 Resolution

The term *resolution* is often used in conjunction with digital sensing devices denoting the ability of the device to distinguish between discrete, individual signal levels. As

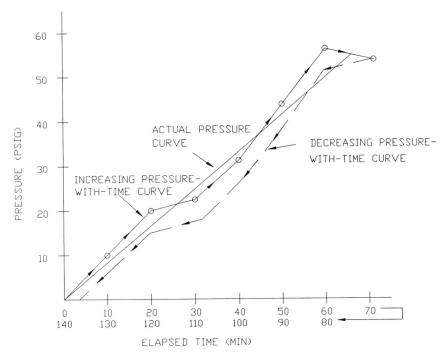

Figure 1-15   Example of hysteresis.

an example, let's say that we wanted to convert an analog voltage level to a digital signal. To do this, an analog-to-digital (A/D) converter circuit must be used. Let's assume that the output of this converter has the capability of handling 8 bits of binary data. This means that for a given input analog voltage span of, say, 40 V (representing a range of, say, 10–50 V dc), we would be able to convert this voltage into 28 parts at the A/D's output. The resolution of this circuit would then be

$$\text{resolution} = \frac{\text{input signal span}}{\text{output bits}} \tag{1-33}$$

In this case,

$$\text{resolution} = \frac{40 \text{ V}}{256 \text{ bits}}$$

$$= 0.156 \text{ V/bit}$$

In other words, we would have the capability of reading the input voltage with our A/D converter to the nearest 0.156 V.

Because of the several methods used in expressing the performance characteristics of a transducer, and because of the apparent similarities of some of the definitions of these expressions, another example problem is presented here to help separate one characteristic from another.

Example 1-28

A particular photocell transducer produces an analog output voltage that is proportional to the light intensity to which it is exposed. A light flow rate of 0.02 lumen (lm) is incident on the transducer to produce an output voltage of 0.376 V dc according to its attached digital display. (A calibration check made earlier shows that for the same light intensity the actual output voltage should have been 0.381 V dc.) The voltage range of the digital display on the transducer is from 0 to 1.5 V dc. It was found through experimentation that a change of 0.001 lm produced a change in output voltage of 0.052 V. This reading was recorded when the input light flow rate was 0.019 lm. It was further noted at this illumination value that the smallest change in voltage that could be detected with the existing digital display was 0.001 V.

Determine the following characteristics for the transducer: (a) the sensitivity, (b) static error, (c) percent accuracy, (d) responsiveness, (e) reproducibility, (f) range, and (g) span.

Solution: (a) To determine the sensitivity [refer to eq. (1-29)]:

$$\text{sensitivity} = \frac{0.052 \text{ lm}}{0.001 \text{ V}}$$

$$= 52 \text{ lm/V}$$

Note that in this case since the output indication was given in volts rather than in divisions, degrees, or some other mechanical measuring unit as dictated by eq. (1-29), the smallest unit of measurement available in this instance was the digital display's smallest display unit capability, 0.001 V.

(b) To determine the static error [refer to eq. (1-30)]:

$$\text{static error} = 0.376 \text{ V} - 0.381 \text{ V}$$

$$= -0.005 \text{ V}$$

(c) To determine percent accuracy [refer to eq. (1-31)]:

$$\% \text{ accuracy} = \frac{0.376 - 0.381}{0.381} \times 100$$

$$= -1.31\%$$

(d) To determine responsiveness [refer to eq. (1-32)]:

$$\text{responsiveness} = \frac{0.001}{0.019} \times 100$$

$$= 5.26\%$$

(e) To determine reproducibility: Because reproducibility is a *quality* possessed by a transducer or instrument, there is no method by which it can be calculated. However, one attribute of reproducibility is found in the percent accuracy figure that was calculated above. This figure may be used to obtain some idea of the transducer's reproducibility.

(f) To determine range (see Section 1-9.6): The range was already stated as being 0 to 1.5 V dc.

(g) To determine span (see Section 1-9.7):

$$\text{span} = 1.5 \text{ V} - 0 \text{ V}$$
$$= 1.5 \text{ V dc}$$

1-10 SUMMARY

To understand how electrical transducers function, it is necessary to have a firm foundation established in physics. We have attempted to make this establishment in Chapter 1 along with defining transducers and sensors, together with how they are categorized for reference purposes. In the chapters that follow, it may be necessary to refer back to the appropriate sections in this chapter to refresh yourself with the necessary physics to understand the characteristics of the measurands in question.

REVIEW QUESTIONS

1-1. Explain the difference between a sensor and a transducer.

1-2. Explain the difference between an analog signal and a digital signal. Give two examples of instruments or devices that utilize each.

1-3. What are the advantages and disadvantages of an analog instrument?

1-4. Explain what is meant by the term *measurand*. List the basic measurands as explained in the text.

1-5. Explain the difference between the electromagnetic waveform of light and the particle form of light.

1-6. List the most often encountered methods used for sensing with electrical transducers and list the primary measurand detected by each.

1-7. Explain the difference between a primary measurand and a secondary measurand.

1-8. What is meant by the term *sensitivity*?

1-9. Define the term *sensitivity* for a transducer.

1-10. Define the term *accuracy*.

PROBLEMS

The following problems are based on the theories and principles discussed in this chapter.

1-11. A sounding device that is part of an automatic focusing device on a camera emits a sound that reflects off an object and the echo received by a microphone adjacent to the emitter. The total travel time of the sound wave is 0.0422 s. The air's temperature is 20°C. Determine the object's distance from the microphone.

1-12. A robot is controlled by a digital controlling system. If a digital sensor is used to measure the extension of the robot's arm, and this sensor can handle a 16-bit word, what can the robot's designer expect regarding tolerance in placing the arm's location if the arm has a maximum extension of 12.70 ft?

1-13. Calculate the energy content of a photon in blue light. What is the percent increase or decrease in energy compared to a photon in red light?

1-14. Calculate the volumetric expansion of an aluminum mechanical arm that is used as a temperature sensor attached to the wiper arm of an indicating potentiometer if the arm is 4.000 in. in length, 0.750 in. wide, and 0.557 in. high. A temperature increase of 47.8°C is being measured.

1-15. If a 10-m^3 canister contains 4.7 mol of nitrogen, and the canister is heated to 37°C, find the pressure inside the canister needed to install a properly sized pressure transducer. (See the discussion immediately following Example 1-13.)

1-16. Determine the size of orifice needed to install in a pipe transporting water if a pressure-sensing transducer used to sense the pressure drop across the orifice has a maximum range of 0.25 m. Water's flowrate is 10 m/s. Assume a K value of 0.61.

1-17. In designing an accelerometer containing a known mass for its sensor (referred to as a *seismic mass*), what acceleration amount is being sensed when a force of 1.34 N is applied and the mass is 155 g?

1-18. If a transmitted radio wave has a frequency of 152.781 MHz, determine its wavelength. Also, if a beam of light from a laser has a wavelength of 7233 Å, determine its frequency.

1-19. Calculate the increase in signal strength, in dB, of a transmitted telemetry signal at a receiving transducer if the transmitter's oscillator voltage of 12.5 V is increased by a factor of 2.5. (Assume the transmitter to be a linear-responding device.)

1-20. Convert a shaft's rotation of 3750 rpm to rev/s; to deg/min. If the shaft's rotation were to be decreased to 3260 rpm over a period of 10 s, what would the angular acceleration be, expressed in rev/s^2?

REFERENCES

BUCHSBAUM, WALTER H., *Practical Electronic Reference Data*. Englewood Cliffs, NJ: Prentice Hall, 1980.

FINK, DONALD G., *Electronics Engineers' Handbook*. New York: McGraw-Hill, 1975.

O'HIGGINS, PATRICK J., *Basic Instrumentation*. New York: McGraw-Hill, 1966.

SEARS, FRANCIS W., MARK W. ZEMANSKY, and HUGH D. YOUNG, *College Physics*. Reading, MA: Addison-Wesley, 1977.

2

DATA TRANSMISSION METHODS

CHAPTER OBJECTIVES

1. To understand the various signal transmission types produced by sensors that form a transducer.
2. To understand the advantages and disadvantages of each transmission type.
3. To study the various kinds of transducer performance data.

2-1 TYPES OF SIGNAL TRANSMISSIONS USED AND HOW THEY ARE RECEIVED

Up to this point we continue to treat the transducer as a black box having an input and an output. We have already discussed the various kinds of inputs a transducer can have; we now want to look at the kinds of outputs we can have.

Transducers must be designed to interface with certain devices, such as displays and other circuits, and because they must also be able to transmit signals over a variety of communications paths (wires, space, fiber optics), not to mention the fact that these devices must also be able to cope with various types of signal noises (i.e., interference) encountered during data transmissions, it is important to understand the characteristics of various signal transmission types that are available. There are certainly advantages and disadvantages associated with each. In Section 1-3 we looked at an overview of the various types of transducer outputs available. In Chapter 2 we want to go into much more detail to increase our understanding of these transmission methods. We want to answer the question most often asked about transducer devices: "Why can't I simply just hook up the output of the transducer to some sort of data receiver regardless of the circuits involved and transmission type being used, and be done with it?"

2-1.1 Analog Current and Voltage Signals

In the early days of transducers, analog data transmissions were the only practical and fairly well understood means of sending signals. (One notable exception was Samuel Morse's invention in 1859 of the telegraph sender and receiver utilizing a digital code for communications.) Immediately following the industrial revolution in the United States around the beginning of this century, individuals began experimenting with methods of sending voltages and currents through wires from strategically placed sensors to remotely placed recording stations. The magnitude of the current flow or the magnitude of voltages being transmitted were directly correlated with the magnitude of the quantity (i.e., the measurand) being detected and measured. Many of the same methods are still being applied in industry and are very effective. We now discuss these systems in detail to obtain a better understanding.

4- to 20-mA current loop system (and others)

The current transmission method is one of the oldest of all existing methods used for sending transducer output information to a remote location. Because of the magnitude of currents being transmitted and because of electrical circuit noise limitations, a transmitting distance of approximately 2500 ft is considered the maximum reliable distance before requiring additional amplification for this type of system.

Figure 2-1 shows a general representation of a 4- to 20-mA current loop system. We see a power supply whose current output is varied in the range 4 to 20 mA, depending on the varying impedance created by the transducer. Often, the power supply is built right into the transducer housing itself to form a single integrated package, as shown in Figure 2-1. The transducer also contains, of course, a sensor whose output response is varied by the measurand. It is this fluctuating output response that is converted to the varying impedance by the transducer's onboard circuitry. The current flow can be converted very simply to an output voltage by inserting a resistor into the loop as shown in Figure 2-1. The 4-mA current represents the minimum measurand amount (often a zero value), whereas the 20-mA value represents the maximum measurand value. The reason it is desirable to have a current flow, in this case 4 mA, represent a zero measurand measurement is so that the receiving or measuring circuit may continue receiving operating power despite no signal being present. Although this is not obvious in Figure 2-1, the resistor is just part of an indicating circuit whose operation is dependent on the current and voltage supplied by the power supply.

Actually, there are other popular analog systems that are used in addition to the 4- to 20-mA system. There are also 10- to 50-mA systems and 1- to 5-mA systems

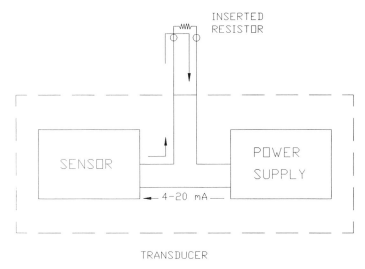

Figure 2-1   4- to 20-mA current loop system.

in use. They function just like the 4- to 20-mA systems; however, the 4- to 20-mA system is probably the most popular one in use at present. However, regardless of the current range being used, there must be a circuit provision onboard the transducer to allow for adjustment of the upper and lower current limits so that the proper span and range are suitable for the measurand span and range. In addition, it is highly desirable to have a transducer that has a very linear output (Figure 2-2). Then it becomes very easy to develop an algorithm using linear components (such as resistors, capacitors, or inductors) such that a simple multiplier circuit can be used to convert the developed current values into actual measurand values. Let's look at an example here using the information given in Figure 2-2.

Example 2-1

Let's say that we will want 1 V (dc) of output from our converter to represent 1°C. Therefore, our converted output (Figure 2-3) will be voltage, whereas our input signal to the converter coming from the transducer's output will be the 4- to 20-mA current signal. The converted output would then be calculated as follows. We would first divide the transducer's output change by its input change as if we were determining the sensitivity of the transducer [see Section 1-9.1, specifically eq. (1-29)]:

$$\frac{\text{output change}}{\text{input change}} = \frac{60 \text{ V dc}}{16 \text{ mA}} \quad \begin{array}{l} \leftarrow 80°C - 20°C \\ \leftarrow 20 \text{ mA} - 4 \text{ mA} \end{array}$$

$$= 3.75 \text{ V dc/mA}$$

Since V dc/mA = kilohms (from Ohm's law), we can, instead, give the answer as 3750 Ω. What this means is this: All we have to do is to install a 3750-Ω resistance (this would have to be a low-tolerance resistor, probably ±1%, since this is not a "standard value" resistor) into our current loop and then measure the occurring voltage drops. The value of each recorded drop would then have the same numerical value as the

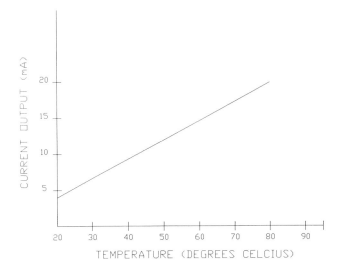

Figure 2-2   Typical output response of a linear transducer.

Figure 2-3   Block diagram showing converter for converting a 4- 20-ma signal from a temperature transducer into an output voltage.

temperature value being measured. The newly installed resistor is now our "multiplication" device since $I \times R = V$.

Admittedly, Example 2-1 is the simplest of examples one might encounter in transducer design, but it does demonstrate a vital principle of unit conversion.

To receive a current analog signal it is only necessary to use a milliammeter that is calibrated within the current range of the transducer. Then you must use a calibration chart to convert the milliammeter readings to the measurand units of measure. A much better method, however, is to use a milliammeter whose scale has been directly marked out in the measurand units of measurement, thereby eliminating the need for conversion charts. In either case, however, the transducer must be designed so that the user can make span and range adjustments of its current output for calibration purposes.

Variable-voltage output systems

In addition to using variable current amounts to represent the variable measurand, the measurand information can be translated into variable voltages. In other words, the sensor's output is converted to a range of voltage values, typically 0 to 1 or 0 to

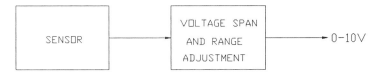

Figure 2-4   Circuit for adjusting output range and span of transducer.

10 V, before being transmitted to the receiving site for interpretation. Figure 2-4 demonstrates this process in simplified form for a transducer device. Similar to the current loop systems, the 0 value of the 0 to 1 or 0 to 10 voltage range typically represents the 0 value of the measurand, whereas the higher voltage value of either of these two voltage ranges represents the upper range limit of the measurand. Also, as in the cases of the current loop systems, adjustment provisions must be present onboard the voltage transducer to allow for proper span and range adjustments so that they can represent the span and range of the measurand.

Reception of analog voltage signals requires a voltmeter with the appropriate input impedance and range to match that of the transducer. Again, as in the case of the analog current loop system, a voltmeter with a direct readout scale calibrated in the measurand units is far handier and easier to use than one requiring conversion charts.

2-1.2 Digital Signals

The generation of digital signals by transducers is a fairly recent innovation in transducer design. There are advantages to this method over the analog methods just described, which we discuss in Section 2-2. The advent of the computer and microprocessor-based process control circuits has had a very large influence on the developing of the digital transducer. Obviously, the fact that a transducer has a digital capability makes that transducer a natural choice for interfacing with a microprocessor chip.

Basically, a digital signal is nothing more than an electrical pulse or a series of electrical pulses that have been modulated with data according to some scheme. These pulses have a finite width, usually measured in micro- or milliseconds, and a pulse height usually measured in millivolts or volts. Some of these modulation schemes can become rather elaborate. The most frequently encountered ones are the ones we discuss below.

Pulse amplitude modulation

As the name implies, *pulse amplitude modulation* (PAM) consists of varying heights or amplitudes of pulses, each pulse height representing a portion of transducer data (Figure 2-5). In those cases where the data are in analog form coming from the sensor, some circuit provision must be made to convert these data into modulated digital data. As an example, a measurand quantity being sensed by a transducer may range from a value of 0 to 10 (0 to 10 in., 0 to 10 dyn, 0 to 10 psi, etc.). This range

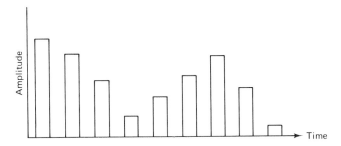

Figure 2-5   Pulse amplitude modulation signal. (From James R. Carstens, *Automatic Control Systems and Components,* copyright 1990, p. 359. Reprinted by permission of Prentice Hall, Englewood Cliffs, NJ.)

may be represented by the transducer's output pulses, whose heights range from 0 to 10 mV (certainly the most straightforward scale to use) or through some other scaled or multiple quantity, such as 0 to 20 mV, 100 to 110 mV, or 0 to 1 V.

To receive a PAM signal, either an oscilloscope can be used or, more conveniently, a receiving circuit that interprets the incoming changing pulse amplitudes and supplies the user with a digital or analog readout display may be used. Two very simple demodulating circuits are shown in Figure 2-6. In these circuits we see a diode that basically acts as an AM detector in an AM radio receiver. The diode "strips" off the amplitude variations (which form an ac signal component) of the incoming digital pulses and converts these variations to a smooth-flowing analog output. Resistor R_1 and capacitors C_1 and C_2 help in this smoothing operation and also act as a low-pass filtering network.

Pulse width modulation

Again, as the name of this modulation scheme implies, the width of each pulse is modulated in accordance with the measurand being detected (Figure 2-7). The amplitude of the pulses remain the same. The receiver or demodulator designed to receive *pulse width modulation* (PWM) must have the capability of measuring the duration of the received pulses and converting that information back into the proper data form. Another name for this type of modulation is *pulse duration modulation* (PDM).

Figure 2-8 shows a block diagram of a typical demodulator used for PWM. The incoming PDM signal is sent to both an integrating circuit and a synch circuit. The integrating circuit sums the total area underneath each pulse to form the trapezoidal forms at 2 in Figure 2-8. The integration is controlled by the incoming signal pulses at 1 through the synchronizing circuit. This circuit also controls the sample-and-hold circuit times. The sample-and-hold circuit then creates the forms at 3, which are processed by a low-pass filtering circuit that "fills in" and smooths the stair-step waveforms coming from the sample-and-hold circuit. The resultant waveform is seen at 4. Here we see a smoothed continuous analog representation of the PWM signal. This waveform should then be a close representation of the original measurand.

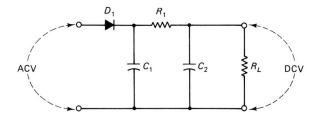

D_1 = Diode
$R_1 C_1 C_2$ = PI Filtering Network for Reducing AC Components Occurring at Output
R_L = Output Load

(a) Half-wave Demodulator

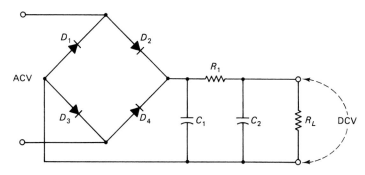

$D_1 D_2 D_3 D_4$ = Diodes Arranged in Full-wave Bridge Configuration
$R_1 C_1 C_2$ = PI Filtering Network
R_L = Output Load

(b) Full-wave Bridge-type Demodulator

Figure 2-6 Simple demodulator circuits for a PAM signal. (From James R. Carstens, *Automatic Control Systems and Components,* copyright 1990, p. 131. Reprinted by permission of Prentice Hall, Englewood Cliffs, NJ.)

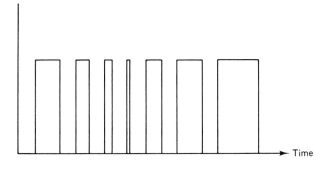

Time

Figure 2-7 Pulse width modulation signal. (From James R. Carstens, *Automatic Control Systems and Components,* copyright 1990, p. 360. Reprinted by permission of Prentice Hall, Englewood Cliffs, NJ.)

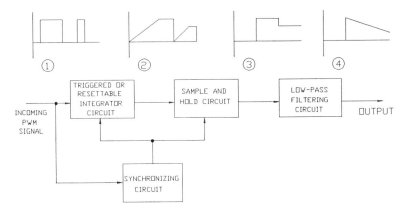

Figure 2-8 Typical demodulator scheme for a PWM signal.

Pulse position modulation

This type of pulse modulation is shown in Figure 2-9. The position of each pulse is measured in units of time relative to a fixed reference mark. The time span is in proportion to the data's magnitude. Figure 2-9 also shows the equivalent analog signal which has created the *pulse position modulation* (PPM).

The receiver for this kind of modulation must have the capability to measure the time differential between each pulse and its reference and then to translate this

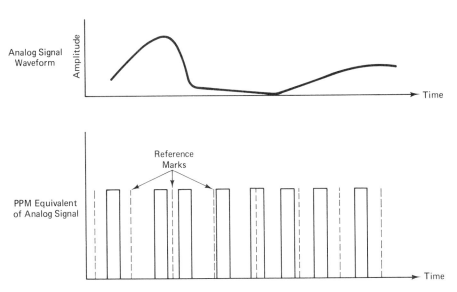

Figure 2-9 Pulse position modulation signal. (From James R. Carstens, *Automatic Control Systems and Components*, copyright 1990, p. 360. Reprinted by permission of Prentice Hall, Englewood Cliffs, NJ.)

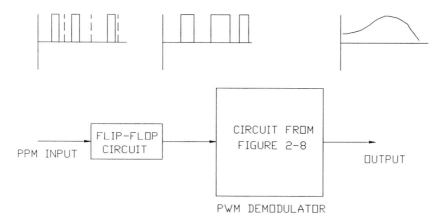

Figure 2-10   Demodulator for PWM modulation signal.

information into data. Figure 2-10 shows such a demodulating scheme. Essentially, this type of demodulator uses a digital flip-flop circuit (either leading-edge or trailing-edge triggering can be used) using the incoming pulse to first trigger the flip-flop and then the reference pulse to retrigger the flip-flop to form one complete pulse. Of course, the time span between the signal pulse and the reference pulse will determine the pulse width leaving the flip-flop. From that point in the demodulator, all that needs to be done is to use a PWM demodulator circuit to continue with the overall demodulation of the PPM signal to produce the final demodulated analog signal.

Pulse frequency modulation

Pulse frequency modulation is a type of modulation where the pulse frequency, that is, the number of pulses per second, are varied in relationship to the measurand. Figure 2-11 shows what a *pulse frequency modulation* (PFM) signal looks like. Notice that the heights and widths of the pulses remain constant; only the frequency changes.

To receive and demodulate a PFM signal a frequency-to-DC current or voltage converter circuit is used. This circuit is shown in Figure 2-12. Basically, a timer chip (a 555 chip is shown) is wired as a one-shot multivibrator. A Schmitt trigger circuit is used to condition incoming variable-frequency waveforms so that rectangular waveforms are received by the multivibrator. The incoming frequency rates of the rectangular waves vary the on/off ratio or duty cycle of the multivibrator in a linear manner (the lower the frequency, the lower the dc output current; the higher the frequency, the higher the output current; etc.) while C_1 and the 0- to 1-mA meter act as integrators to smooth out the varying dc currents. The FET acts as a constant current source in the meter circuit; this makes the calibration current for the meter (adjusted by R_2) relatively independent from the circuit's 5-V supply. This particular

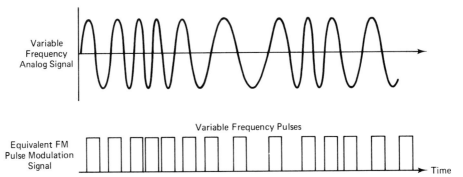

Figure 2-11 Pulse frequency modulation signal. (From James R. Carstens, *Automatic Control Systems and Components,* copyright 1990, p. 358. Reprinted by permission of Prentice Hall, Englewood Cliffs, NJ.)

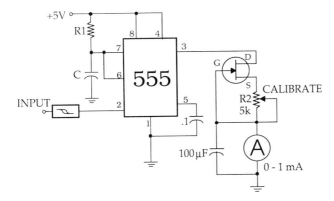

Figure 2-12 Demodulator for PFM signal. (From Howard M. Berlin, *the 555 Timer Applications Sourcebook with Experiments,* copyright 1976, p. 53. Reprinted by permission of Howard W. Sams, Inc., Indianapolis, NJ.)

converter can be calibrated so that the meter will indicate frequencies in the range 0 to 1000 Hz (i.e., 0 to 1 mA on the meter's scale).

This is a very simple and rather crude frequency-to-dc current converter being illustrated, but it does show the basic principle involved for making PFM conversions.

Pulse code modulation

Of all the pulse modulation schemes that we have discussed in this section, this particular scheme is the one modulation type most often used today. It has the most variations, too. One form of *pulse code modulation* (PCM) is shown in Figure 2-13. Here we have a group of pulses that combine to form a binary-coded group for the purpose of representing an *alphanumeric byte*. (An alphanumeric byte is a group of pulses or "bits" whose decimal numerical or alphabetical value is determined by the coding scheme being used. See the discussion below on the ASCII coding system.) Out of the group of four pulses shown in Figure 2-13 (note that the second pulse from the left is missing), the binary number 1011, or decimal 11, is formed. This grouping

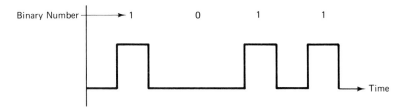

Figure 2-13   Form of pulse code modulation. (From James R. Carstens, *Automatic Control Systems and Components,* copyright 1990, p. 361. Reprinted by permission of Prentice Hall, Englewood Cliffs, NJ.)

system is based on what is referred to as the *8–4–2–1 binary-coded decimal* (BCD) system. Which pulses show up in which of four possible locations within the grouping determines the value of the decimal number. Additional groups of four binary bits can be formed to create larger-magnitude decimal-valued numbers. Table 2-1 is a truth table for the 8–4–2–1 numbering system. A truth table is simply a list of all the desired combinations that one would want to have for a coding or decoding scheme using a given group of symbols, in this case, using the binary bits of 0 and 1. As an example, if we wanted to represent the decimal number 471 using the 8–4–2–1 code, we would write the binary equivalent as 0100 0111 0001. By noting which columns have 1's occurring in them, and then adding the weighted values of those columns together, we can produce the equivalent decimal number value. It is interesting to note that in Table 2-1 we used only 10 of the possible 16 states or combinations that can exist using four columns of binary digits. The remaining six unused states (the states beyond 9) are called "forbidden" states. Should any of the forbidden binary numbers crop up within a transducer output data transmission (e.g., the binary number 1101), these would be considered erroneous data and would automatically be disallowed.

There are many other BCD systems in use in industry, each having its own

TABLE 2-1 TRUTH TABLE FOR THE 8–4–2–1 BINARY-CODED SYSTEM

Decimal number	Weighting code			
	8	4	2	1
0	0	0	0	0
1	0	0	0	1
2	0	0	1	0
3	0	0	1	1
4	0	1	0	0
5	0	1	0	1
6	0	1	1	0
7	0	1	1	1
8	1	0	0	0
9	1	0	0	1

TABLE 2-2 FIVE BIT BINARY-CODED-DECIMAL (BCD) CODES

Decimal	2-out-of-5	63210	Johnson code
0	00011	00110	00000
1	00101	00011	00001
2	00110	00101	00011
3	01001	01001	00111
4	01010	01010	01111
5	01100	01100	11111
6	10001	10001	11110
7	10010	10010	11100
8	10100	10100	11000
9	11000	11000	10000

particular advantage and disadvantage for usage. (Be sure to check the References at the end of this chapter for further information on BCD systems.) Let's now discuss some of the other popular codes used by transducer manufacturers. Table 2-2 lists some of the more common 5-bit codes, and Table 2-3 lists some of the more-than-5-bit codes that are used.

The 2-out-of-5 code has an interesting built-in *parity check* provision. Parity checking is a method used by a special digital circuit built within the transducer that counts all the 1's and 0's to make certain that there is the required total number or that there is the required odd or even number of 1's and 0's present for that particular character. For instance, in the 2-out-of-5 code there are always two 1's. The same is true for the 63210 code. We can add parity checking to the code in Table 2-1 by adding another column of bits. For even-parity checking we can add a bit to those binary numbers required to generate an even number of bits. This is shown in Table 2-4.

Another method used for checking data validity is to use those codes that are *self-complementing*. Table 2-5 contains a self-complementing code, the 5–1–1–1–1 code. To understand how self-complementing codes work, let's first look at the decimal number 3 in Table 2-5 for an example. The *9's complement* of 3 is 6, because

TABLE 2-3 COMMON MORE-THAN-5-BIT CODES

Decimal	50 43210	9876543210
0	01 00001	0000000001
1	01 00010	0000000010
2	01 00100	0000000100
3	01 01000	0000001000
4	01 10000	0000010000
5	10 00001	0000100000
6	10 00010	0001000000
7	10 00100	0010000000
8	10 01000	0100000000
9	10 10000	1000000000

TABLE 2-4   USING EVEN-PARITY CHECKING WITH THE 8–4–2–1 BCD CODE

Decimal	8	4	2	1	Even-parity check bit
0	0	0	0	0	0
1	0	0	0	1	1
2	0	0	1	0	1
3	0	0	1	1	0
4	0	1	0	0	1
5	0	1	0	1	0
6	0	1	1	0	0
7	0	1	1	1	1
8	1	0	0	0	1
9	1	0	0	1	0

the definition of a 9's-complemented number is simply that number subtracted from 9. (If we were interested in finding the 10's complement of 3, we would have subtracted 3 from 10 to get 7.) Now that we have established the 9's complement of 3 to be 6, look at the BCD codes for these two numbers. Notice that the BCD codes are the *1's complement* of one another. (The 1's complement of a binary number is obtained simply by changing all the 1's to 0's and all the 0's to 1's.) Choosing any decimal number from Table 2-5 will produce the same results.

Because of this self-complementing feature found in some BCD codes, a very simple and rapid check can be made on the validity of the data being generated and transmitted by transducer circuits.

Another very popular form of PCM is the *ASCII code*. The letters ASCII (pronounced "askey") stand for "American Standard Code for Information Interchange." Although this code is seldom used in the lesser, expensive transducer outputs because of its complexity, it is often found in some of the very sophisticated and larger transducer systems. The primary intent of the ASCII system is to be used for communications purposes, where data are exchanged between data terminals or computers. Table 2-6 shows the truth table for this particular code. This code is

TABLE 2-5   SELF-COMPLEMENTING 5–1–1–1–1 BCD CODE

Decimal	51111
0	00000
1	00001
2	00011
3	00111
4	01111
5	10000
6	11000
7	11100
8	11110
9	11111

TABLE 2-6 ASCII EIGHT-BIT CODE TRUTH TABLE

BIT 7:				0	0	0	0	1	1	1	1
BIT 6:				0	0 1	1	1	0	0	1	1
BIT 5:				0		0	1	0	1	0	1
BIT 4:	BIT 3:	BIT 2:	BIT 1:								
0	0	0	0	NULL	DC$_0$	b	0	@	P	↑	↑
0	0	0	1	SOM	DC$_1$	!	1	A	Q		
0	0	1	0	EOA	DC$_2$	"	2	B	R		
0	0	1	1	EOM	DC$_3$	#	3	C	S		
0	1	0	0	EOT	DC$_4$ (STOP)	$	4	D	T		U
0	1	0	1	WRU	ERR	%	5	E	U		N A S S I G N E D
0	1	1	0	RU	SYNC	&	6	F	V	U N A S S I G N E D	
0	1	1	1	BELL	LEM	' (APOS)	7	G	W		
1	0	0	0	FE$_0$	S$_0$	(	8	H	X		
1	0	0	1	HT / SK	S$_1$	)	9	I	Y		
1	0	1	0	LF	S$_2$	*	:	J	Z		
1	0	1	1	V$_{TAB}$	S$_3$	+	;	K	[		
1	1	0	0	FF	S$_4$	, (COMMA)	<	L	\		ACK
1	1	0	1	CR	S$_5$	−	=	M	]		ACK (1)
1	1	1	0	SO	S$_6$	.	>	N	↑		ESC
1	1	1	1	SI	S$_7$	/	?	O	←		DEL

Legend:

NULL	Null/Idle
SOM	Start of Message
EOA	End of Address
EOM	End of Message
EOT	End of Transmission
WRU	"Who Are You ?"
RU	"Are You . . . ?"
BELL	Audible Signal
FE$_0$	Format Effector
HT	Horizontal Tabulation
SK	Skip (Punched Card)
LF	Line Feed
V$_{TAB}$	Vertical Tabulation
FF	Form Feed
CR	Carriage Return
SO	Shift Out
SI	Shift In
DC$_0$	Device Control Reserved for Data Link Escape
DC$_1$–DC$_3$	Device Control
DC$_4$ (Stop)	Device Control (Stop)
ERR	Error
SYNC	Synchronous Idle
LEM	Logical End of Media
S$_0$–S$_7$	Separator (Information)
b	Word Separator (Space, Normally Nonprinting)
<	Less Than
>	Greater Than
↑	Up Arrow (Exponentiation)
←	Left Arrow (Implies/ Replaced By)
\	Reverse Slant
ACK	Acknowledge
(1)	Unassigned Control
ESC	Escape
DEL	Delete/Idle

Example: Character "R" is represented by 0100101

Source: James R. Carstens, *Automatic Control Systems and Components,* copyright 1990, p. 364. Reprinted by permission of Prentice Hall, Englewood Cliffs, NJ.)

known as an 8-bit code, whereas only 7 bits are shown in the truth table. This is because the eighth bit is used as a parity-check bit for the other seven.

2-1.3 Carrier Signal Modulation

Carrier signal modulation techniques used by transducers resemble very closely the modulation techniques used for long-distance radio communications. Perhaps the

Figure 2-14   Frequency-modulated waveform. (From James R. Carstens, *Automatic Control Systems and Components,* copyright 1990, p. 354. Reprinted by permission of Prentice Hall, Englewood Cliffs, NJ.)

most often one used for transducer outputs is *frequency modulation* (FM). Figure 2-14 shows what this particular signal looks like. The sensor onboard the transducer is usually the responsible element for creating the sinusoidal waveform. This, in turn, is due to some mechanical rotating member that is part of a frequency generator. The frequency then varies due to the changing of speed of this rotating member, thus creating the frequency-modulated waveform. At the receiver site the receiver then interprets the frequency change amount as the varying magnitude changes in the measurand data.

There are other means of transmitting data through modulated carrier signals as already mentioned above, but when these other means are used it becomes a question of: Where do the transducer hardware boundaries stop and the hardware comprising a full-fledged radio transmitter begin? In the case of the FM technique just described, there is relatively little problem in the packaging of this particular transducer so that it meets our transducer definition criteria as laid down in Chapter 1. However, any of the other techniques, such as amplitude modulation, single- or double-sideband modulation (to mention just a few for those of you familiar with radio transmission techniques) will probably require a separate circuit module external to the transducer, leaving the transducer hardware itself intact; therefore, we do not include these other transmission techniques in our transducer discussions here.

2-2 ADVANTAGES AND DISADVANTAGES OF EACH SIGNAL TYPE

Now that we have discussed all of the transducer signal output types that we plan to cover in this book, we look at the advantages and disadvantages of each.

Advantages and disadvantages of analog current and voltage signals

Analog current and voltage signals have the distinct advantage of being the easiest signals to process and understand. Probably the least amount of hardware is required for the transmitting and receiving of these particular signal types since many sensors are naturally compatible with analog currents and voltages.

Analog current and voltage signals suffer from two major disadvantages: (1) the signals can be transmitted only over relatively short distances, usually no more than 2500 ft or so, without requiring additional signal boosting and repeating (Fig-

Figure 2-15 Analog
telephone-line amplifier.

ure 2-15); and (2) analog current and voltage signals are very susceptible to electrical
noise, either naturally occurring or human-made. Human-made noise is especially
rich in amplitude-modulated signals, which unfortunately mix very well and become
attached to the desired signals that are being transmitted and received. Because of
this, it becomes very difficult to filter out or otherwise remove the unwanted noise
(Figure 2-16). As we will learn in a moment, digital signals have the distinct advantage of being easier to "scrub" of unwanted signal noise.

Figure 2-16 Example of AM-type "noise" affecting a TV picture.

Advantages and disadvantages of digital signals

Some forms of digital signals are readily adaptable to interfacing with computer circuitry. This reason alone has accounted for the great popularity in using this type of signal for transducer work. In addition, as pointed out in the preceding paragraph, digital signals enjoy some immunity to many types of electrical noises. The one possible exception is PAM; we will discuss this further in the paragraph concerning PAM signals below. Otherwise, it is relatively easy to design filtering circuits for the purpose of scrubbing out unwanted analog noise.

There is one big disadvantage in using digital signals for transmission purposes. Although they are relatively impervious to noise, they themselves generate their own fair share of noise that infects other circuits. Since digital pulses are basically square-like waves, these waves contain an infinite number of sine-wave frequencies. (This can be proven through the Fourier expansion of certain mathematical functions; any good integral calculus text will show how this is done.) Mathematically, these sine waves add together to make up the square pulses. This means that each sine wave has the potential to be transmitted or radiated from the wire conductor to cause interference with other radio-sensitive circuits. Unfortunately, this can become a serious drawback to the utilization of digital signals for many data transmission systems. This is especially true for very high data rate transmissions.

Now that we have discussed the general advantages and disadvantages of digital signals, let's analyze the characteristics of the five transducer digital modulation methods that we have discussed.

Advantages and disadvantages of pulse amplitude modulation

Transducers that use PAM, unfortunately, have little of the noise immunity and easy-scrubbing qualities of the other digital signals. This is due to the similarities of PAM signals and ordinary AM carrier transmissions used in radio communications. AM demodulation suffers from poor signal-to-noise differentiation; that is, it has a difficult time telling the difference between data and noise under conditions where the data signal is weak and the background noise is quite strong.

A major advantage in using PAM transducers is their relatively inexpensiveness compared to other types. The modulation circuitry is easy to design and fabricate, as is the receiver or demodulator.

Advantages and disadvantages of pulse width modulation

An advantage of PWM is that it has very good noise immunity properties when compared to analog systems. As pointed out earlier, most human-made and naturally occurring noise tend to be of an amplitude-modulated variety. Consequently, electrical noise contains very little PWM interference.

A major drawback to using PWM, and this is true for any pulse-generating system, is the generating of other unwanted frequencies that interfere with other communications services. This was discussed earlier concerning the theory of square-wave generation. For this reason adequate radio-frequency shielding is a must; all circuitry and transmission lines must contain shielding and proper grounding techniques used.

Another disadvantage to PWM compared to analog signal generation is that pulse-coded systems in general may cost more to design and manufacture. In addition, somewhat more sophisticated demodulating techniques are usually required to decode data than in analog systems.

Advantages and disadvantages of pulse position modulation

PPM has very good noise-immunity qualities and is very similar to PWM in many respects concerning this quality. Perhaps one of the greatest advantages of PPM is that of having *multiplexing* capabilities. Multiplexing is the "mixing" of several different data transmissions from different sources on a single, shared transmission line. Other coded signal forms also allow multiplexing, but PPM and PWM are perhaps the easiest to do. Using PPM as an example, each transducer in a multiplexing setup would have its own particular reference marker and data spacing from that marker. When decoded by the demodulator, a particular range of PWM signals is generated. Each transducer then has its own distinctive pulse width range containing its data (Figure 2-17). Pulse-width-measuring circuits at the receiving site are then easily able to differentiate between the various ranges of pulse widths contained within the multiplexed signal.

A disadvantage of a PPM system is the complexity of hardware needed for demodulating the data signals. However, as integrated circuitry becomes less expensive and more compact, the argument for this being a disadvantage fades.

Advantages and disadvantages of pulse frequency modulation

PFM can be a naturally generating code with many transducers that use mechanical rotating members (Figure 2-18) for reacting with their measurands. Because of this, the amount of design hardware for that transducer can be reduced substantially. Also, PFM has very good noise immunity and can be filtered rather easily.

A disadvantage of PFM is that it does not lend itself very well to the time multiplexing of multiple signal channels. In other words, if several transducers were to share the same transmission line simultaneously for the transporting of all their data, PFM would not work. Data multiplexing depends on having to know the positions of each pulse within the waveform; PFM does not allow this.

Advantages and disadvantages of pulse code modulation

The single major advantage of using PCM is that, by its nature, it allows for provisions for parity error or other error-checking routines. Because of this, PCM can be transmitted over very long distances with virtually error-free capabilities. None of the other coding schemes have these kinds of provisions. Another advantage is the ability to encode alphanumeric commentary along with the data being generated. As a result, considerably more information can be displayed at the transducer's receiving site, making the data easier to interpret.

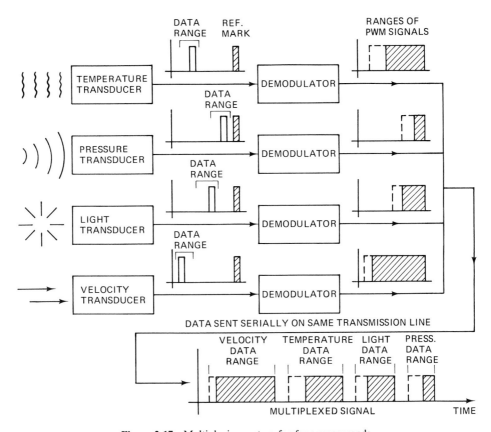

Figure 2-17   Multiplexing system for four measurands.

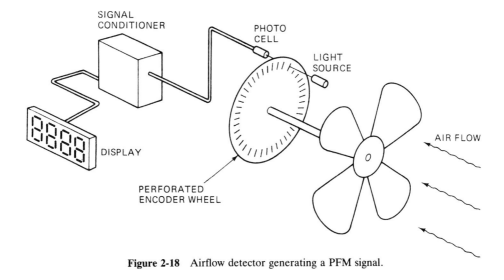

Figure 2-18   Airflow detector generating a PFM signal.

The only disadvantage of PCM is that it requires more encoding hardware onboard the transducer to enable circuitry to convert sensor data into the required PCM information. This additional hardware increases the cost of the transducer. Similarly, at the receiving site, decoding hardware and possible software is also required.

Other digital coding schemes used in transducers

As you can probably guess, there are many, many other digital codes that could be used for encoding and decoding data in a transducer setup. No attempt has been made here to list and describe them all. Only a few representative codes have been described. Commonly, however, modifications are made to the ones we have discussed to make them function better and more reliably. Rather than describe these modifications now, it will be better to discuss them with the appropriate sensors in succeeding chapters. These modifications are usually associated only with certain sensing methods.

2-3 MODIFYING THE SENSOR TO TRANSMIT THE DESIRED SIGNAL TYPE (SIGNAL CONDITIONING)

As it turns out, due to their construction and operating characteristics, many sensors have the capability to output only one kind of signal form, as we will discover in the next several chapters. As is often the case in transducer design, what the sensor is capable of outputting and what is actually desired by the user may be two different things. As a result, many transducers contain modifying circuits or converters to alter the sensor output to the desired form. However, if the transducer has been purchased and the user still wishes to modify its output signal, any modifications usually must be done external to the transducer housing itself.

Table 2-7 lists the modifying circuit needed to convert a transducer output from

TABLE 2-7 CONVERSION CIRCUIT NEEDED TO CONVERT FROM ONE TYPE OF TRANSDUCER OUTPUT TO ANOTHER

To convert:	To:	Use:
Analog voltage or current	BCD	A/D converter
Analog voltage or current	Frequency-varying pulses	Voltage-controlled oscillator
Digital (BCD or Gray code)	Analog voltage or current	D/A converter (plus Gray–BCD converter)
Frequency-varying pulses	Analog voltage or current	Integrating circuit (see Fig. 2-12)
Frequency-varying pulses	BCD	Frequency counter plus scale converter

one form to another. In addition to the conversions above, we may also want to be able to convert a sensor's output into any one of the several digital pulse schemes that we discussed in Section 2-1. However, before we do this, let's discuss the conversion techniques mentioned in Table 2-7.

2-3.1 Converting Analog Voltage or Current to BCD

Analog-to-digital converters are available as integrated circuit modules (Figure 2-19), and consequently, very few additional discrete components are needed to construct the circuit needed to perform this conversion. There are many popular methods used in making A/D conversions in industry.

VCO and frequency counter method

One of the easiest methods to understand is shown in Figure 2-20. Here we see an analog voltage (or this could be a result of an analog current converted to a voltage through a resistance drop) being fed to a voltage-to-frequency converter (whose basis of operation we have yet to explain, but we will do so below). This particular circuit converts the changing voltage levels into proportional changes in frequency pulses. That is, the higher the voltage levels, the higher the frequencies produced; the lower the voltage levels, the lower the frequencies produced. These proportional changes in frequencies are then sent to a frequency or binary counter, where timed units of counting are controlled by an adjustable timing gate. The timing can be

Figure 2-19 A/D converter.

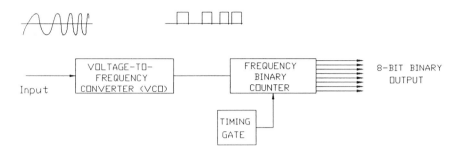

Figure 2-20 VCO used to convert an analog signal to a frequency-varied signal.

selected to produce whatever values are desired to coincide with the input voltage levels; these values then appear at the binary data output terminals.

This method of A/D conversion does have some serious drawbacks, one of which is in accuracy and another in conversion speed. However, because of its simplicity it is popular in many of the smaller transducer processing systems, such as those found in hand-held or portable units.

Flash converter method

Another popular and very accurate A/D conversion technique is the flash converter shown in Figure 2-21. "Flash" comes from the fact that this particular conversion method is extremely fast, typically measuring in the nanosecond region. In this method we see a voltage divider network comprised of a string of precision resistors, each creating a predetermined amount of voltage drop for one of the two inputs to an attached comparator. (Note the supply voltage at the top of the divider network.) The analog signal to be converted is wired to each of the other inputs of each comparator. When a particular comparator goes high due to a matching input voltage, that output voltage is sent to a digital encoder circuit that creates a corresponding binary output signal. The network pictured in Figure 2-21 is capable of generating a 3-bit binary word. In general, you need $2^n - 1$ comparators to create n bits with this system.

2-3.2 Converting Analog Signals to Frequency-Varying Pulses Using the Voltage-Controlled Oscillator

The voltage-controlled oscillator is nothing more than an oscillator whose output frequency is proportional to and is controlled by the input signal voltage. The operating frequencies are typically as low as tens of hertz and can be as high as perhaps tens of thousands of hertz. Figure 2-22 shows a typical circuit. In this circuit we see a current creating a voltage drop across R. This voltage is amplified linearly by an op amp having a gain of $1 \times 10^6 \ \Omega \div 1 \times 10^4 \ \Omega$ or 100. This amplified voltage is sent to pin 5 of a 555 timer chip, where it modulates the frequency of an oscillator circuit whose base frequency is determined by C_1 and R_1.

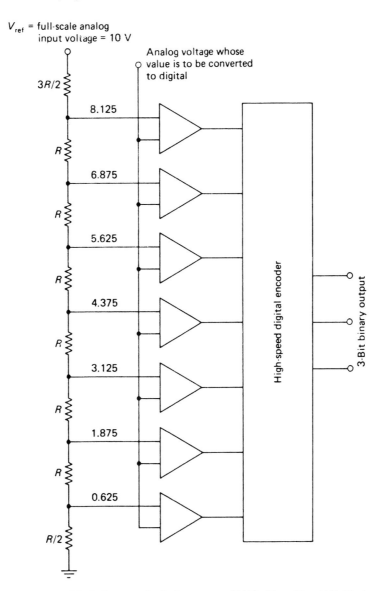

Figure 2-21 Block diagram of a flash converter (3-bit). (From Ronald P. Hunter, *Automated Process Control Systems,* copyright 1987, p. 325. Reprinted by permission of Prentice Hall, Englewood Cliffs, NJ.)

2-3.3 Converting Digital (BCD or Gray Code) to an Analog Signal

We will save the Gray code conversion techniques until later (Chapter 18), where we will have an opportunity to explain what the Gray code is used for. For the

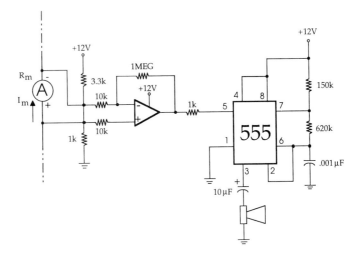

Figure 2-22 Schematic of a frequency converter converting frequency to a variable-frequency output. (From Howard M. Berlin, *The 555 Timer Applications Source-book with Experiments,* copyright 1976, p. 111. Reprinted by permission of Howard W. Sams, Inc., Indianapolis, IN.)

present, however, we will discuss how to convert BCD information into analog voltages.

As in the case of A/D conversion, there are integrated circuits (ICs) available that will perform a D/A conversion all housed into one component. Let's take a look at the circuit in Figure 2-23. We see a binary-weighted resistance ladder with each resistor having a relative resistance value, as shown. Note that these values are relative to the feedback resistor across op amp 6. Op amps 1 through 5 act as amplifiers such that when a binary bit appears at its input from the flip-flop register, the op amp's output triggers an electronic switch, causing the gain of op amp 6 to change accordingly for the supplied reference voltage. The output voltage of op amp 6 will be a summation of whatever switches were activated by the digital signals arriving at its input from the other individual op amps.

2-3.4 Converting Frequency-Varying Pulses to Analog Voltages

This conversion process was described earlier in our discussion of pulse frequency modulation (Section 2-1.2). Also, be sure to refer to Figure 2-12 in that discussion.

2-3.5 Converting Frequency-Varying Pulses to BCD

Figure 2-24 shows a block diagram layout of a circuit that converts frequency-varying pulses into a BCD signal. The figure shows a binary counter outputting a three-decimal equivalent BCD code. Actually, the counter can be purchased on a single

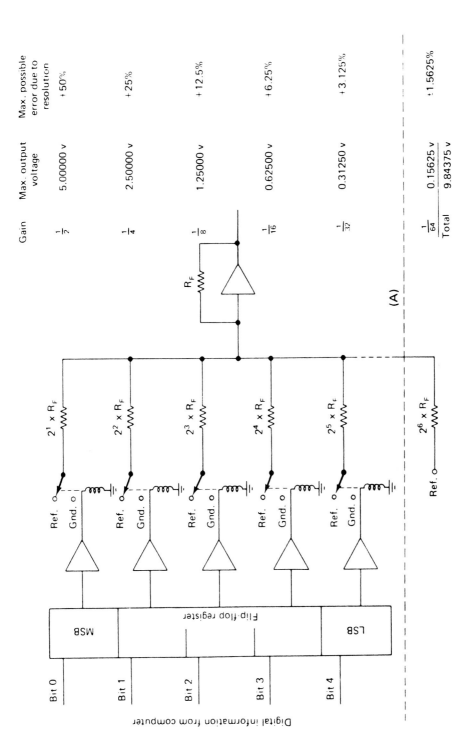

Figure 2-23 Block diagram of a binary-weighted ladder-type D/A converter. (From Ronald P. Hunter, *Automated Process Control Systems*, copyright 1987, p. 309. Reprinted by permission of Prentice Hall, Englewood Cliffs, NJ.)

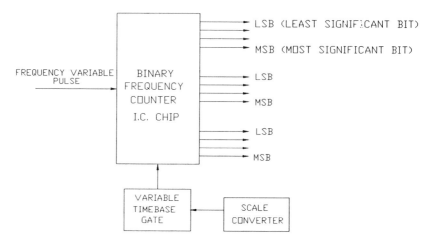

Figure 2-24   IC chip that converts a variable-frequency pulse signal to BCD signal.

IC chip along with a crystal-controlled gating circuit. The scale converter seen wired
to the gate is nothing more than a circuit that varies the time base of the gate so that
the number of cycles being counted each second by the counter can be varied. In
this manner any desired output scale can be obtained to match the transducer's
original measurand values.

2-4 DECIPHERING TRANSDUCER DATA AND PUTTING THEM TO WORK

Assuming at this point that we now understand how signals are modified at their
transducer's output and how they are demodulated or received at the receiving
station or location, we must now understand how these data can be put to practical
use. Transducers may be put to work performing either or all of the following duties:
(1) measurement, (2) display, (3) recording, and (4) control. Each of these functions
is discussed below.

2-4.1 Measurement

The simplest and easiest way to receive transducer data signals is merely to hook up
a recording or display instrument such as a voltmeter, oscilloscope, or ammeter to
the transducer's output and record the readings. Obviously, the instruments used
must be capable of responding adequately to the output's signal quantity. If you do
use this method, it will be necessary to have some sort of calibration curve available
to convert the instrument's readings into the proper measurand amounts. Many
times this calibration information is supplied by the transducer's manufacturer, but
in many other cases you will have to generate the calibration information yourself.
If this turns out to be the case, and chances are pretty good that it will be, you must

FLOW TECHNOLOGY, INC.
4250 East Broadway Rd.
Phoenix, Arizona 85040

RUN WITH PICKOFF HORIZONTAL AND NOMENCLATURE UP.
JOB # 7550

Customer MICHIGAN TECH. UNIV.

GENERAL TRANSDUCER
and
GAS FLOW CALIBRATION
DATA SHEET

Model No. FTM-U10-6?? Size Omniflo Serial No. 8502275 End Fitting 1/8" NPT

Pickoff Type MAG Pickoff P/N 30105 Bearing Type JEWEL

Orifice Size (Omniflo Only) Up .37 inches Down .37 inches

Calibration Method WET TEST METER Calibration Fluid Air

Calibration Pressure ATMOSPHERIC Calibration Fluid Temp. 76

Run No.	FLOW RATE FT³/MIN	Output Frequency (Hz)	PULSES/FT³ "K" Factor
1	1.0084	671	39924.63
2	.7742	500	38749.68
3	.6936	441	38148.79
4	.6417	404	37774.66
5	.5854	367	37615.31
6	.5042	311	37009.12
7	.4317	259	35997.22
8	.3636	214	35313.53
9	.2703	147	32630.41
10	.0857	29	20301.02
11			
12			
13			
14			
15			
16			
17			

Voltage Output @ 30 Hz 20 millivolts

Calibration by Don Nicholas Certified by [illegible]

Calibration Date 11/14/77

Figure 2-25 Typical calibration chart for a transducer. (Courtesy of Flow Technology, Inc., Phoenix, AR.)

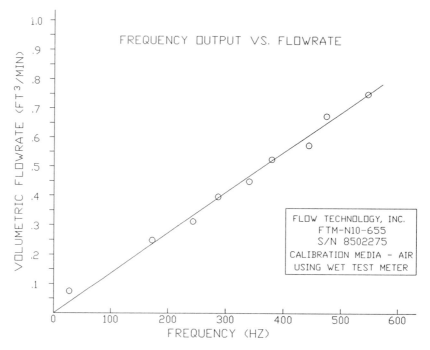

Figure 2-26 Conversion graph for converting frequency to flowrate based on data
in Figure 2-25.

then be able to calibrate the transducer using some reliably known standard for
accuracy comparison purposes.

Figure 2-25 shows a typical calibration chart for a transducer. This particular
chart is for a transducer that responds to fluid flow in a pipe. The transducer
transmits a variable frequency that varies directly (but not proportionally in this
case) to the fluid flowing through it. The volumetric flow rate can then be interpreted
by reading the transducer's output using a frequency counter since the calibration
curve makes these data available. These data are plotted as a curve in Figure 2-26.

2-4.2 Display

In portable or hand-carried instruments that require the use of transducers, some
sort of direct readout display is usually built into the instrument, making hand
conversion of transducer output data unnecessary. This display can be either an
analog meter (Figure 2-27) or a digital readout display such as the one shown in
Figure 2-28. In either case the display usually is not an integral part of the transducer
itself but rather is housed in a separate unit, although there is no firm rule governing
this.

Figure 2-27 Typical analog display on a measuring device. (REPRODUCED WITH THE PERMISSION OF OMEGA ENGINEERING, INC.)

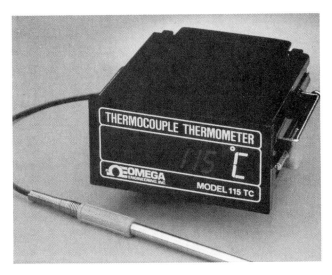

Figure 2-28 Typical digital display on a measuring device. (REPRODUCED WITH THE PERMISSION OF OMEGA ENGINEERING, INC.)

2-4.3 Recording

Transducers may also have their outputs directed to the graphics display of a computer for direct readout or they may even drive the output of a chart recording system. In the case of the computer, appropriate software is needed to interpret the transducers' data and to make any appropriate conversions and preparations necessary for display purposes.

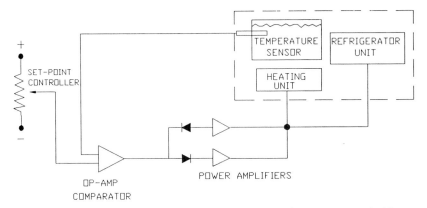

Figure 2-29   Temperature controller used for controlling temperature inside a liquid-filled vat.

2-4.4 Control

In many instances the transducer signals may be coupled directly into a system for control purposes. Transducers are used extensively for process control applications where a measurand is sensed, a control signal is generated, and some controlling action is taken on the measurand to control its behavior. A good example is a temperature controller used for monitoring and controlling the temperature of a liquid-filled vat similar to the one shown in Figure 2-29. The sensing transducer would be housed inside the vat. The sensor would then send its temperature signal to a comparator circuit, where the voltage data would be compared to the desired set-point controller temperature (i.e., the thermostat) voltage. The difference between these two voltages would then result in an error signal that would, after proper amplification, drive either the vat's heating unit or refrigeration unit in a proportional manner. The greater the error signal's positive voltage value, the hotter the heating unit would become; the greater the error signal's negative voltage value, the more rapidly the vat would be cooled by its refrigeration unit.

Up to this point in our exploration of transducers we have discussed, in very general terms, the physics needed to understand sensors. We have discussed the various measurands, the methods used for evaluating transducers, and the means to transmit the generated data along with how these data are received and used. We have completed our treatment of the transducer as a black box. We are now ready to delve into these boxes to see what is taking place and how they manage to do what they do. We start this investigation beginning with Chapter 3.

REVIEW QUESTIONS

2-1. Explain 4- to 20-mA analog standard data transmission.

2-2. Explain the difference between PWM and PAM.

2-3. Why is AM not used for data transmission?

2-4. Why are FM transmissions not as prone to electrical noise interference as AM transmissions?

2-5. Why can an AM demodulator not detect a PWM signal? (*Hint:* Can an AM detector "hear" a CW signal? Why or why not?)

2-6. Explain in detail how a PFM signal is demodulated. Use a block diagram if necessary.

2-7. What advantage is there in using a Gray code for a rotary encoder versus using an 8–4–2–1 BCD code?

2-8. Explain what is meant by the term *parity checking*.

2-9. What is a voltage-controlled oscillator? Give an example of how one is used.

2-10. What is a flash converter? What are their advantages?

PROBLEMS

2-11. A particular transducer, when used in conjunction with a certain power supply, produces an output current between 5 and 50 mA. The transducer has a range of 0 to 150 psi. What is the pressure being measured for a current reading of 38.5 mA from this transducer?

2-12. A transducer purchased from a catalog is capable of indicating length measurements between 10 and 150 mm. If its output range is 1 to 4 V dc, what sort of voltage scale resolution is needed (i.e., what is the smallest voltage increment that has to be read) in order to read to the nearest millimeter?

2-13. Determine the sensitivity of the sensing device whose response curve is depicted in Figure 2-2.

2-14. Draw a complete flow diagram of a PPM demodulator that uses a PWM demodulation scheme. (*Hint:* refer to Figure 2-10.) Explain the functions of each block that you decide to include for each circuit portion, using waveforms where necessary.

2-15. Refer to Table 2-4. Determine the odd-parity check values for each of the ten 8–4–2–1 BCD codes listed. Explain how an odd- or even-parity check system can be devised to check the validity of binary data emitting from a transducer. Would this system be foolproof? Explain your reasons.

REFERENCES

CARSTENS, JAMES R., *Automatic Control Systems and Components*. Englewood Cliffs, NJ: Prentice-Hall, 1990.

DUNCAN, FRANK R., *Electronic Communications Systems*. Boston: Breton, 1987.

LEACH, DONALD P., and ALBERT PAUL MALVINO, *Digital Principles and Applications*. New York: McGraw-Hill, 1981.

MILLER, GARY M., *Modern Electronic Communication*. Englewood Cliffs, NJ: Prentice-Hall, 1988.

The Radio Amateur's Handbook, 60th ed. Newington, CT: American Radio Relay League, 1983.

VERGERS, CHARLES A., *Handbook of Electrical Noise: Measurement and Technology*. Blue Ridge Summit, PA: TAB Books, 1979.

3

SONIC SENSORS

CHAPTER OBJECTIVES

1. To understand how sound is quantified.
2. To understand how microphones work.
3. To study sonic detection methods.

3-1 INTRODUCTION

A general review of the physics of sound was given in Section 1-5.3. We will review additional properties of sound as we begin discussing specific sensors.

Sonic sensing, as it is referred to in this book, refers to any type of sensor that has been specifically designed to respond to sound, whether that sound is subaudible (i.e., below about 16 Hz, which is the lower limit of the unimpaired human ear), audible (between 16 Hz and 20 kHz, which is considered the hearing range of the unimpaired human ear), or superaudible (above 20 kHz). The term *ultrasonic* has in recent years replaced the expression "superaudible." Sound is comprised of periodically spaced fronts of modified atmospheric pressure, each front being either slightly above or below the existing ambient atmospheric pressure. These fronts can easily be detected using a flexible diaphragm or other free-moving member of fairly broad surface area designed so that it is free to vibrate. The approaching wavefronts then impinge on this diaphragm, causing movement or vibration to take place. As it vibrates, however, some method has to be employed to sense the diaphragm's movements as it responds to these fronts. It is these methods of detection that create new categories of sound-sensing microphone transducer.

In general, then, we can state that the purpose of the microphone transducer is to convert sound to an electrical signal that is proportional to that sound's intensity. This type of sensor is used primarily for the use of audio reproduction applications and is used extensively in the radio, recording, and television industries.

Another category of sonic detection is one that is used not so much for the detection of the sound measurand but rather for the measurement of distance and motion in water. This category of sound detection is called *sonar detection*. *Sonar* is an acronym meaning *sound navigation and ranging*. Sonar was developed and used extensively during World War II for the purpose of detecting enemy submarines.

3-2 COMMONLY SENSED MEASURANDS

The measurands most often sensed using sonic sensors are the following: (1) sound, (2) position (i.e., distance), (3) motion (i.e., velocity), and (4) flow (usually liquids). Sound, the primary measurand, is detected by using what is commonly referred to as a *microphone*. It is the transducer with which most of us are most familiar. The other three secondary measurands are detected by using some form of sonar device. It has been only in the last few years that much research and development has been done in the area of sonar detection, especially in the areas of motion and flow detection. We investigate next all four of these areas, beginning with the detection of sound.

3-3 HOW SOUND INTENSITY IS MEASURED

Before getting into the sensing methods used for sound detection, we must determine how sound levels are measured. To begin with, sound intensity is usually expressed in decibels by making a comparison to a standard reference intensity. The amount or magnitude of this intensity is stated as a sound pressure level given in decibels and defined as

$$dB = 20 \log \frac{p}{p_0} \qquad (3-1)$$

where dB = sound intensity expressed in decibels (dB)

  p = sound pressure front created by a sound source (expressed in the same units of pressure as the reference source: microbars, lb/in^2, dyn/cm^2, etc.)

  p_0 = sound pressure created by reference source, usually stated as 0.0002 μbar or 0.0002 dyn/cm^2

Example 3-1

It was found that a power hand tool generated a sound pressure of 0.030 μbar. Convert this measurement into decibels by comparing the sound to a standard sound source.

Solution: Using the standard sound source value of 0.0002 μbar and eq. (3-1), we obtain

$$dB = 20 \log\frac{P}{P_0} = 20 \log\frac{0.0300}{0.0002} = 20 \log(150)$$

$$= 20 \times 2.176$$

$$= 43.5 \text{ dB}$$

This means that a sound pressure of 0.0300 μbar generates a sound with a loudness value of 43.5 dB, compared to a standard sound source of 0.0002 μbar.

The standard reference pressure most often used for making sound comparisons is 0.0002 dyn/cm² (or 0.02 μbar). This amount of pressure change is just barely detectable by the average human ear and is equivalent to 0 dB as produced by a 1000-Hz signal. In other words, a pressure change of this magnitude will just produce an audible-level change in hearing. Now that we have defined sound intensity, let's look at the variety of ways that it is used to detect sound.

3-4 AUDIO REPRODUCTION: THE VARIABLE-RESISTANCE MICROPHONE

The variable-resistance microphone, often referred to as a *carbon microphone*, is pictured in Figure 3-1. The carbon microphone is probably the oldest method of sound detection in use today. The method was developed and perfected by Thomas Edison sometime before 1900.

3-4.1 Theory of Operation

The variable-resistance microphone gets its name from the fact that its electrical dc resistance changes in accordance with incoming sound waves that impinge on an edge-supported diaphragm. A schematic of this type of system is shown in Figure 3-2. The sound waves, which are initially captured by some sort of acoustical enclosure that concentrates the sound onto a diaphragm, cause the diaphragm to vibrate. This vibrating motion is transferred to an attached insulated container or "button" containing loosely packed granules of powdered carbon. Attached to the carbon container are two electrodes through which a small dc current is passed. This current is supplied by a battery (*E* in Figure 3-2). Because of the vibrations now transferred to the granules, the granules are caused to compress and decompress, causing, in turn, a proportional increase and decrease of current flow through the battery's circuit. This varying current flow is an analog representation of the sound

Figure 3-1 Carbon microphone.

waves impinging on the microphone's diaphragm. If a load such as a resistor, or more typically, a telephone receiver or amplifier, is placed in series with the battery and microphone, this analog current would produce a proportional voltage drop across that load. In the case of the telephone receiver, the varying voltages or current would be converted back into an accurate sound reproduction of the original sound. Figure 3-3 shows a cross-sectional view of a modern carbon microphone. It is important to compare this illustration with the schematic shown in Figure 3-2 so that all the important features of this microphone can be identified.

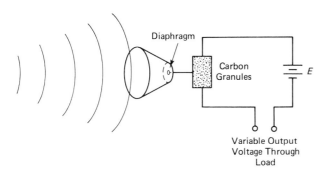

Figure 3-2 How the carbon microphone works. (From James R. Carstens, *Automatic Control Systems and Components,* copyright 1990, p. 96. Reprinted by permission of Prentice Hall, Englewood Cliffs, NJ.)

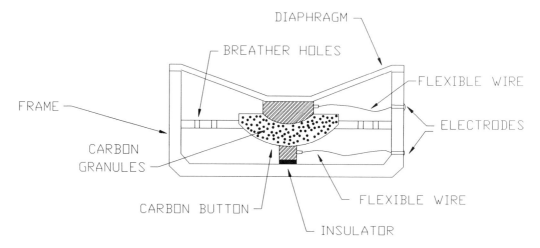

Figure 3-3 Cross-sectional view of a carbon microphone.

3-4.2 Characteristics

The carbon microphone was extremely popular in World War II for communications applications. It was easy to manufacture, inexpensive, and it had acceptable sound characteristics despite its simplicity. Also, the electrical signal it produced required little amplification because of the direct variation of resistance in the battery line, therefore producing a high-signal-level output. It is because of this last characteristic that the Bell Telephone Company used this type of microphone in so many of their earlier telephone models.

The microphone's limited audio-frequency response, a habit of generating electrical noise or interference, and audio distortion are all undesirable features compared to other microphone types. It was these faults that encouraged development of other microphone types.

The carbon microphone is a self-amplifying device. That is, it can control relatively large amounts of current flowing through its granular carbon structure with relatively little audio power input at its diaphragm. It is this property that has made the device simple to install, understand, and use.

Figure 3-4(a) shows a typical frequency response curve that you can expect from a carbon microphone. Notice, first, the nonlinearity of the curve. Second, the response range is only from about 100 to 4000 Hz. Third, the position or attitude of the transducer adversely affects the response characteristics. If the microphone were tilted at a 45° angle and a 90° angle, the curves in Figure 3-4(b) and (c) would probably occur. Obviously, tilting the microphone alters its frequency response and will cause significant alterations in gain where the gain varies by more than 3 dB as the frequency changes.

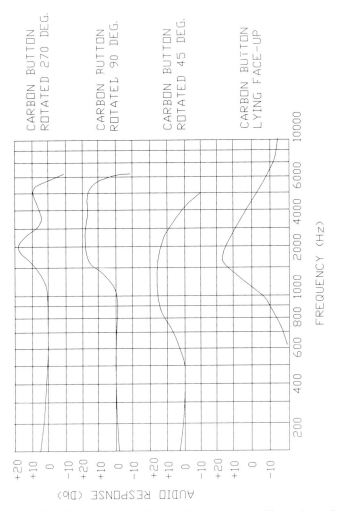

Figure 3-4 How the carbon microphone varies its response with angular position.

To understand the reason for this third undesirable characteristic, we have to look again at its mechanical construction. In the carbon granule area shown in Figure 3-3, the chamber holding the granules is filled to over 90% of its volume (although this is not evident in the illustration). This means that the granules are somewhat loosely packed and free to move about (Figure 3-5). Depending on the orientation of the chamber, a significant change will be made in the amount of contact between the two electrodes and the carbon granules; these changes in contact area bring about the various changes in resistance and gain response. As a result of this peculiarity of carbon microphones, they should not be tilted more than 45° from their vertical orientation.

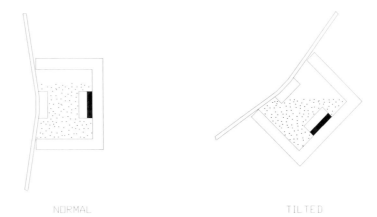

NORMAL   TILTED

Figure 3-5   How the carbon chamber varies with angular tilt of microphone.

3-4.3 Typical Construction

The carbon used in the carbon chamber is pure anthracite carbon. This material must be entirely moisture free and free of impurities. Unfortunately, after extended use, the granules tend to wear down and become "gassy." This is due primarily to abrasion created by vibration and because of minute arcing between individual granules. It is the latter problem that is responsible for the characteristic "bacon frying" sound that can often be heard in the audio background of a carbon microphone.

The carbon chamber is usually constructed of one convex and one concave electrode to form the curved chamber depicted in Figure 3-5. Through experimentation it was found that this style of chamber minimized granule packing, therefore reducing nonlinearity and consequently, the audio distortion.

The electrode or contact surfaces within the carbon chamber are made from a carbonized metal material. Gold, silver, and platinum were all used at one time but obviously increased the expense of the microphone. It was finally determined that the best operation was achieved using carbon electrodes, or more recently, carbonized metal. Their performance was even superior to that of the more expensive gold contacts.

3-4.4 Practical Applications

The largest users of carbon microphones today are the telephone companies, which use them as the transmitting element in telephone hand sets. The limited frequency response of carbon microphones lends them very well to use in speech transmission, which is typically in the range of 200 to 7000 Hz (although for telephone transmission a range of 300 to 3000 Hz is often the criterion used). Any frequencies outside this range would be greatly attenuated or would not be transmitted at all, thereby creating a desirable naturally occurring noise-canceling circuit.

3-5 AUDIO REPRODUCTION: THE MOVING-IRON MICROPHONE

The moving-iron microphone, sometimes referred to as the *variable-reluctance microphone*, is shown schematically in Figure 3-6. This style of microphone was very popular many years ago but has, for the most part, been replaced by other types. The variable-reluctance microphone was first used by Alexander Graham Bell. It was easily fabricated and worked quite well for his purposes. This type of microphone was used by the Bell Telephone Company in various forms, but in later years the company settled on the carbon microphone for use in the majority of its telephone manufacturing.

3-5.1 Theory of Operation

To understand the theory of operation, let's look first at Figure 3-7(a). We see a typical magnetic circuit, or electromagnet, comprised of an iron path, a sheet steel path that is wound with a coil having N number of turns, and an air gap. The coil is supplied with a voltage, E, causing a current, I, to flow through it. As a result of this current supply a flux, Φ, is set up within the electromagnet as shown. The strength of this flux is determined by the individual "resistances" or reluctances that

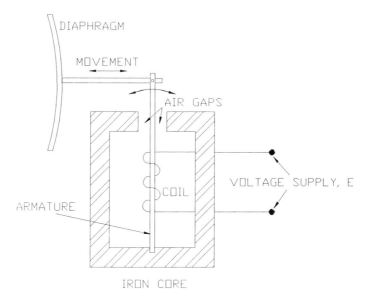

Figure 3-6   Schematic of moving-iron microphone.

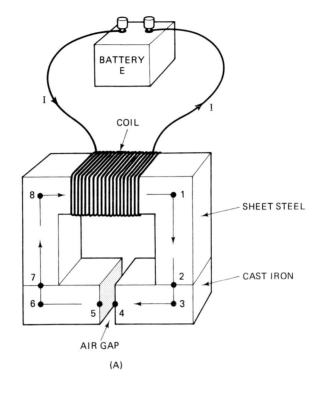

COIL

SHEET STEEL

CAST IRON

AIR GAP

(A)

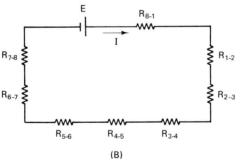

(B)

Figure 3-7 Magnetic circuit with its dc analogy.

occur between each of the numbered locations around the magnetic circuit. The greatest reluctance is experienced between points 4 and 5, the air gap.

Figure 3-7(b) shows the dc circuit analogy for the magnetic circuit shown in Figure 3-7(a). Applying Ampère's circuital law for magnetic circuits (which is identical to his circuital law for electric circuits), one obtains

$$NI = H_{8\text{-}1}\,l_{8\text{-}1} + H_{1\text{-}2}\,l_{1\text{-}2} + H_{2\text{-}3}\,l_{2\text{-}3} + H_{3\text{-}4}\,l_{3\text{-}4} + H_{4\text{-}5}\,l_{4\text{-}5}$$

$$+\ H_{5\text{-}6}\,l_{5\text{-}6} + H_{6\text{-}7}\,l_{6\text{-}7} + H_{7\text{-}8}\,l_{7\text{-}8} \tag{3-2}$$

where NI = number of turns in coil times the current in coil in amperes (ampere–turns or At); this is also called the magnetomotive force (MMF)

H = magnetizing force per unit length of the conductor or magnetic path (At/m); see the note below

l = path length (m)

Note: To find H, it is first necessary to find the flux density, B, expressed in webers/m^2. This is found by noting that $B = \Phi/A$, where A is the cross-sectional area of the conductor through which Φ is flowing. Once Φ is found, a B–H curve for the material in question can be used to find H. These curves can be obtained from reference books on magnetic circuits. A typical curve is shown in Figure 3-8. Solving for I, we get

$$I = \frac{(H_{8\text{-}1}\, l_{8\text{-}1} + H_{1\text{-}2}\, l_{1\text{-}2} + H_{2\text{-}3}\, l_{2\text{-}3} + H_{3\text{-}4}\, l_{3\text{-}4} + H_{4\text{-}5}\, l_{4\text{-}5} + H_{5\text{-}6}\, l_{5\text{-}6} + H_{6\text{-}7}\, l_{6\text{-}7} + H_{7\text{-}8}\, l_{7\text{-}8})}{N}$$

$$(3\text{-}3)$$

As can be seen from eq. (3-3), the magnitude of I in the coil can be changed by varying any of its "$H \times l$" or reluctance values. As pointed out earlier, these values are the individual reluctance values between the numbered locations on the magnetic circuit shown in Figure 3-7(a).

We now apply what we have learned in the magnetic circuit just discussed to the circuit shown in Figure 3-6. We note, first, the moving diaphragm attached to the flexible armature around which a coil is wound. The coil is wired to a voltage source, E. As was the case of our original magnetic circuit in Figure 3-7(a), the voltage source causes a current to flow in the windings of the coil. However, unlike the previous circuit, the coil is now constructed around a flexible armature whose base is attached to the main iron or magnetic circuit. The voltage across the coil sets up a flux that circulates through the iron core, across the two air gaps, and back to the coil. In addition, the iron ends at the air gaps become polarized, acting as magnets. The polarities of these ends will be determined by the flux direction inside the iron circuit. Note, too, that there are two paths of circulation, parallel to each other. The largest reluctance in either circuit will be that of the air gap. Since this gap will vary according to the position of the armature, the reluctance will vary proportionally. This, in turn, will cause a proportional change to be reflected in the current being supplied to the armature's coil. Sound that enters the diaphragm will create the armature movements needed to produce fluctuations in the coil's current. As a result, the current flowing in the power supply will have an ac component representing the output signal superimposed on the dc current supply.

3-5.2 Operating Characteristics

The typical frequency response curve for a moving-iron microphone is somewhat restricted and quite similar to that of the carbon microphone. However, unlike the carbon microphone, the moving-iron microphone is quite stable. In other words, it

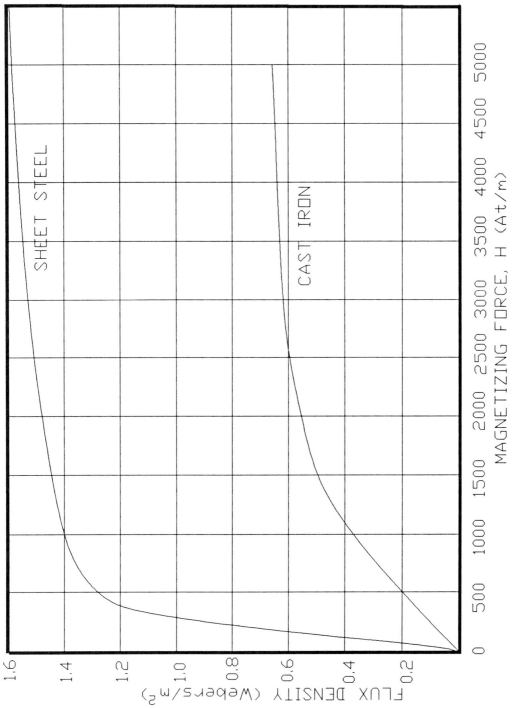

Figure 3-8 Typical *B–H* curves.

will perform equally well in any position. In addition to this, it can operate under high-vibration conditions. However, one drawback to the microphone is that its impedance output is highly reactive, due to the armature's coil. This could present a problem in some electrical circuits if proper circuit modifications are not made.

3-5.3 Typical Construction

Probably the most common style of construction for the moving-iron microphone is that shown in Figure 3-9. The armature is in the shape of a ring that has the diaphragm mounted inside. The outer edge of the ring is held rigidly by a surrounding iron core, whereas the inner ring portion is allowed to flex with the diaphragm. Mounted beneath the armature is a circular coil with its supplied voltage source. The amount of current flowing in the coil is then varied by the diaphragm movement at the air gaps, the movement, of course, being created by the sound.

3-5.4 Practical Applications

Because of the way the magnetic circuitry functions in the moving-iron or reluctance microphone, you can reverse the operation of this system and make it into an audio speaker system. Instead of modulating the current flow inside the armature coil

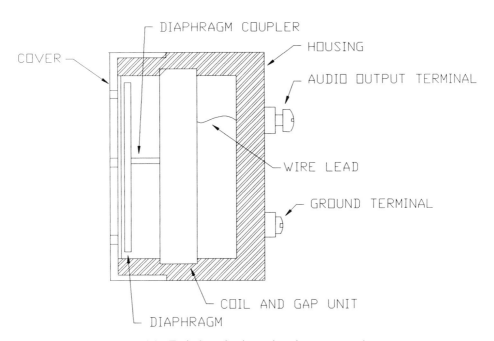

Figure 3-9 Typical moving-iron microphone construction.

as you would normally do with the variable-reluctance microphone, an audio-modulated voltage is supplied to the coil instead, causing variable magnetic fields to be set up across the iron circuit paths at the air gaps. These variable fields would then cause the diaphragm to respond to the varying attraction of the magnetic field, vibrating at the desired audio-frequency rates, causing, in turn, a re-creation of the sounds detected originally. It is this reversible transducer nature that has made the moving-iron microphone popular in many two-way intercom and telephone system designs.

3-6 AUDIO REPRODUCTION: THE MOVING-COIL MICROPHONE

In many respects, the moving-coil microphone looks and acts very much like the moving-iron microphone. Another name for this type of microphone is *dynamic microphone*. Like the moving-iron microphone, this one can also be reversed to act as a speaker system. We now investigate its theory of operation.

3-6.1 Theory of Operation

The dynamic microphone works on the theory of an electrical generator. To illustrate this, look at Figure 3-10 to review our generator theory. The right-hand rule reminds us that if we point our thumb in the direction of travel of a moving wire inside a

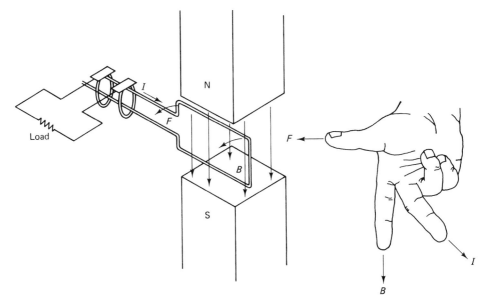

Figure 3-10   Review of electrical generator theory. (From James R. Carstens, *Automatic Control Systems and Components,* copyright 1990, p. 23. Reprinted by permission of Prentice Hall, Englewood Cliffs, NJ.)

magnetic field (F) and point our index finger in the direction of the magnetic flux (B), the current (I) generated within the wire would travel in the direction of the middle finger. Figure 3-10 shows a simplified schematic of the generator while illustrating the principle of the right-hand rule.

If we now place a moving coil of wire into a magnetic flux field, the amount of generated voltage would be determined by Faraday's law, which states that

$$e = N\frac{d\Phi}{dt} \tag{3-4}$$

where e = induced voltage (V)

$\quad\quad N$ = number of turns of wire in coil

$\quad\quad d\Phi$ = increment of change in flux, Φ

$\quad\quad dt$ = increment of change in time, t

A schematic of a moving-coil microphone is shown in Figure 3-11. The coil, which is wound around an armature, is also attached to a diaphragm. The coil itself is hooked to a load, such as an audio amplifier; in the figure a resistor is shown to simulate this load. Attached to the coil is a diaphragm, so that as the diaphragm vibrates in response to the sound pressure, the coil will also vibrate. Surrounding the coil is a magnetic field produced by a large permanent magnet. As the coil vibrates, the turns of wire in the coil cut the lines of magnetic flux, causing a current to become induced within the coil. The amount of the induced current will be in direct proportion to the velocity of the vibrations, as implied in eq. (3-4). (Remember that the derivative of a quantity written with respect to time is an expression of the velocity of that quantity.)

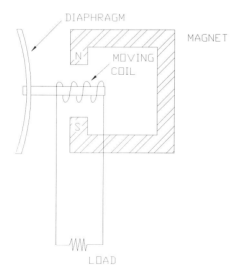

Figure 3-11 Schematic of a moving coil or dynamic microphone.

3-6.2 Operating Characteristics

Figure 3-12 shows a typical frequency response curve for a moving-coil microphone. Comparing this curve to that of Figure 3-4 for the carbon microphone, you can see the improvement in frequency response. Dynamic microphones can have response curves that extend well beyond 20,000 Hz, exceeding the upper range limit of the human ear. It is important to remember, though, that some reshaping of the response curve is possible through design of the acoustical cavity containing the microphone's diaphragm. For the most part, however, the comparison just made between these two response curves is valid.

      One of the benefits of using the moving-coil microphone is that it is a self-generating transducer. That is, it does not require a power supply for its operation, as is the case with some of the other styles. Power supplies can add considerable weight to a microphone, especially a hand-held model. Consequently, the dynamic microphone is comparatively light in weight. Because the output of a dynamic microphone is from a coil, its characteristic output impedance is highly reactive inductively, much like the moving-iron microphone.

3-6.3 Typical Construction

Figure 3-13 is a photograph of several dynamic microphone styles, each style depending on its application. Because of the suitability of the dynamic microphone to cover such large spans of frequencies, it is used in a wide variety of applications. It has

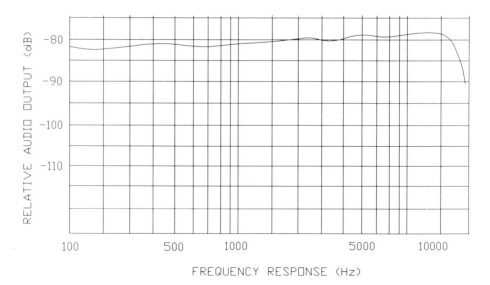

Figure 3-12   Frequency response curve for a moving-coil microphone.

Figure 3-13   Typical dynamic microphone.

replaced the moving-iron microphone mainly in applications where the microphone doubles as a speaker phone. In addition, while the dynamic microphone is not considered to have extraordinary sensitivity, it is more sensitive than the others we have discussed so far. Because of its sensitivity, the size of this type of microphone has been reduced tremendously over the years. It no longer requires the large acoustical cavities associated with older microphones.

3-6.4 Practical Applications

We have already mentioned the fact that moving-coil microphones have relatively good sensitivity. Dynamic microphones are used in many radio and television studios for the transmission of studio programming information. Figure 3-14 shows such an application in a radio communications setup. We see a popular two-way intercom system in Figure 3-15, where the dynamic microphone doubles as a speaker to allow for the two-way exchange of audio information between remote locations in a building or with a moving vehicle.

3-7 AUDIO REPRODUCTION: THE ELECTROSTATIC MICROPHONE

The electrostatic microphone, otherwise known as a *condenser* or *capacitor microphone*, was invented in the mid-1920s by E. C. Wente. This type of microphone was a tremendous improvement over the carbon and moving-iron microphones in existence at the time. It was also found that the capacitor microphone was capable of

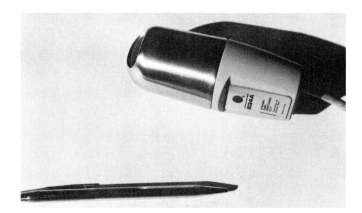

Figure 3-14   Example of studio microphone.

generating very stable frequency response curves that were quite repeatable and dependable. This made the microphone capable of being a precision calibration device for both studio and laboratory use, something that the other microphone styles were, for the most part, incapable of doing.

3-7.1 Theory of Operation

As the name implies, a capacitor microphone's means of operation is centered around the function of a capacitor. One plate of the capacitor is formed by the speech

Figure 3-15   Two-way intercom system. (Courtesy of Heath Co., Benton Harbor, MI.)

diaphragm, while the other plate is the back plate of the microphone, forming a parallel-plate capacitor (Figure 3-16). In operation the capacitor contains an electrostatic charge between its plates developed by an onboard power supply, typically in the range of 50 to 100 V dc. As the diaphragm vibrates due to sound pressure, the distance between the two capacitor plates will vary, producing a variable-capacitance output. In general, eq. (3-5) shows the relationship that exists between capacitance, C, plate separation, d, plate area, A, and the dielectric constant, ϵ.

$$C = \frac{A\epsilon}{d} \qquad (3\text{-}5)$$

where C = capacitance [farads (F)]

   A = plate area (m²)

   ϵ = dielectric constant (8.85×10^{-12} F/m)

   d = distance (m)

Having established now that capacitance, C, varies inversely with the distance of plate separation, d, eq. (3-6) shows the relationship that exists between the capacitance and the voltage existing between the plates for a given charge on the capacitor's plates:

$$V = \frac{Q}{C} \qquad (3\text{-}6)$$

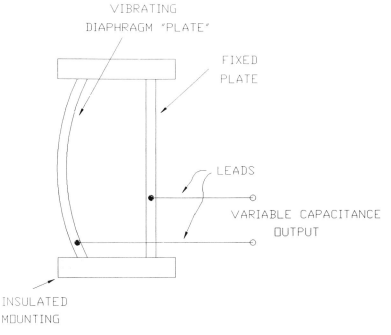

Figure 3-16   How the capacitor microphone works.

where V = voltage between plates (V)

Q = electrical charge (C)

C = capacitance (F)

We can now see that as the distance varies between the sensing diaphragm and the back plate, the output voltage of the capacitor will also vary in proportion to this distance.

A special type of capacitor microphone has been developed to overcome the necessity of having an onboard power supply for supplying the needed charge between the two capacitive plates. This type of microphone utilizes a "precharged" plastic membrane coating for the one stationary plate, a charge that stays intact for very long periods of time (typically, years). The diaphragm is also plastic coated, so that the two combined coatings, together with the airspace in between, form the dielectric material for the capacitor's plates. This type of microphone is called an *electret microphone*. Figure 3-17 shows the construction of this special type of microphone.

3-7.2 Operating Characteristics

Figure 3-18 shows a typical frequency response curve for the capacitor microphone. Note the extensiveness of the response curve and how flat the curve appears. This is typical of this microphone type and has given the capacitor microphone a very good reputation for audio fidelity.

The electrical output of the capacitor microphone is characterized by having

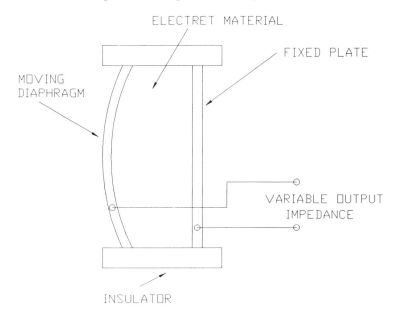

Figure 3-17 Schematic of an electret condenser microphone.

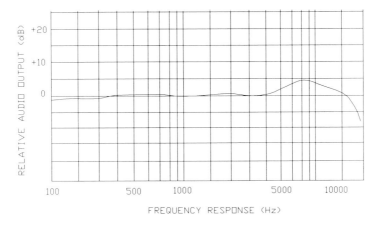

Figure 3-18   Frequency response curve for a capacitor microphone.

a nominal high impedance—much higher than the others that we have discussed so far. The term *nominal* is important here because obviously the actual impedances will vary greatly according to the sound frequencies. In the case of the capacitor microphone, these impedances could range upward into the thousands of ohms, whereas in many other microphone types, the impedances would be in the tens of ohms only. The capacitor in this microphone is rated at about 50 pF, so when you calculate the impedance (or reactance, assuming little or no resistance associated with the capacitance) at a nominal frequency of, say, 500 Hz, the impedance turns out to be over 6 MΩ!

Because of this high-impedance characteristic, the capacitor microphone presents special problems as far as output wiring to an amplifier is concerned. Most amplifiers have relatively low input impedances. But because of the high output impedance of the capacitor microphone, it is necessary to use amplifiers that utilize a field-effect transistor (FET) input stage for proper matching.

Another problem associated with this microphone (and this also has to do with its very high output impedance) is cable noise. If the microphone's high-impedance output is transmitted any appreciable distance to the amplifier system (say, more than 20 ft or so), a problem with *triboelectricity* occurs with the high-impedance cable being used. This type of cable has a habit of developing static charges between its insulated conductors. These charges create unwanted electrical noise (hissing and popping) that is superimposed on the audio signals coming from the microphone. As a result, it becomes very difficult to remove or to filter out these noises.

3-7.3 Typical Construction

A typical capacitor microphone is shown in Figure 3-19. This particular model shows an onboard preamplifier with its FET input stage. This gets around the problem of transmitting audio signals through an otherwise high-impedance circuit. With the

Figure 3-19 Typical capacitor microphone.

preamplifier shown, it is not unusual to have cable runs to the main amplifier unit of 150 ft or more.

Some of the earlier capacitor microphones contained high-voltage-generating power supplies that produced the charge needed to create variable output voltages. However, difficulty with interference created by the high switching frequencies of these supplies made the microphone difficult to use.

Typical operating voltages used to maintain the necessary static charge between the capacitor plates of this microphone are in the range of 50 to 100 V dc. In recent years, though, certain designs have been developed that use much lower voltage values.

The diaphragm is constructed out of a specially treated foil, or in some cases, plastic material less than $\frac{1}{2}$ mil in thickness. In many cases the foil is coated with a thin gold or aluminum surface to prevent corrosion and to form a highly conductive surface for one of the two capacitor surfaces. The spacing between the diaphragm or forward plate and the rear plate is generally about 1 mil.

3-7.4 Practical Applications

Because of the predictability of the performance characteristics of the capacitor microphone, it has found much use in the laboratory for calibration standardizing. Also, this microphone has been used extensively in commercial communications, such as in television and radio. One of the more interesting applications of this type of microphone is in the production of remote cordless microphones. Because of its variable capacitance output, this type of microphone is a natural for becoming part of an *LC* network in a resonant circuit. It becomes very easy to frequency-modulate a small built-in radio transmitter that can transmit an FM signal up to several hundred yards away. Figure 3-20 shows a circuit for such a system. The microphone becomes an integral part of the oscillator circuit. The radiated power of the transmitter is quite small, usually under about 100 mW. This is to prevent the microphone's user from receiving dangerous radio-frequency burns from the antenna if he or she comes into accidental contact with it. The antenna is also kept purposely short to prevent interference with nearby service frequencies. Typical frequencies used for remote microphone broadcasting are in the region of 108 MHz or higher. This further reduces any likelihood of interference.

3-8 AUDIO REPRODUCTION: THE PIEZOELECTRIC MICROPHONE

Like the moving-coil and moving-iron microphones, the piezoelectric microphone requires no external power supply for its operation. The functioning of the microphone depends on a piezoelectric substance, that is, a substance that generates electrical signals in proportion to the amount of mechanical impulses applied to it. The theory of operation is discussed in the next section.

3-8.1 Theory of Operation

To understand how the piezoelectric microphone works, we must first understand what has to take place in generating piezoelectricity. Certain naturally produced crystals and several human-made chemically produced crystals possess piezoelectric properties. Certain forms of quartz found in nature display this unique property, as does the mineral tourmaline. (Tourmaline is usually found containing aluminum and boron silicate together with other metallic elements combined with it, such as iron or magnesium.) Other natural crystalline piezoelectric substances are the chemicals lithium sulfate, ammonium dihydrogen phosphate, and Rochelle salts. (Discovered in La Rochelle, France, this salt, which is potassium sodium tartrate, is frequently used in medicine as a laxative. Figure 3-21 demonstrates this electricity-generating

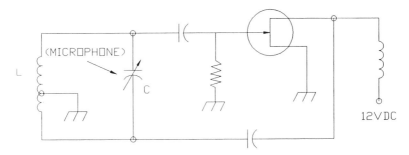

Figure 3-20   Schematic for a small compact FM transmitter.

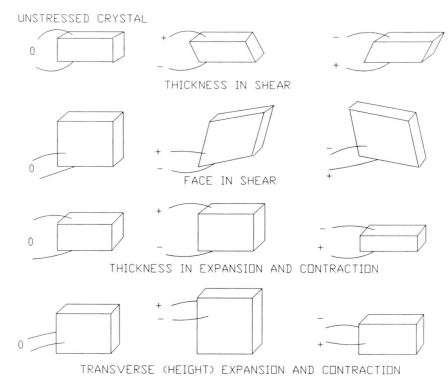

UNSTRESSED CRYSTAL

THICKNESS IN SHEAR

FACE IN SHEAR

THICKNESS IN EXPANSION AND CONTRACTION

TRANSVERSE (HEIGHT) EXPANSION AND CONTRACTION

Figure 3-21   How piezoelectric crystals generate voltage through either compres

property. As shown, a voltage can be generated through either a compressive load or a shearing load. The amount of output charge depends largely on the piezoelectric substance used and the geometrical shape or cut of the crystal itself. Table 3-1 lists the magnitude of electrical charge output per force input that can be expected from these materials.

The interesting thing about piezoelectric materials is that their properties are reversible. That is, in addition to being able to generate electricity by physical deformation, the materials themselves deform automatically when subjected

TABLE 3-1   CHARGE/FORCE RATIOS OF CERTAIN
PIEZOELECTRIC MATERIALS

Ammonium dihydrogen phosphate (ADP)	48×10^{-12} C/N
Lead zircontate–barium titanate (PZT)	$150–741 \times 10^{-12}$ C/N[a]
Lithium sulfate	16×10^{-12} C/N
Quartz	2.3×10^{-12} C/N
Rochelle salts (at 30°C)	550×10^{-12} C/N
Tourmaline	1.9×10^{-12} C/N

[a] This is a human-made ceramic material having a very wide range of charge/force response. See the discussion in Section 3-8.2.

to a voltage. However, we are interested only in their first-discussed property—generating electricity for a microphone application.

A piezoelectric microphone is shown schematically in Figure 3-22. We see a diaphragm attached to a piezoelectric cartridge by means of a rigid link. The cartridge has leads attached to its two opposite faces by means of electrodes. The electrodes are either glued to the crystal slab surfaces by means of a conductive cement, or the electrodes are attached to "pressure" plates with the slab sandwiched between them. As the diaphragm vibrates due to sound pressure waves, the force each wave creates is transmitted to the crystal. The crystal, in return, generates a proportional voltage between its electrodes due to the charge generated.

3-8.2 Operating Characteristics

Up to this point we have not discussed any of the environmental limitations of the various types of microphones, such as limitations in humidity or temperature. With the piezoelectric microphone, however, it becomes necessary to address a rather significant problem that Rochelle salt crystal microphones have with regard to moderately high temperatures.

At a certain temperature level a piezoelectric substance loses its piezoelectric characteristics. This temperature is called the *Curie point*. For most piezoelectric substances the Curie point is quite high, above 100°C. For crystals made from Rochelle salts, however, this temperature is only 45°C (113°F). Manufacturers of microphones used in mobile radio communications prior to the late 1950s found that their Rochelle salt microphones were failing at an alarming rate. Users were leaving

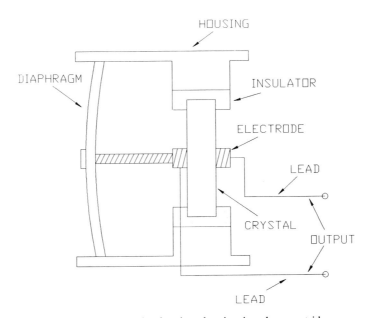

Figure 3-22 Schematic of a piezoelectric microphone cartridge.

the microphones in their cars with the windows rolled up. Unfortunately, the heat of the summer sun at midday quickly destroyed these microphones.

Because of this problem and because piezoelectric materials are used for sensing applications other than sound (applications that require very high temperature environments), work was done to develop a more temperature-resistant substance. The result was the development around 1958 of the *ceramic cartridge microphone*. Piezoelectric ceramics were found to have Curie points in excess of 190°C, and in many cases in excess of 300°C. One of the most commonly used ceramics to cure the problem of the burned-out microphone was lead (Pb) zirconate–barium titanate (PZT). With its extraordinarily high Curie point, PZT could withstand extremely high temperatures.

Two apparent drawbacks to the use of a ceramic microphone in preference to a natural crystal microphone relate to its frequency response and sensitivity. Ceramic microphones initially lacked both the natural sound of crystal microphones and their sensitivity. However, that has all been changed. It would be very difficult now to tell a ceramic microphone from a natural crystal microphone. Tremendous improvements have been made in both their frequency response and their sensitivity.

3-8.3 Typical Construction and Practical Applications

Figure 3-23 shows a typical ceramic microphone cartridge in use today. Because of its durability, ceramic has replaced virtually all natural piezoelectric cartridges. One inherent advantage that ceramic microphones have over other types is cost: They are very inexpensive to produce. Also, they can be made extremely small in size.

3-9 AUDIO REPRODUCTION: THE RIBBON MICROPHONE

From the standpoint of fidelity, the ribbon microphone is considered one of the best in the audio industry. Because of the diaphragm's lightweight construction and consequent low mass, it has the ability to respond to a broad frequency range. This

Figure 3-23 Typical ceramic microphone cartridges.

causes the ribbon microphone to have a very flat and broad response curve. We will see now why this is the case.

3-9.1 Theory of Operation

To understand how the ribbon microphone works, let's look at Figure 3-24. In this illustration we see a ribbon made of aluminum that is suspended between the poles of magnets. Typically, this ribbon is only a few ten thousandths of an inch thick, roughly 1 in. in length, and perhaps $\frac{1}{4}$ in. in width at the very most. Sound waves impinging on the ribbon will, of course, cause the ribbon to vibrate. At this point the ribbon microphone functions quite similarly to the moving-coil microphone. That is, an electromotive force (EMF) is generated across the diaphragm's terminals (in the case of the moving-coil microphone, an EMF was generated across the terminals of the coil) due to the movement of the diaphragm cutting across the lines of magnetic flux. Admittedly, the EMF that is generated is quite small compared to the coil; nevertheless, the amount is very detectable and usable.

3-9.2 Operating Characteristics

The sound quality of the ribbon microphone is excellent. The microphone has a very flat, broad frequency response range, which makes it ideal for studio audio work. However, its one major disadvantage is that it is susceptible to low frequencies created by vibration and wind noise. As a result, it is used only indoors.

Looking again at Figure 3-24, you can see that the dc resistance between the

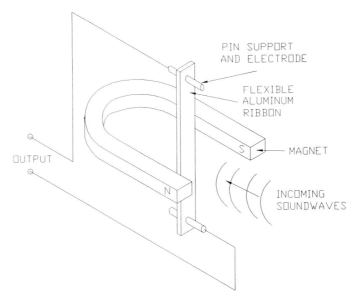

Figure 3-24   How the ribbon microphone works.

diaphragm's two terminals would be quite low, typically much less than 1 Ω. This means that the characteristic impedance is at least this value; therefore, to have this microphone work in conjunction with an amplifier whose input resistance is in the range of 30 to 1000 Ω or higher, some sort of impedance-matching network is needed. Impedance transformers are sometimes used for this purpose, but care must be taken not to affect the microphone's frequency response when using them.

3-9.3 Typical Construction and Practical Applications

Figure 3-25 shows a typical ribbon microphone used for studio audio work in the recording industry and in radio and television. The ribbon microphone is rarely used for other applications.

3-10 SONAR DETECTION

Sonar, a form of sonic detection involving frequencies from about 1 Hz to as high as 10 MHz, was developed by the United States during World War II for the purpose of detecting enemy submarines. When you hear the term used, it will be in the context of underwater detection.

3-10.1 Theory of Operation

The principle of sonar detection is rather simple (the implementation is the difficult part). A high-frequency sound is emitted from a transmitter toward the direction of a suspected unseen target. The transmitted sound reflects off the surface of the

Figure 3-25 Studio ribbon microphone.

target, forming an echo that is detected by sensitive audio receivers. The target's range can then be calculated by measuring the time interval between the sound reflection and detection of the echo.

Figure 3-26 shows a diagram of a typical sonar installation aboard a military ship sending out sound waves that strike an object, creating return echoes. A similar application is used in the operation of depth sounders used by fishermen for the purpose of locating schools of fish and keeping track of the bottom of the waterway. Let's take a look now at the transmitter and receiver used in these applications.

The transmitter consists of a "speaker" device capable of emitting several tens or hundreds of watts of sound power. This system can also act as a receiving device, depending on the type of transmitting element used, or the transmitter and receiver can be two separate devices. In one application the transmitter consists of a piezoelectric material such as a ceramic element or crystal element. When subjected to a voltage the piezoelectric material is caused to change its physical size at a rate equal to the voltage's frequency and duration. In other words, the piezoelectric sensor oscillates at this frequency and the energy produced is "acoustically coupled" to the water, much as an ordinary loudspeaker is "coupled" to the air for sound transmission. The acoustic coupling is done by submerging the oscillating piezoelectric transducer into the water and then using specially designed sound baffling to concentrate the sound source in one direction. This type of transmitter can also be used as a sonar receiver system, in that the transmitting diaphragm can act as a microphone or hydrophone for receiving returning sound echoes. Figure 3-27 illustrates a typical transmitting and receiving pattern for a sonar system. The sound baffling referred to above is responsible for the highly directional characteristics seen in this pattern.

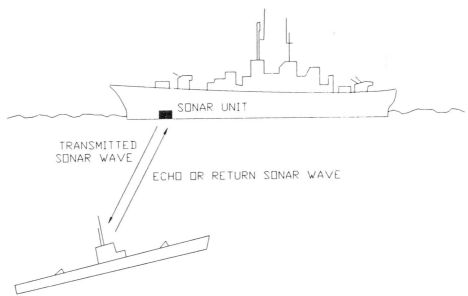

Figure 3-26   How sonar works onboard a military vessel.

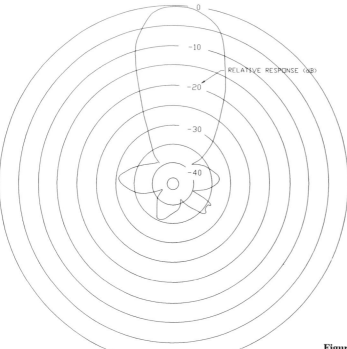

Figure 3-27 Transmitting and receiving pattern for a sonar onboard a ship.

Another popular form of sonar transmitter is the dynamic coil and diaphragm. Again, this is very similar to a dynamic loudspeaker in that a coil suspended inside a magnetic field is energized with a signal voltage. In this case, however, the voltage's frequency and pulse duration is such that a water-submerged diaphragm is made to oscillate rather than an air-suspended voice cone. Like the piezoelectric system described above, the moving-coil or dynamic coil system can also serve as a hydrophone for receiving return signals.

Sonar usually deals with the reception of very weak signals superimposed on a background of spurious noise signals. These noise signals come from at least three major sources: (1) self-noise (noise from platform or vehicle where the sonar is installed, i.e., noise from the engine, pumps, propellers, etc.), (2) ambient noise (noise from waves, ice motion, fish, human-made noise, etc.), and (3) reverberation noise (noise from spurious reflections of the transmitted signal, i.e., reflections from the bottom, surface, or scattered layers of varying water densities due to temperature variations, etc.). It is extremely difficult to design a transducer that can distinguish between background noise and the desired echo or sound source. Figure 3-28 demonstrates this problem. In Figure 3-28(a) we see a graphical representation of three desirable signals we would like to detect. However, when noise is added to our signals we could get the pattern shown in Figure 3-28(b). This illustration could be from an oscilloscope pattern actually received from the output of a hydrophone. It

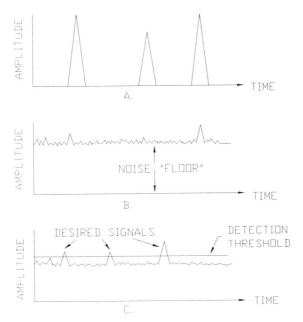

Figure 3-28 (a) Desirable signals; (b) noise added to the desired signals; (c) setting the detection threshold to detect only the desired signals.

is obvious from what we see here that it would be very difficult to distinguish the desired signals from the noise.

One method frequently used with sonar transducers for separating signals from noise is to incorporate a threshold setting circuit. This circuit is set so that any signals "peaking" above the background noise at a certain predetermined voltage level will be singled out for identification. This is illustrated in Figure 3-28(c)—an oversimplification of the actual method used but depicting essentially what is done. In reality, a rather complicated statistical analysis is made of the noise patterns, and based on these data, the probability of what is noise is subtracted from the signal-plus-noise signals, leaving only the signal amplitudes desired. The threshold level is then set to indicate any signals occurring at or above this calculated level.

To determine the range of an object that has reflected a signal back to the sonar's receiver, it is first necessary to know the signal's velocity as it travels through water. Having obtained that piece of information, we can then use the equation

$$d = 0.5vt \qquad (3\text{-}7)$$

where d = distance (m, ft, etc.)

v = propagation velocity of sound (m/s, ft/s, etc.)

t = time (s)

The factor 0.5 in eq. (3-7) is necessary to halve the results, since the distance over which the sound travels represents a double path. That is, the incident or outgoing path of the transmitted signal, plus the reflected path of travel of the incoming reflected signal, make up a path that is double the distance that separates the sound source and the reflecting surface.

3-10.2 Operating Characteristics

The accuracy of a sonar system is based to a great extent on its ability to measure accurately the time intervals between emitted and received sound signals. With the modern highly regulated and precisely controlled timing circuits, this is not a problem. The system's accuracy also depends on the user's ability to program in the sonic velocity of the water involved in the particular application.

3-10.3 Typical Construction and Practical Applications

Figure 3-29 shows a popular application of a sonar device—as a depth finder to locate fish and keep track of underwater hazards for boating purposes. This particular unit transmits at a frequency of 200 kHz with a maximum power of 150 W rms. It can receive and transmit signals to a depth of 1500 ft.

Sonar devices are generally divided into two major groups: passive sonar devices and active devices. Passive sonar devices are "listening only" devices. Their function is to receive radiated acoustic energy only. Applications include surveillance and early warning detection devices. Active systems contain transmitting capabilities in addition to receiving capabilities. The depth finder is an example of this application.

Figure 3-29 Depth finder used for navigation and fishing. (Courtesy of Heath Co., Benton Harbor, MI.)

3-11 ULTRASONIC DETECTION

Ultrasonic detection deals with the detection of sonic energy that is beyond the upper end of the audio spectrum. In other words, ultrasonic sound is of higher frequency than 20,000 Hz. As a matter of fact, it is not unusual to have frequencies as high as 5 MHz used in an ultrasonic system.

Ultrasonic detection is used extensively in medicine for diagnostic purposes. The most recent advancement in this application is in the production of ultrasonic imaging, which produces results similar to those of x-rays but without the hazards of x-rays. Ultrasonics are also used in industry and in consumer electronics for the measurement of distance and velocity, much like sonar. We discuss these applications in detail in the sections that follow.

3-11.1 Theory of Operation

In principle, ultrasonic devices are similar in operation to sonar systems. These devices contain a sonic transmitter and a receiver. Ultrasound is used in applications dealing with the detection of very small objects, usually located in media other than water, although this is not always the case. The reason for the very high frequencies being used for object detection as opposed to the lower frequencies used in sonar is due to the reflection efficiency of the object itself. Shorter wavelengths (i.e., higher frequencies) bounce off smaller surface areas of objects far more efficiently than do the longer-wavelength sound waves. We have somewhat the same situation in radio-antenna reflector design, where the size of reflector is dependent on the wavelength being transmitted or received. Specifically, the size of object being detected by an ultrasonic signal must be at least one wavelength in length or width to be detected effectively.

The construction of a typical ultrasonic transducer is shown in Figure 3-30.

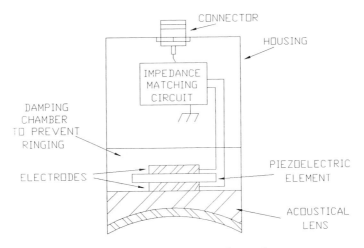

Figure 3-30   Typical ultrasonic transducer.

There are six basic components in this device: (1) the piezoelectric element, (2) an impedance-matching circuit, (3) the connector, (4) the damping chamber, (5) housing, and (6) the acoustical lens or sonic focuser. The sonic focuser is not present in all transducer designs.

The piezoelectric element is made to resonate at a particular frequency determined by its geometry. An ac voltage near or at the element's resonant frequency is supplied to the element, causing it to vibrate at its resonant frequency. This voltage is usually applied in a single pulse rather than in continuous pulses. Therefore, to prevent "ringing" from occurring following a transmitted pulse, a damping chamber adjacent to the element absorbs any transient energy following a pulse transmission.

The operation of some ultrasonic transducer systems closely resembles that of a sonar system. Figure 3-31 shows the theory of operation of how an ultrasonic transducer is used for the detection of flaws in a workpiece. Because of the known geometry of the workpiece, the echo pattern is known ahead of time. Any flaws in the workpiece will add additional echoes to these patterns. This detection technique is sometimes referred to as the *pulse echo ultrasonic method*.

Acoustical lensing

The *acoustical lens* or *sonic focuser* is used to concentrate the radiated energy from the piezoelectric element to the target area. Depending on the shape of this lens, the pattern of sound energy radiated will travel along certain predictable paths. Figure 3-32 illustrates this behavior. The *spot focus* is produced by using what is referred to as a *spherical lens*; the *line focus* is produced by a *conical lens*.

The len's function is much like that of an optical lens. As a matter of fact, sound

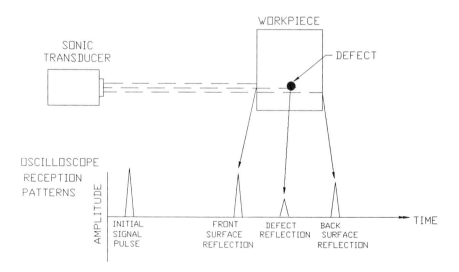

Figure 3-31 Detecting flaws in a test part using ultrasonic detection.

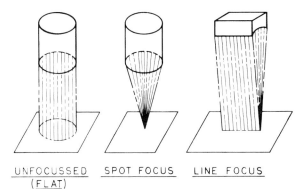

UNFOCUSSED   SPOT FOCUS   LINE FOCUS
(FLAT)

Figure 3-32 Focusing patterns of acoustical lenses. (From Kenneth H. Beck, *Methods for Ultrasonic Inspection of Tubes On-Mill and Off,* from paper presented at Tube Asia Conf., Singapore, 1982, p. 7. Reprinted by permission of TAC Technical Instrument Corp., Trenton, NJ.)

waves at very short wavelengths behave very much like light waves. In the case of a spherically shaped sonic focusing unit on the sonic transducer, sound from the transducer is radiated from a concave surface, much like a narrow beam of light being reflected from a concave mirror. The exact shape and size of the radiation pattern can be determined by means of Huygens' principle, developed by Dutch astronomer and mathematician Christian Huygens (1629–1695). Each point on the concave surface acts as a radiating source, causing the individual "source beams" to converge at a distant intersection point called the *focus* or *focal point*. Figure 3-33 illustrates this phenomenon. Only a very few of an infinite number of these radiating point sources are shown here. Notice, however, that each radiant beam is perpendicular to the concave surface at the radiating point, thereby creating the focal point. The exact location of the focal point in relation to the radiating source may be calculated using the following equation from optics:

$$FL = \frac{r}{1 - (c_1/c_2)} \qquad (3\text{-}8)$$

where FL = focal length, measured from center of radiating lens to focal point (m)

   r = radius of sphere used to create concave lens surface (m)

   c_1 = velocity of sound inside lens (m/s)

   c_2 = velocity of sound in target media (m/s)

Note: c_2 must be larger than c_1 in order for the equation to determine a usable location for the focal point.

Figure 3-34 shows actually measured radiation patterns for a conical lens and for a spherical lens, both mounted in an ultrasonic transducer operating at 3 MHz or higher. It would seem from the results shown in this figure that compared to the spherical acoustical lens the lens on the right has a narrower focusing ability over a longer distance in steel.

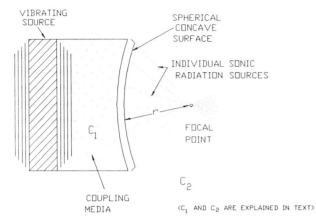

Figure 3-33 How sonic focusing works on a spherical concave surface.

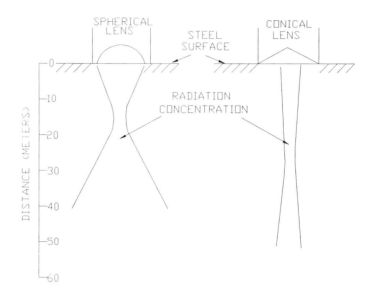

Figure 3-34 Radiation concentration patterns for a spherical and conical lens.

Acoustic intensity

The term *acoustic intensity* refers to the amount of power concentration per unit area of an object. In the case of a sonic transmission, a transducer will be capable of transmitting just so much sonic power over a given area. The unit of measurement is the watt per square meter (W/m^2). The idea of using a focusing system such as the one just described is to concentrate and therefore increase the sonic power intensity at a given location. With the increased intensity, a stronger echo can result.

Comparing acoustic intensities

In Section 1-5.3 we discussed the logarithmic characteristics of sound. Equation (3-1) allowed us to make a comparison between the sound pressures generated between two different sound sources and to express that comparison in decibels. We can perform the same comparison between two different intensities, I_1 and I_2, and similarly, produce an expression showing a comparison ratio in decibels:

$$dB = 10 \log \frac{I_2}{I_1} \tag{3-9}$$

where dB = decibels

$\quad\quad I_1$ = intensity of first sound source (W/m^2)

$\quad\quad I_2$ = intensity of second sound source (W/m^2)

Acoustic impedance

The *acoustic impedance* of any sound propagating system is defined as the product of the density of the medium through which the sound is being propagated and the velocity of the sound. In other words,

$$Z_a = \rho c \tag{3-10}$$

where Z_a = acoustic impedance (kg/m^2-s)

$\quad\quad \rho$ = mass density of medium (kg/m^3)

$\quad\quad c$ = velocity of sound in medium (m/s)

Example 3-2

Find the acoustic impedance for water. The velocity of sound in fresh water is approximately 1.49×10^3 m/s.

Solution: The mass density of water is 1 g/cm^3 or 1×10^3 kg/m^3. Therefore, applying eq. (3-10), we obtain

$$Z_a = \rho c = (1 \times 10^3 \text{ kg/m}^3)(1.49 \times 10^3 \text{ m/s})$$

$$= 1.49 \times 10^6 \text{ kg/m}^2\text{-s}$$

Equation (3-10) may be rewritten by substituting eq. (1-6) in Section 1-5.3 for c to obtain

$$Z_a = \sqrt{Y\rho} \tag{3-11}$$

where Z_a = acoustic impedance (kg/m^2-s)

$\quad\quad Y$ = modulus of elasticity (or Young's modulus) of medium (N/m^2)

$\quad\quad \rho$ = mass density of medium (kg/m^3)

Example 3-3

Determine the acoustic impedance for aluminum using eq. (3-11).

Solution: Referring to Table 1-3, we see that the modulus of elasticity for aluminum is 7.3×10^{10} N/m^2. (In Table 1-3 the symbol Y is used for Young's modulus, which is the same as the modulus of elasticity.) According to the table, the mass density of aluminum is 2643 kg/m^3. Therefore, according to eq. (3-11),

$$Z_a = \sqrt{Y\rho} = \sqrt{(7.3 \times 10^{10} \text{ N/m}^2)(2643 \text{ kg/m}^3)}$$

$$= \sqrt{1.93 \times 10^{14} \text{ N-kg/m}^5}$$

$$= 1.39 \times 10^7 \text{ kg/m}^2\text{-s}$$

It is important to know what the acoustic impedance is for a certain substance because it enables the ultrasonic transducer user to match the impedance of the transducer being used to that of the medium being used. This matching then allows the maximum power to be transferred from the transducer to the medium and back again. Recall from electrical theory that electrical impedance has very similar behavior.

In the event that there is an impedance mismatch between the ultrasonic sensor and the medium through which the sound is traveling, an impedance-matching transformer can be devised to effectively create an impedance match. If the impedance for the transformer is expressed as Z_{xf}, then

$$Z_{xf} = \sqrt{Z_t Z_m} \tag{3-12}$$

where Z_{xf} = impedance of the matching transformer (kg/m$_2$-s)

$\quad\quad Z_t$ = impedance of transducer (kg/m^2-s)

$\quad\quad Z_m$ = impedance of medium (kg/m^2-s)

Having solved for Z_{xf}, it becomes a matter of finding a combination of mass density and sound propagation in eq. (3-10) that will produce a practical material that can be used for the impedance-transforming material. Or, you could look for a combination of the modulus of elasticity and mass density, as asked for in eq. (3-11), to find a practical impedance-matching material. Either approach will give the same results.

Example 3-4

In Examples 3-2 and 3-3 we found the acoustical impedance for water and for aluminum. Now find the impedance for the matching acoustical transformer, whose output transmitter is made from aluminum, that will efficiently couple a sonar device to water. Select a material that would serve as this impedance coupler.

Solution: First, we must determine the coupling impedance for the transformer by using eq. (3-12):

$$Z_{xf} = \sqrt{Z_t Z_m} = \sqrt{(1.39 \times 10^7) \text{ kg/m}^2\text{-s} (1.49 \times 10^6) \text{ kg/m}^2\text{-s}}$$

$$= \sqrt{2.071 \times 10^{13} \text{ kg}^2/\text{m}^4\text{-s}^2}$$

$$= 4.55 \times 10^6 \text{ kg/m-s}$$

TABLE 3-2 ACOUSTIC DATA FOR SEVERAL COMMON MATERIALS

Material	Sound velocity (m/s)	Mass density (kg/m)	Acoustic impedance, Z (kg/m-s)
Air	332	1.281	425.3
Aluminum	5102	2643	13.48×10^6
Brass	3499	8553	29.93×10^6
Copper	3557	8906	31.68×10^6
Hydrogen	1269	0.090	114.2
Iron	5000	7100	35.50×10^6
Clay, ceramic	3000–5000	1500–2500	4.5×10^6–12.5×10^6
Water	1461	998	1.46×10^6
Wood	3048–4572	480–800	1.5×10^6–3.7×10^6

Having found the value for the matching impedance, we must now locate a material whose $\rho \times c$ value [eq. (3-10)] approximates the impedance value above. Table 3-2 lists the acoustic impedance values for several materials. According to this table, an impedance transformer constructed of a ceramic-clay material would probably be suitable for our application.

Ultrasonic scattering

An undesirable occurrence in propagating ultrasonic energy is the scattering of this energy due to objects lying in the path of the target. Figure 3-35 shows what happens to the energy when a particle or object is struck. Each obstruction causes a portion of the incident or transmitted energy to develop spherical wavefronts around each obstructing particle, a consequence of Huygens' principle. Less energy is then received by the target, causing weaker reflections back to the ultrasonic receiver.

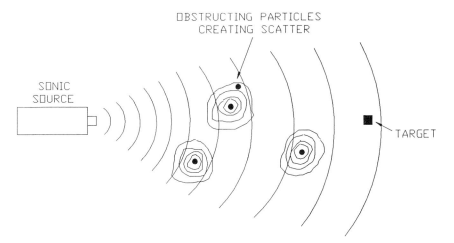

Figure 3-35 Particles creating scatter in a sonic field.

Utilizing the Doppler shift

The velocities of moving fluids in pipes can be measured remotely by using an ultrasonic transducer and the Doppler effect. The only requirement is that the fluid contain higher-density material than that of the transporting fluid. The higher-density material will then create the necessary echoes for the ultrasonic receiver to sense.

Figure 3-36 shows a circuit used for Doppler sensing. Since we are looking for a change in frequency only, we can ignore the measuring of time using pulsed waves, and instead, transmit a continuous signal. By comparing the original signal with the echo signal received, any change in frequency can easily be noted. In the system depicted in Figure 3-36, the incoming echo frequency is mixed or heterodyned with the original outgoing signal. Any difference in frequencies between these two signals will then produce a proportional output voltage which is sent to a velocity output indicator. If there is no change in frequency, implying no measurable velocity, no output voltage will result.

Ultrasonic imaging

The precise distance-measuring capabilities of ultrasonic transducers can be utilized in a very interesting manner. By scanning a target with a highly collimated ultrasonic beam, similar to the scanning raster of a television screen, it is possible to develop an accurate representation of the target's relief features. In Figure 3-37 we see such an object being scanned by an ultrasonic scanning system. The resultant image is produced on a CRT screen, where the various distances being detected by the scanner produce varying proportional shadings within the developed image.

3-11.2 Characteristics of Sonic Detectors

Distance-measuring accuracy using sonic or ultrasonic waves is quite good. Positional accuracies of ±0.001 in. are not unusual in some applications. In the case of

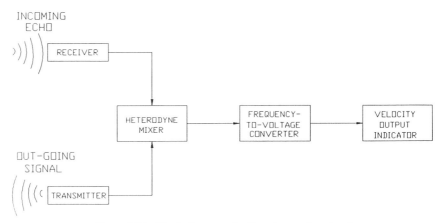

Figure 3-36 Block diagram of a Doppler sensing circuit.

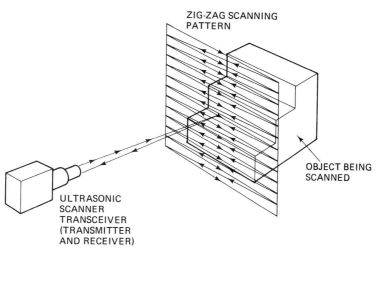

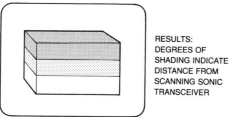

Figure 3-37 Object being scanned by an ultrasonic scanner.

sonar where distances of several kilometers are involved, accuracies of less than ±1% are common. The only major drawback to using sound waves for measuring is that they are slow (limited by the velocity of propagation); also, extraneous noise can be a problem. Otherwise, sonic and ultrasonic transducer systems are very reliable and relatively inexpensive to manufacture and to operate.

3-11.3 Practical Applications

As mentioned in an earlier section, ultrasonic transducers are used extensively in medical research. They are presently being used in ultrasonic imaging, where x-rays were used formerly. There are two great advantages to the use of ultrasonic images rather than x-rays. One is that no harmful radiation is produced that could have future harmful tissue-damaging effects on the body. Second, by varying the intensity of the transmitted ultrasonic wave, it is possible to obtain different depths of image penetration. In this manner it is possible to obtain images of organs that would normally be transparent to x-rays.

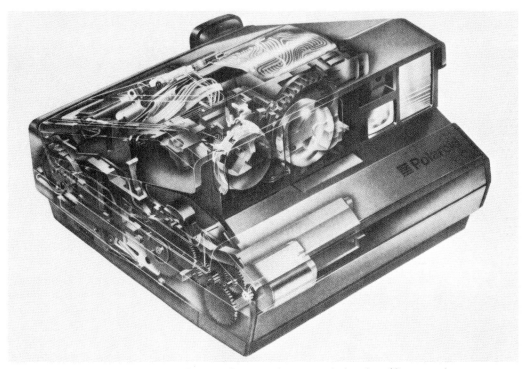

Figure 3-38 Camera using a sonic system for automatic focusing. (Courtesy of Polaroid Corp.)

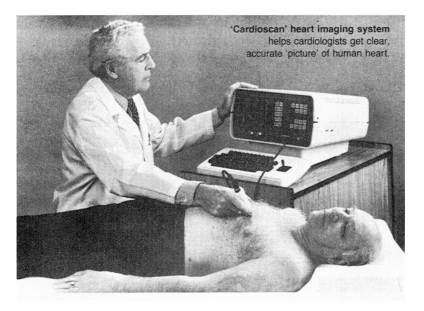

'Cardioscan' heart imaging system helps cardiologists get clear, accurate 'picture' of human heart.

Figure 3-39 Using ultrasound in place of x-rays to peer inside the human body. (Courtesy of Design News, Apr. 7th, 1986.) Imaging System Helps Cardiologist get the Picture." pg. 45.

A practical application of sonar was shown in Figure 3-29: a popular depth finder used by boaters and fishermen. The Polaroid Camera Company has developed a very successful ultrasonic ranging device that is used in many of their cameras for automatic focusing. In the camera illustrated in Figure 3-38, the ranging device can measure the distance between the camera lens and the photographic subject to within approximately ±5% of the actual distance. A typical hospital installation of ultrasonic scanning is shown in Figure 3-39.

REVIEW QUESTIONS

3-1. Explain how the dynamic microphone works. Why is it called a self-generating transducer?

3-2. To operate properly, why does an electret microphone require an onboard charge?

3-3. Why does the capacitor microphone have a high input impedance, whereas a dynamic microphone has a low input impedance?

3-4. What does *triboelectricity* mean? Why does it occur primarily with capacitor microphones?

3-5. Why have cartridges made with Rochelle salts been replaced with ceramic piezoelectric cartridges in microphones? Why is PZT a popular material for manufacturing piezoelectric sensors?

3-6. Explain how the ribbon microphone works.

3-7. In sonar imaging or small-object detection, why are higher-frequency waves used rather than lower-frequency waves?

3-8. Explain the functions of the six basic components that comprise an ultrasonic transducer.

3-9. What is acoustical impedance? How is it determined?

3-10. What is a major drawback in using sound for ranging purposes? What are some advantages in using sound for this purpose?

PROBLEMS

3-11. Design a sonic ranging system, using block diagrams only, that can measure distances of several hundred feet and is temperature compensated.

3-12. Calculate the acoustical impedance for air. What would the matching impedance be for a sonic ranging device whose sonic emitter is made of iron?

3-13. Calculate the Doppler shift at a receiver site of a sonar signal in water whose transmitted frequency is 250 Hz and whose transmitter is traveling toward the receiver at a velocity of 6.7 m/s.

3-14. Explain how you would set up a device to measure the density changes in liquids and solids in a testing facility by using a sounding device and precise timing equipment.

REFERENCES

ALLOCCA, JOHN A., and ALLEN STUART, *Transducers: Theory and Operation*. Reston, VA: Reston, 1984.

BIBER, C., S. ELLIN, E. SHENK, and J. STEMPECK, *The Polaroid Ultrasonic Ranging System*. Audio Engineering Society, New York, NY, 1980.

NORTON, HARRY N., *Sensor and Analyzer Handbook*. Englewood Cliffs, NJ: Prentice Hall, 1982.

4

MAGNETIC SENSORS

CHAPTER OBJECTIVES

1. To understand the theory and operation of the following magnetic field-sensing devices:

   a. The magnetic reed switch
   b. The Hall effect transistor
   c. The inductive pickup coil
   d. The vibrating wire transducer

4-1 INTRODUCTION

In this chapter we study those sensors that either use or interact with a magnetic field to cause the sensor to operate. This subject covers quite a wide range of sensing devices, so we will study only the more commonly used ones frequently encountered in industry.

The magnetic field referred to above is one that is usually produced by a permanent magnet. This magnet is often mounted externally to the transducer body and associated somehow with the measurand. A typical example is given below. We discuss the various configurations in which magnetic sensors are found; additionally, we will see how and why these configurations all work.

4-2 COMMONLY SENSED MEASURANDS

The measurands most often sensed using magnetic sensors are (1) position, (2) motion, and (3) flow. In all three cases the sensing is contactless, which is very desirable. However, as we will find out later in our discussions, there is a drawback to this kind of sensing produced by what is known as *viscous damping*. Viscous damping is an interaction between a magnetic field and a moving member reacting to that field and causing the member to have its motion impeded, so to speak, due to the attractive forces involved.

As we discuss the various types of magnetic sensor applications we will see how each of the three measurands—position, motion, and flow—is detected.

4-3 MAGNETIC REED SENSOR

The magnetic reed sensor is perhaps the simplest of all magnetic sensors in its construction. It came into existence around the mid-1960s and has been manufactured in a variety of forms. The thing to bear in mind about this type of sensor is the fact that it is a digital device. In other words, it is either in one state or another (i.e., either electrically on or off, as in a relay contact). Its theory of operation and various physical configurations are discussed below.

4-3.1 Theory of Operation

Figure 4-1 illustrates a typical magnetic reed sensor. Usually, the sensor is constructed inside an evacuated glass bulb or one that contains an inert gas such as nitrogen. Inside the envelope is a set of electrical contacts in which one side of the contacts, which is movable, is made from a ferrous metal. The other contact half, or stationary side, is made of a nonferrous material. The movable contact or reed is "oil-canned"; that is, it is constructed so there is a built-in snap action that takes place when the contact is moved slightly.

In the example shown in Figure 4-1 there are actually two sets of contacts

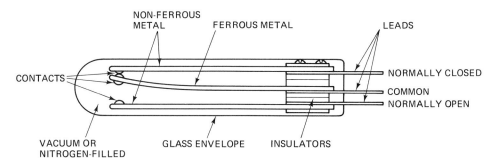

Figure 4-1 Magnetic reed sensor.

shown. The outer two contacts are nonferrous. Only the middle "common" contact is ferrous. In this design, then, one set of contacts is normally "closed" in the nonenergized state, whereas the other set is normally "open" in the nonenergized state. When a small magnet is brought near the glass envelope of the reed switch, the common contact is attracted toward the magnet, causing it to snap into its new position against the other contact. When the magnet's magnetic field is removed, the middle reed snaps back to its original position against the other contact. The reed switch in this example is being used as a proximity switch.

4-3.2 Operating Characteristics

The triggering distance between the switch and the magnet will vary according to the strength of the magnet's magnetic field. Typically, this distance is between 5 and 25 mm.

The contacts of the reed switch are not made to handle very large currents or voltages. Maximum voltages are usually no greater than 60 to 70 V, although some manufacturers have rated their contacts for 125 V ac. Such high voltages for the small contacts being used are usually not recommended, however.

Current ratings are normally no greater than 0.5 A at 50 V dc (i.e., 25 W). Any larger amounts of current and voltage will cause excessive arcing and pitting of the contacts.

As long as the power ratings of the contacts are not exceeded, the life span of the reed switch is extremely long. Typically, a reed switch can withstand at least 10 million switch closures before failing. A hundred million cycles is not uncommon with these devices.

As for specifying magnetic field intensities needed to cause switch closures with the reed switch, these data are not often published by the manufacturer. In reality, it is generally assumed that the user will, with some experimentation, determine the size of magnet needed and the spacing to cause the switch to trip.

4-3.3 Physical Configurations

Figure 4-2 shows a typical example of a manufactured reed switch. The size and shape can easily be configured by the manufacturer for many types of specialized applications.

4-3.4 Practical Applications

As already mentioned in Section 4-3.2, one of the primary functions of the reed switch is to act as a proximity switch, making simple contact openings and closures. The one popular aspect of using reed switches for this kind of application is that they are so easy to install onto existing equipment or structures. Let's look at some examples.

Figure 4-3 shows another possibility for installation, in which it is desired to

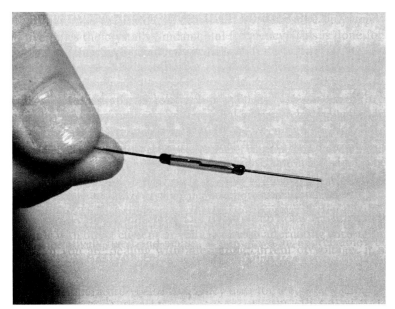

Figure 4-2 Typical reed switches.

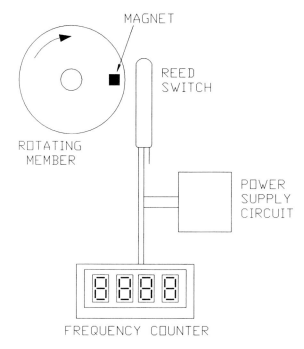

Figure 4-3 Monitoring a rotating machine member using a reed switch.

know the rpm of a rotating machine member. A magnet is mounted (glued) onto the rotating member and a reed switch mounted in close proximity to the magnet on a stationary portion of the machine. Each passage of the magnet would then trip the switch's contacts, thereby indicating one complete revolution.

4-4 HALL EFFECT TRANSISTOR

The *Hall effect* is a phenomenon that was predicted for the transistor soon after its development in the late 1950s. The discovery that the current-carrying behavior of a conductor can be altered when it is immersed in a magnetic field became evident many years before. The Hall effect was named in honor of the American physicist Edwin Hall (1855–1938). We explore the reason for the existence of this effect in our next discussion.

4-4.1 Theory of Operation

In Figure 4-4 we see an illustration depicting an electrical conductor subjected to a magnetic field. The magnetic field is at right angles to a current flow that has already been established within the conductor. One electron, which is a member of this current, is shown. Its velocity of travel is v. Note that the direction of travel is opposite the current direction following the convention that current flow is always

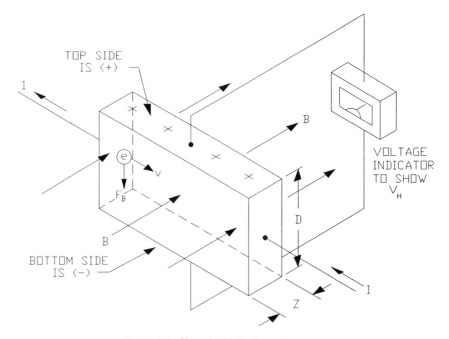

Figure 4-4   How the Hall effect takes place.

in the opposite direction of electron flow. According to the right-hand rule for determining the direction of force experienced by an electron traveling through a magnetic field (i.e., the extended fingers of the hand point in the direction of the magnetic flux, the thumb points in the opposite direction of electron travel, and the palm of the hand faces the direction of the exerted force), a force F_L (called the Lorentz force) will be generated on the electron as shown. This means that all the electrons traveling as shown in the slab of conductor will congregate toward the bottom of the conductor. As a result, the bottom edge of the conductor will become negatively charged while the upper edge becomes positively charged. In other words, an EMF will develop across the width of the slab that is in proportion to the amount of flux, current, and resultant charge separation, d.

The amount of generated voltage due to the Hall effect, V_H, can be calculated using the relationship

$$V_H = avd \qquad (4\text{-}1)$$

where V_H = Hall effect voltage (V)

a = flux density of magnetic field [Wb/m or tesla (T)]

v = velocity of traveling charge (m/s)

d = height or width of conductor (m)

To find the velocity, v, of the electrons we have to go to physics, which states that

$$v = \frac{I}{Ane} \qquad (4\text{-}2)$$

where v = velocity of traveling charge (m/s)

I = current (A)

A = cross-sectional area of conductor (m^2)

n = number of electrons per unit volume of conductor [i.e., the electron density (number of electrons/m^3)]

e = charge per electron (1.6×10^{-19} C)

Remember that current, I, is charge per unit time. This is the reason that the velocity term, which is time dependent, is developed from the expression above where time units are not apparent. Note, too, that the cross-sectional area of the conductor, A, is $z \times d$. Then eq. (4-2) can be rewritten as

$$v = \frac{I}{zdne} \qquad (4\text{-}3)$$

Substituting Eq. (4-3) for v in eq. (4-1), we get

$$V_H = \frac{\beta I}{zne} \qquad (4\text{-}4)$$

where V_H = Hall effect voltage (V)

   β = flux density of magnetic field (Wb/m or T)

   I = current flowing through conductor (A)

   z = thickness of conductor (m)

   n = number of electrons per unit volume (number of electrons/m³)

   e = electron charge (1.6×10^{-19} C)

Finally, often what is done in Eq. (4-4) is to assign a constant, the Hall effect constant, K_H, to the variables $1/n$ and $1/e$ so that eq. (4-4) becomes

$$V_H = \frac{K_H \beta I}{z} \qquad (4\text{-}5)$$

where V_H = Hall effect voltage (V)

   K_H = Hall effect constant (m³/number of electrons-C)

   β = flux density of magnetic field (Wb/m or T)

   I = current flowing through conductor (A)

   z = thickness of conductor (m)

Up to this point we have been using a conductor to illustrate the behavior of the Hall effect. Actually, semiconducting material is used to manufacture Hall effect devices, but the explanation of how electrons are deflected at right angles to the magnetic flux remains the same. The only difference is that with semiconducting materials you are working with charge carriers and holes instead of electrons. Let's look at an example.

Example 4-1

In Figure 4-5 we see a piece of semiconducting material being subjected to a magnetic field of 1.8 Wb/m. The dimensions of the piece are given. A supply current of 350 mA is circulating through it in the direction shown and the K_H value is 1.6×10^{-4} m³/C. What is the generated Hall effect voltage?

Solution: Because of the flux and current directions shown, and according to the right-hand rule stated earlier, the charge electrons would drift to the top of the Hall effect slab. Therefore, according to eq. (4-5),

$$V_H = \frac{(1.6 \times 10^{-4} \text{ m}^3/\text{C})(1.8 \text{ Wb/m})(0.35 \text{ A})}{2.5 \times 10^{-3} \text{ m}}$$

$$= 0.403 \times 10^{-1} \text{ Wb/s}$$

$$= 40.3 \text{ mV}$$

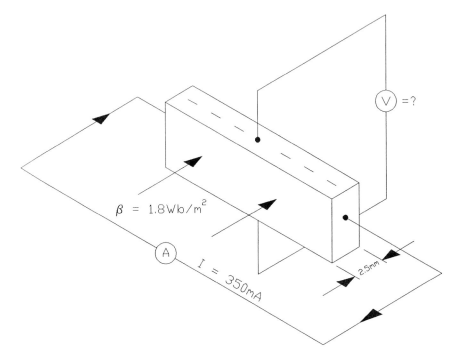

Figure 4-5   Illustration for Example 4-1.

4-4.2 Operating Characteristics

Hall effect devices may be manufactured from either p-type or n-type semiconducting materials. The only difference between the way these two materials behave is in
the internal flow direction of electrons. Figure 4-6 shows that when using n-type
material, you are dealing with a flow of electrons whereas when working with p-type
material, you are working with a flow of hole carriers. For all practical purposes these
are the same as positively charged particles, or positrons, that flow in the opposite
direction from the n-type electrons. Consequently, the outward results of using these
two types of material are identical; only the polarities of the biasing currents are
reversed. The Hall effect sensor can be used as either a proportional sensing device
or as a proximity sensor.

4-4.3 Typical Construction

Hall devices can be manufactured to fit into a variety of packages. Figure 4-7 shows
some of the packages commonly used. These are housings normally associated with
transistors and other solid-state devices. In many cases it is very difficult to single
out the Hall device within a transducer system because of the integral design
packaging that is often used.

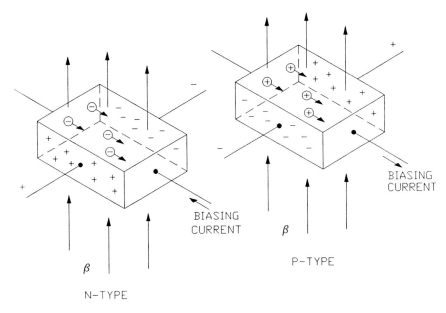

Figure 4-6   How p-type and n-type materials react to form Hall effect sensors.

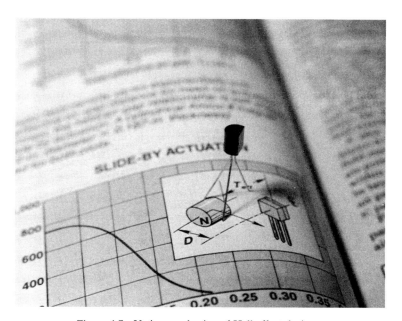

Figure 4-7   Various packaging of Hall effect devices.

4-4.4 Practical Applications

The amount of voltage generated by the Hall effect is dependent on the velocity of the charged particles flowing through the conductor, and the medical profession has taken advantage of this fact in the form of a blood flow monitor. Figure 4-8 shows how this is done. Tiny voltage probes are placed on either side of the blood vessel, while the poles of an electromagnet are placed perpendicular to these probes, again on either side of the vessel. The amount of voltage generated due to the Hall effect created by the ions within the blood supply is a direct indication of a person's blood flow.

Figure 4-9 shows a Hall device being used as a wattmeter. Current from the load is used to create the magnetic flux, whose amount will vary proportionally with the power consumption of the load. The control current for operating the Hall device is tapped from the load current supply.

Hall devices are frequently used for linear displacement transducers. Figure 4-10 shows two types of applications. In part A, two magnetic fields are set up so that one opposes the other. With the Hall detector in the middle, the two magnetic fields nullify each other, resulting in no output from the device. Movement to either side of center causes a voltage to develop. The polarity of this voltage depends on which side of center the device is located.

Figure 4-11 shows a Hall device being used as a tachometer. A rotating gear causes a nearby magnetic field to become redirected toward one of its teeth as it rotates past the Hall detector. The redirection, in turn, causes the field to sweep

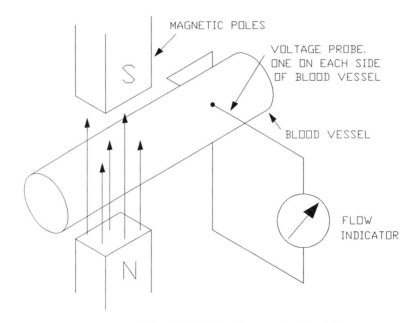

Figure 4-8 Utilizing the Hall effect in measuring blood flow.

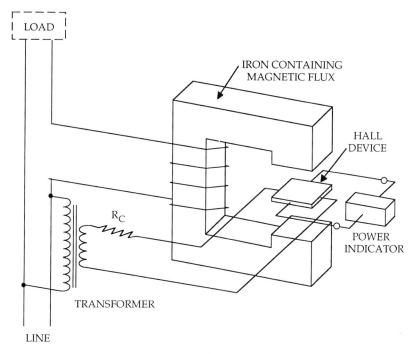

Figure 4-9 Using a Hall effect device in a wattmeter application.

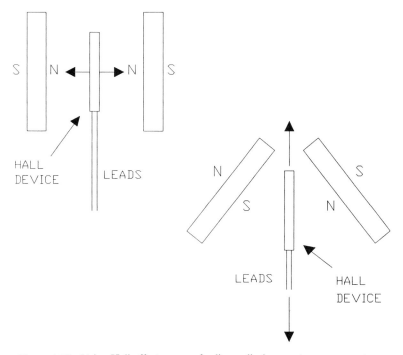

Figure 4-10 Using Hall effect sensors for linear displacement measurements.

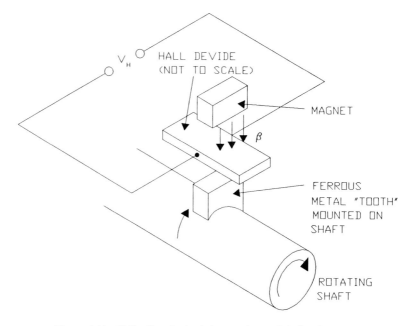

Figure 4-11   Hall effect device being used as a digital tachometer.

through the device, thereby generating a Hall voltage. The rotational accuracy of this system is $\pm\frac{1}{2}$ revolution. This very same system is easily adaptable to a flow-monitoring system for fluids in which the gear is replaced by a rotating turbine and its blades. The blades produce the magnetic field deflections, which periodically activate a Hall device, much as the gear teeth do in the tachometer.

4-5 INDUCTIVE PICKUP COIL

An inductive pickup coil is a coil that generates a current proportional to the change in flux of a magnetic field. Basically, this coil acts as an electrical generator, producing voltage according to Faraday's law. This type of sensor is used to sense the presence of a magnetic field and when used by itself, is used primarily as a proximity sensor. However, when used in conjunction with signal-conditioning circuits, the inductive coil can be used as an analog or digital device.

4-5.1 Theory of Operation

Faraday's law states that a voltage, e, will be induced across a conductor in a form of a coil having N turns of wire. This relationship was discussed in Section 3-6.1, where we discussed the moving-coil microphone; the equation is being repeated here for clarity. Faraday's law states the following relationship:

$$e = N \frac{d\Phi}{dt}$$
(4-6)

where e = induced voltage (V)

N = number of turns of wire in coil

$d\Phi$ = change in magnetic flux (Wb)

dt = change in time

In Figure 4-12 we see a conductor moving at right angles to a magnetic flux field produced by the poles of a magnet. The amount of voltage, e, induced into the conductor is dependent on the number of lines of flux being cut in a given instant of time. The greater number of conductors passing through this magnetic field, the greater the amount of induced voltage. Also, the greater the change in flux of this coil ($d\Phi/dt$), the larger e becomes. The important thing to note here is that for a voltage to be induced into the coil, there must be relative motion between the coil and the magnetic field. The absence of motion causes the $d\Phi/dt$ expression to equal zero, thereby causing the entire expression in eq. (4-6) to become zero.

The pickup head used in a recording and playback audio system is a magnetic induction coil whose construction is similar to that of the design shown in Figure 4-13. Note the added core in the coil's construction. The purpose of the core is to concentrate or localize the magnetic flux "sensitivity" of the coil. In this particular application the pickup head must be able to respond to very small areas of magnetism

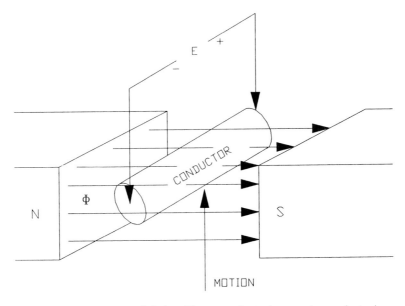

Figure 4-12 How current is induced into a conductor by a moving conductor in a magnetic field.

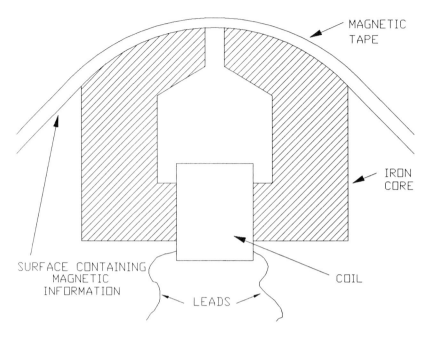

Figure 4-13   Using an induction coil in a magnetic tape audio system.

on a moving tape. Since it is too difficult to wind very small coils for this application, a core having low reluctance is used instead to concentrate the pickup area. This has the advantage of increasing the resolution or data density that can be previously recorded on the tape.

The read and write heads on a data storage disk device for a computer operate very similarly to the system described above. The only difference is that in a computer, you are dealing with receiving recorded digital data from the media in the form of magnetic pulses rather than in the form of variable-strength analog magnetic signals.

4-5.2 Operating Characteristics

Because the inductive coil functions only when there is a change in magnetic flux, it is relatively easy to use this type of device for a proximity sensor. However, for it to function in this role, the magnetic flux and coil must experience relative motion at right angles to each other for optimum voltage generation to take place. Relative motion at any angle other than 90° will produce correspondingly smaller values of e according to the sine of that angle. A typical installation as a proximity switch would be virtually identical to that of the Hall effect detector shown installed in Figure 4-10.

Two very important quantities associated with magnetic induction coils are their sensitivity to magnetic fields and their frequency response. Sensitivity has to do with the coil's ability to detect weak magnetic fields and is defined as the ratio of the change in the coil's output signal to a change in the magnetic field at the coil's input producing the change in the output. The units associated with this sensitivity value are V-m/Wb or V/T.

Frequency response for a magnetic induction coil refers to the coil's ability to respond to a particular range of magnetic signal changes per second. The unit of measure is the hertz but is rapidly being replaced by the unit *bits per second* (bps) in digital applications, where the concern is with *data rate* capability rather than frequency response.

One of the limiting factors that determine the frequency response of a coil is its *hysteresis,* a retention property in which the coil's magnetic core retains magnetism from a previous magnetic signal. This, in effect, desensitizes or biases the next-received magnetic signal, resulting in distortion or a false signal.

4-5.3 Practical Applications

The most frequent application for the magnetic induction coil is in the manufacture of the reading and writing heads used in data handling. These heads detect the very small magnetic pulses or variable magnetic areas implanted on magnetic media during a recording session. ("Magnetic media" refers to the material used in the manufacturing of recording tape, cassettes, floppy disks, and hard drives on computers, videocassettes, and so on. This material is comprised of iron oxide or other easily magnetized particles suspended on a plastic tape. Refer to Figure 4-16.) Figure 4-14 is a photograph of a typical recording head used in the common cassette player. Figure 4-15 is a photograph of a "read/write" head on a floppy disk drive in a computer. In both cases the magnetic media passing beneath the pickup head has been magnetized with an electromagnetic recording head or "writing head," very similar in construction to the pickup or reading head. Figure 4-16 illustrates how all of this is done.

Figure 4-14 Pickup head on a cassette player system.

Figure 4-15   Read/write head on a floppy drive in a PC. Heads are visible in center of photo.

4-6 VIBRATING-WIRE TRANSDUCER

Various transducers have been designed around the mechanical vibration propertie of stretched members inside an unknown magnetic field. The following transduce description is such a device. It is used for monitoring magnetic fields in general wher it is desirable to know if any fluctuations are taking place within that field.

4-6.1 Theory of Operation

In Figure 4-17 we see a stretched wire between the poles of a permanent magnet Its ends are fastened between two stationary posts, *a* and *b*. Because of the wire'

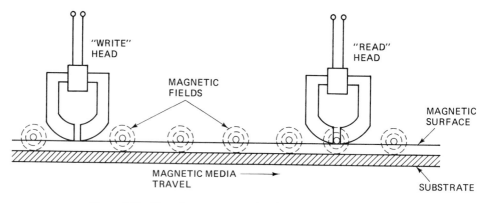

Figure 4-16   How charges are stored and read on magnetic media.

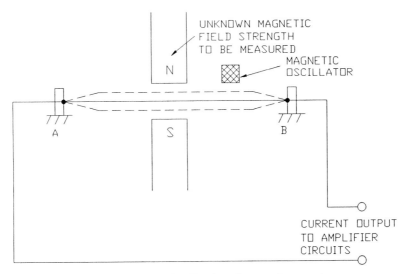

Figure 4-17   How the vibrating wire transducer works.

length, it will have a natural resonant frequency, f_r. This value is determined by the equation

$$f_r = \frac{1}{2L} \sqrt{\frac{T}{m}} \tag{4-7}$$

where f_r = natural resonant frequency (Hz)

   L = length of wire (m)

   T = tension force on wire (N)

   m = mass of wire per unit length (kg/m)

Initially, the wire is made to oscillate by a smaller oscillating magnetic field located near the wire. A small ferromagnetic sleeve or other attachment is installed on the wire so that it will respond to the field. The frequency of oscillation coincides with the natural resonant frequency of the wire. The induced current in the wire will then be in proportion to the larger magnetic field being monitored since the wire's frequency of vibration is held constant.

4-6.2 Practical Applications

This type of transducer can be used in applications where constantly maintained magnetic fields are needed, such as in certain research applications. Using a vibrating wire system for these measurements represents an independent source of measurement, rather than depending on related parameters associated with generating and maintaining the magnetic field.

REVIEW QUESTIONS

4-1. Explain why it is necessary to construct a magnetic reed switch inside a sealed, nitrogen-filled glass container. Construct a circuit in which a magnetic reed switch is used for the purpose of locating position.

4-2. What is meant by the term *viscous damping*?

4-3. Explain how the Hall effect works in a piece of semiconducting material. Show a practical circuit that uses a Hall effect device for proximity sensing.

4-4. Explain how an inductive pickup device works in conjunction with a gear tooth on a gear attached to a rotating shaft to determine the rpm of the shaft.

4-5. Explain how the vibrating-wire transducer works, and describe a practical application for it.

4-6. How would you use a magnetic sensor to measure the flow rate of water being transported in a pipeline?

4-7. Why is or is not a magnetic sensor a good device for analog sensing? Support your answer by explaining how the three most often sensed measurands are sensed.

4-8. Give a reason why magnetic sensing may be used instead of a mechanical proximity switch for certain applications.

4-9. Give two factors that limit the life or usefulness of a magnetic reed switch.

4-10. What advantage is there in using a Hall effect device for proximity sensing rather than a magnetic reed switch? Can you think of a disadvantage?

PROBLEMS

4-11. Calculate the Hall effect voltage generated by a charge traveling at 2.8×10^8 m/s in a conductor having a width of 0.5 cm which is exposed to a magnetic field of 1.93×10^{-7} Wb/m.

4-12. A vibrating wire transducer is made from a piece of No. 26 AWG steel wire of length 13 cm. The tension used to stretch the wire is 84 N, and its mass per length is 0.00127 kg/m. Determine the wire's resonant frequency.

4-13. Determine the tension needed for an 18-cm length of wire, with a mass per length of 0.0025 kg/m, to vibrate in a vibrating wire transducer at a frequency of 315 Hz.

4-14. Using the voltage calculated in Problem 4-1, calculate the Hall effect constant, K_H, associated with the system described in that problem. Assume a current flow of 13.4 μA.

REFERENCES

BLATT, FRANK J., *Principles of Physics*. Boston: Allyn and Bacon, 1986.

BUCHSBAUM, WALTER H., *Practical Electronic Reference Data*. Englewood Cliffs, NJ: Prentice Hall, 1980.

FINK, DONALD G., *Electronics Engineers' Handbook*. New York: McGraw-Hill, 1975.

5

PIEZOELECTRIC SENSORS

CHAPTER OBJECTIVES

1. To understand the theory of piezoelectricity.
2. To study the applications of the radio-frequency oscillator.
3. To understand how shock and vibration are measured.

5-1 INTRODUCTION

We encountered one use of the piezoelectric sensor in Chapter 3, where we discussed the crystal microphone. In that application we were concerned about the conversion of sound waves into electrical signals. However, as we will soon discover in this chapter, this is but one of several other common applications of piezoelectric substances.

5-2 THEORY AND DESIGN OF PIEZOELECTRIC SENSORS

5-2.1 Single-Element Crystal

As mentioned in Section 3-8.1, certain naturally occurring crystalline materials display that piezoelectric phenomenon. Recall from that discussion that a piezoelectric substance generates an electric charge when subjected to a sudden change in stress, such as a sudden blow to the crystal. Conversely, when this substance is exposed to a changing electrical charge, the physical shape of the crystalline substance is altered. Piezoelectric materials can also be produced artificially. Table 3-1 lists both human-made and naturally occurring materials.

Figure 5-1 shows the natural crystal shapes of the more popular piezoelectric crystals used. Notice that an axis coordinate system has been assigned to each crystal shape. The *z*-axis is the longitudinal axis in each case (although in the case of the lithium sulfate crystal, it is somewhat difficult to ascertain which is the longitudinal axis; notice that the *z*-axis in this case is parallel to the lines of intersection of the surfaces comprising the crystal's sides). The *x* and *y* axes can only be identified from the shape of the crystal and are therefore not as easily determined as the *z*-axis.

Figure 5-2 shows how a piezoelectric crystal can be fitted with an electrode plate on opposite faces so that a crystal can accommodate and respond to a particular stress direction. In addition, the output sensitivity of the crystal can be enhanced for certain directions merely by altering the crystal housing's design to allow it to move in that direction, as shown in Figure 5-3. Therefore, in reality, there are only two

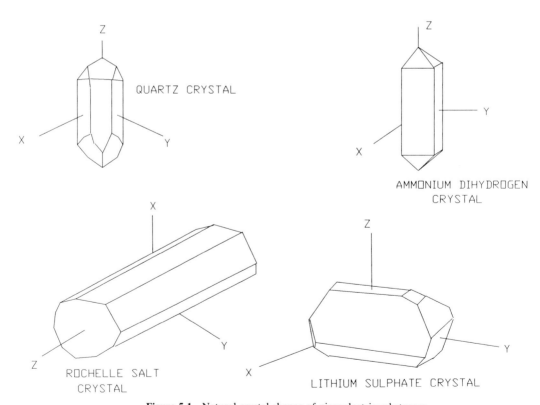

Figure 5-1 Natural crystal shapes of piezoelectric substances.

Figure 5-2   Assembly of a piezoelectric crystal in its holder showing electrode plates used on both sides of the crystal.

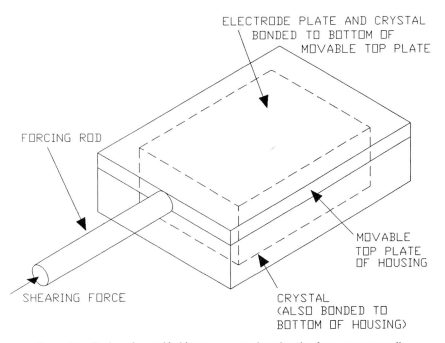

Figure 5-3   Design of crystal holder to accommodate shearing force across crystal's face.

modes of deformation associated with a piezoelectric crystal: shear and compression (or tension).

The crystal can be identified by the type of cut that has been made with a crystal slab cutter. These cutters are designed for cutting very hard crystalline materials and are usually impregnated with diamonds or other extremely tough wear-resistant cutting material. When a cut has been made through a crystal, resulting in a slab whose two major surfaces are perpendicular to, say, the *y*-axis, it is called a *y-cut* crystal. The other axes cuts are identified in a similar manner. It is also possible to cut slabs at angles other than perpendicular to an axis. The angle of cut has a significant effect on the crystal's behavior as a piezoelectric sensor, and crystal manufacturers have spent much time researching this effect.

The type of cut made in a certain crystal also has a dramatic effect on the crystal's output. For instance, a *y*-cut made from lithium sulfate has a very strong thickness expansion characteristic. An *x*-cut made from Rochelle salts has one of the highest outputs known.

In the case of *PZT* (lead zirconate–barium titanate) or "ceramic" crystals, the axes used for positioning and location information are determined by a means somewhat different from that described above. These axes are found by the direction of a dc polarizing field used to polarize the ceramic element after manufacture. This process is called *poling*. The poling direction is determined by the application of a high dc voltage during the crystal's manufacture. This direction goes from plus to minus on the polarizer's electrodes. Once polarized, the axes are identified, usually by a numbering system rather than by letters, so that we have axis 1, axis 2, axis 3, and so on. The directions of the applied stress and the polarization axis are referred to as the *mode* of the ceramic material. The mode is described by the numbered axes. For instance, referring to Figure 5-4, if the polarization direction were along axis 3 and the stress axis were along axis 1, the ceramic material would be operating as a mode 31 type (the first digit referring to the polarizing field or poling direction, the second digit to the stress direction).

It is interesting to note that when an applied dc voltage has the same polarity as that used for creating the poling direction, the element expands in the poling direction and contracts along the perpendicular axes. This is illustrated in Figure 5-5(a). However, when a force in tension is applied in the same direction as the polarizing field in this ceramic crystal, the output voltage generated by the crystal has a polarity which is *opposite* that of the original polarizing voltage [Figure 5-5(c)]. On the other hand, when a signal voltage is applied to an axis perpendicular to the poling voltage axis, shearing takes place within the crystal element, as shown in Figure 5-5(b). The direction of shear is determined by the applied voltage's polarity. Figure 5-5 also shows the relationships between other combinations of poling directions and applied force directions. Also, equations are given showing the mathematical relationships among these variables. Single-crystal slabs that are placed between electrodes as implied in Figure 5-5 are referred to as *sandwiched* crystals.

We have referred to the piezoelectric element's electrodes several times but have not described how these electrodes are constructed. In the case of the older quartz crystals these electrodes were nothing more than plates of copper that were

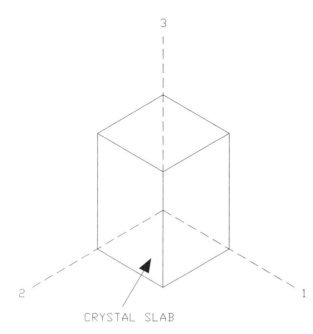

CRYSTAL SLAB

Figure 5-4 Operating modes of a crystal.

spring-loaded so as to press against opposite sides of the crystal's body. In the case of ceramic crystals the electrodes are either plated or sprayed onto the crystal element's body. Graphite and copper are two commonly used materials.

5-2.2 Design Information for Piezoelectric Ceramic Crystals

In designing and analyzing transducers using ceramic piezoelectric crystals, it is necessary to have a good idea ahead of time what sort of performance is to be expected from the crystal being used. Table 5-1 lists some of the nomenclature typically encountered in this design analysis process. Figure 5-6 shows both the *generator mode* and the *motor mode* for cantilevered series-connected or parallel-connected ceramic crystals and their relating equations. In the generator mode case these equations show the relationship existing between the output voltage (V) or charge (Q) and the applied input force (F). In the case of the motor mode, the equations show the relationship between the input voltage (V) and the output deflection (X). (Multiple-element crystal construction is covered in more detail in Section 5.2.3.)

It is important to note that in the cases presented, where output voltage or charge is shown as a function of applied force, F, the time dependency that generated voltage or charge has with respect to a force change in a piezoelectric device (see the discussion in Section 5-3.3) is not evident. This time change is inherent in the charge and voltage coefficient values referred to in the equations given in Figures 5-5 and 5-6. The equations listed in these figures are only approximations, as are the coefficients supplied elsewhere.

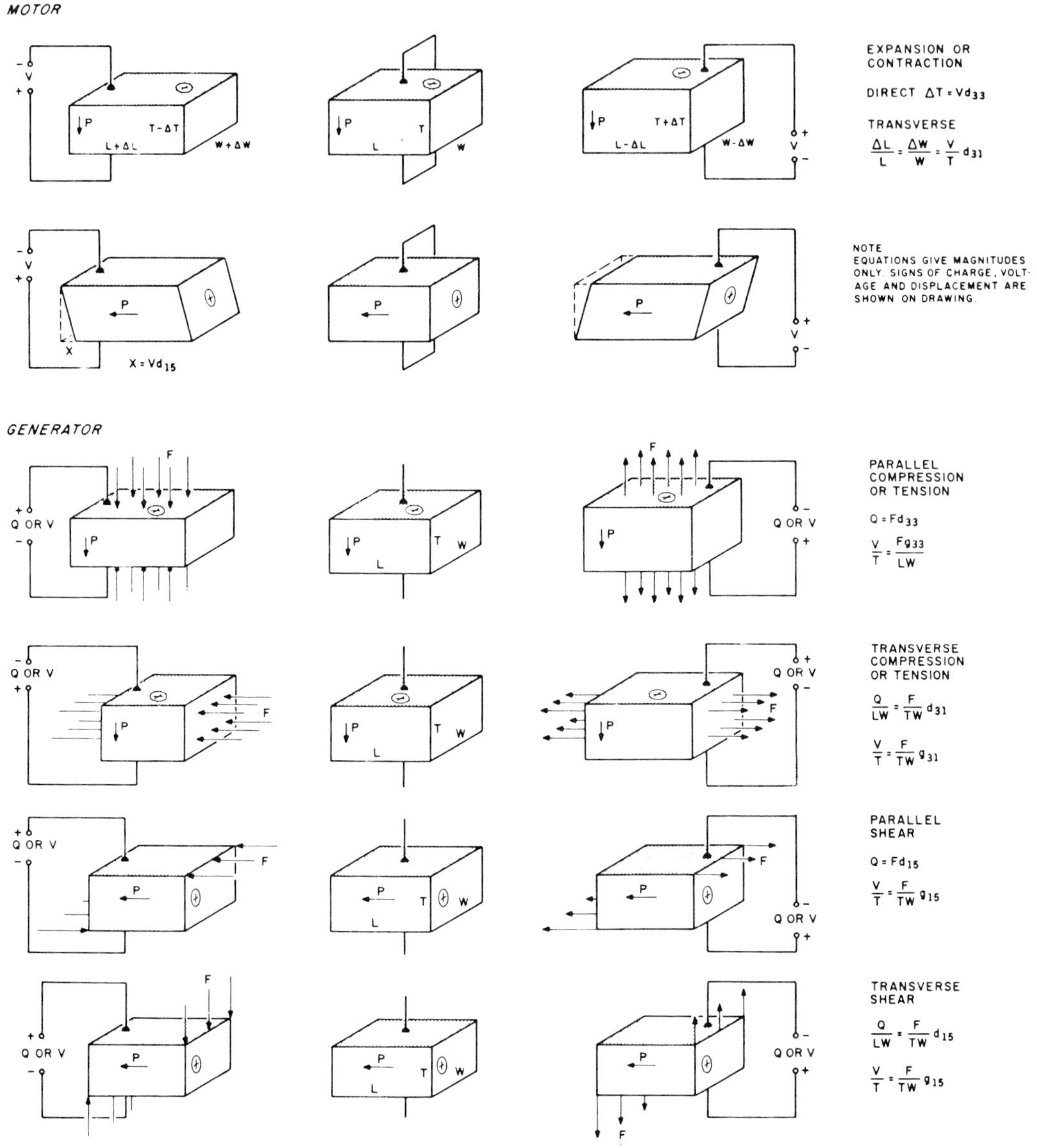

MOTOR

EXPANSION OR CONTRACTION

DIRECT $\Delta T = V d_{33}$

TRANSVERSE

$$\frac{\Delta L}{L} = \frac{\Delta W}{W} = \frac{V}{T} d_{31}$$

$X = V d_{15}$

NOTE
EQUATIONS GIVE MAGNITUDES ONLY. SIGNS OF CHARGE, VOLTAGE AND DISPLACEMENT ARE SHOWN ON DRAWING.

GENERATOR

PARALLEL COMPRESSION OR TENSION

$Q = F d_{33}$

$$\frac{V}{T} = \frac{F g_{33}}{LW}$$

TRANSVERSE COMPRESSION OR TENSION

$$\frac{Q}{LW} = \frac{F}{TW} d_{31}$$

$$\frac{V}{T} = \frac{F}{TW} g_{31}$$

PARALLEL SHEAR

$Q = F d_{15}$

$$\frac{V}{T} = \frac{F}{TW} g_{15}$$

TRANSVERSE SHEAR

$$\frac{Q}{LW} = \frac{F}{TW} d_{15}$$

$$\frac{V}{T} = \frac{F}{TW} g_{15}$$

Figure 5-5 Generator and motor modes of a "sandwiched" crystal showing applied forces and voltages and poling directions. (Courtesy of *Sensors*, North American Technology, Inc., Peterborough, NH, from the article, "Piezoelectric design notes," by Piezo Electric Products, Inc., March 1984, pp. 20–27.)

TABLE 5-1 Symbol Designations for Crystals

V	—Voltage
Q	—Electric charge
C	—Capacitance
F	—Force
T, W, L, & D	—Dimensions: Thickness, width, length, and diameter respectively
ΔT, ΔL & ΔD	—Small changes in dimensions
d_{33}	—Direct charge coefficient
d_{31}	—Transverse charge coefficient
d_{15}	—Shear charge coefficient
g_{33}	—Direct voltage coefficient
g_{31}	—Transverse voltage coefficient
g_{15}	—Shear voltage coefficient
$P\downarrow$	—Direction of the poling axis. The arrow is parallel to the poling electric field, pointing toward the negative poling electrode
k_{33}	—Direct electromechanical coupling coefficient
k_{31}	—Transverse electromechanical coupling coefficient
k_{15}	—Shear electromechanical coupling coefficient
k_p	—Planar electromechanical coupling coefficient
K_3	—Relative dielectric constant measured along the poling axis
K_1	—Relative dielectric constant measured at right angles to the poling axis
ϱ	—Density of ceramic
Y^E_{ij}	—Young's modules measured at constant electric field
Q_m	—Mechanical Q (quality factor)
P_r	—Remanent polarization
E_c	—Coercive field
Z_m	—Impedance at resonance
ε^T_{11}	—Free permittivity
ε^S_{11}	—Clamped permittivity
f_r	—Resonance frequency
f_v	—Antiresonance frequency

Source: Courtesy of *Sensors,* North American Technology, Inc., Peterborough, NH, from the article, ''Piezoelectric design notes,'' by Piezo Electric Products, Inc., March 1984, pp. 20–27.

144

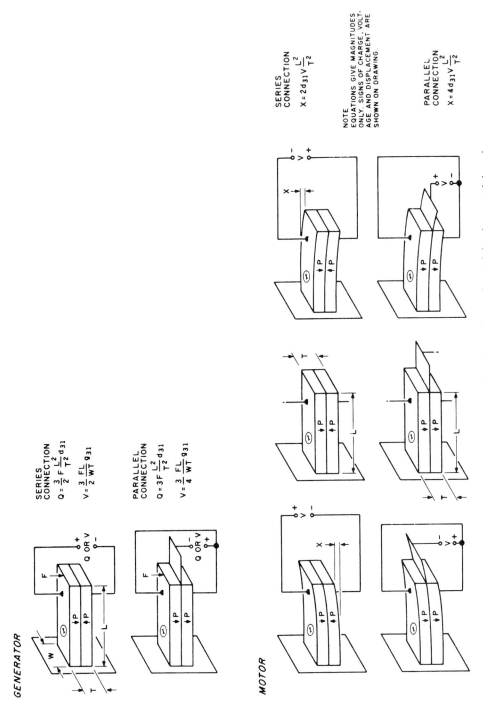

GENERATOR

SERIES
CONNECTION

$$Q = \frac{3}{2} F \frac{L^2}{T^2} d_{31}$$

$$V = \frac{3}{2} \frac{FL}{WT} g_{31}$$

PARALLEL
CONNECTION

$$Q = 3F \frac{L^2}{T^2} d_{31}$$

$$V = \frac{3}{4} \frac{FL}{WT} g_{31}$$

MOTOR

SERIES
CONNECTION

$$X = 2 d_{31} V \frac{L^2}{T^2}$$

NOTE
EQUATIONS GIVE MAGNITUDES
ONLY. SIGNS OF CHARGE, VOLT-
AGE AND DISPLACEMENT ARE
SHOWN ON DRAWING.

PARALLEL
CONNECTION

$$X = 4 d_{31} V \frac{L^2}{T^2}$$

Figure 5-6   The generator and motor modes of a cantilevered crystal showing crystals in series and parallel. (From *Sensors*, North American Technology, Inc., Peterborough, NH, from the article, "Piezoelectric design notes," by Piezo Electric Products, Inc., March 1984, pp. 20–27.)

The following example problems should help you become familiar with the various crystal configurations and the equations used for determining their characteristics.

Example 5-1

Calculate the amount of voltage generated with a cantilevered series-connected ceramic crystal of thickness 3 mm, length 5 mm, and width 20 mm. The force applied to the sensor containing this crystal is 5 N. The force is a direct force, with no associated frequency; it is applied one time only. The transverse voltage coefficient, g_{31}, is -11.1×10^{-3} V-m/N.

Solution: Since we are working with a cantilevered crystal in its generator mode, we refer to Figure 5-6 and find that part (a) seems to depict the crystal description we are working with. Using the equation alongside the diagram, we get

$$V = \frac{3}{2} \frac{FLg_{31}}{WT} \tag{5-1}$$

Inserting the values given in the problem, we get

$$V = \frac{(3)(5 \text{ N})(0.005 \text{ m})(11.1 \times 10^{-3} \text{ V-m/N})}{(2)(0.020 \text{ m})(0.003 \text{ m})}$$

$$= 6.9 \text{ V}$$

Example 5-2

What would the output voltage be if the crystal in Example 5-1 were used in a parallel-connected circuit rather than a series-connected circuit?

Solution: We will now use Figure 5-6(b) instead of part (a) for our system analysis, and will consequently use part (b)'s equation to calculate V.

$$V = \frac{3}{4} \frac{FLg_{31}}{WT} \tag{5-2}$$

$$= \frac{(3)(5 \text{ N})(0.005 \text{ m})(11.1 \times 10^{-3} \text{ V-m/N})}{(4)(0.020 \text{ m})(0.003 \text{ m})}$$

$$= 3.5 \text{ V}$$

We see that the output is only approximately one-half as much as the output in Example 5-1.

Example 5-3

Calculate the amount of voltage generated with the same crystal cut as above (i.e., having the same dimensions) and having the same applied force. However, assume that the crystal is now "sandwiched" (i.e., not cantilevered, but mounted between compression plates), that the force is applied in the same direction as the polarization field, and that both are along axis 3 (i.e., the Y-axis). The direct voltage coefficient, g_{33}, is 26.1×10^{-3} V-m/N.

Solution: Since we are working again with a crystal in its generator mode, but now sandwiched, we refer to Figure 5-5 and find that part (c) seems to describe the crystal design we are dealing with here. Therefore, we rearrange this equation to solve for the voltage, V:

$$V = \frac{Fg_{33}T}{LW} \tag{5-3}$$

Inserting our known given values, we get

$$V = \frac{(5\ \text{N})(26.1 \times 10^{-3}\ \text{V-m/N})(0.003\ \text{m})}{(0.005\ \text{m})(0.020\ \text{m})}$$

$$= 3.9\ \text{V}$$

5-2.3 Multielement Crystal

In an effort to obtain even higher output signals from a piezoelectric sensor, multiple crystal structures or "stacked" systems have been constructed. Figure 5-7 shows two such systems. In part (a) we see a series-stacked system in which two crystals have their electrodes wired such that when two crystal slabs are mounted face to face, only the outer two electrodes are used; the inner-facing electrodes are not used at all. In part (b) of the figure the outer two electrodes are wired together to form one lead and the two inner facing electrodes are wired together to form the other lead. This

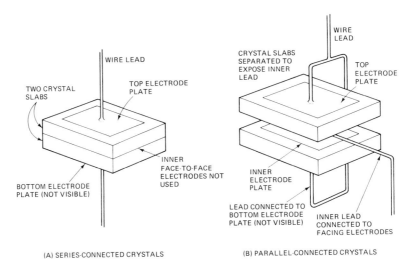

Figure 5-7 Multiple or "stacked" crystal systems.

particular construction is a parallel-stacked system. As in series and parallel voltage circuits in general, the series ceramic elements provide twice the voltage output as the parallel-wired crystal elements for the same input driving force. However, the parallel-wired elements provide twice the output displacement as that of the series elements for the same input voltage. In both cases just described, the two crystal slabs are cemented together with the appropriate electrodes connected and interleaved to form an integral unit (see Examples 5-1 and 5-2).

5-2.4 General Characteristics of Piezoelectric Materials

One serious problem associated with a piezoelectric sensor is its characteristically high output impedance. A hint of this characteristic can be seen while making a qualitative inspection of sorts of a typical crystal output condition. For a given crystal voltage output, very little current output will be observed. Because impedance, Z, and voltage, E, are related by means of the current, I, through a variation of Ohm's law, that is,

$$Z = \frac{E}{I} \tag{5-4}$$

it can be seen that for a given E and an extremely small I, a rather large impedance, Z, might well result. Furthermore, to transfer this high-impedance signal from the crystal to a signal cable, the cable itself must have an impedance that matches the crystal's impedance in order to obtain maximum signal transfer. Unfortunately, high-impedance cable, in the form of coaxial cable, suffers from a condition called *triboelectric noise*. This is a condition unique to this type of cable. Because of the dielectric material used in its construction and because of the possibility of relative motion existing between the cable's dielectric and the outer shield, electrostatic charges may easily develop. The net result is the creation of extraneous electrical noise such as popping or hissing. This is a condition brought on by cable movement as would be experienced in the case of using a piezoelectric crystal for vibration or impact measurements. A particularly troublesome vibrational frequency range for these cables seems to be below about 20 or 30 Hz. As a result, particular care must be used in selecting a low-noise high-impedance cable for this type of application.

Piezoelectric crystals are readily affected by temperature. As a matter of fact, any piezoelectric material will lose its charge generation characteristics if heated to a sufficiently high temperature. The temperature point at which the crystal's piezoelectric quality is lost is called the *Curie point*. This discussion, given in Chapter 3, bears repeating here. Some of the piezoelectric substances listed in Table 3-1, which concerned the discussion of piezoelectric microphones, are repeated in Table 5-2. However, for each of the substance types listed, the Curie points are also given.

As can be seen in Table 5-2, the range of Curie point temperatures is from 45 to 550°C. It is important to be able to identify the Curie point temperature value for the crystal in a given transducer design so that you have some idea what its maximum operating temperature limitations are.

TABLE 5-2   CURIE POINTS FOR CERTAIN PIEZOELECTRIC MATERIALS

Material	Curie point temp. (°C)
ADP	120
PZT	300
Quartz	550
Rochelle salts	45

Crystals are affected by temperature in another way, too. Their frequency of oscillation is directly related to temperature. This characteristic will be explained further in Section 5-3.1.

5-3 APPLICATIONS OF THE PIEZOELECTRIC TRANSDUCER

One company that specializes in the design and construction of the multielement piezoelectric structure is the Vernitron Corporation in Bedford, Ohio. Their trade name for this style of element is Bimorph. Figure 5-8 illustrates applications of this construction type in some products in common use. In the paragraphs that follow we discuss other applications of piezoelectric transducers and how they operate.

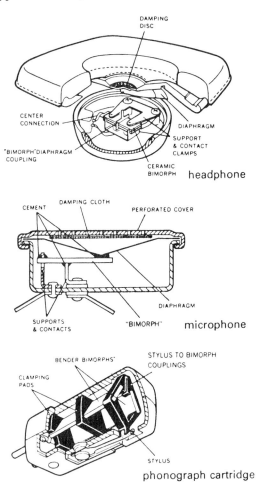

Figure 5-8 Bimorph crystals. (Courtesy of Morgan Matroc, Inc., Vernitron Division, Bedford, OH.)

5-3.1 RF Oscillator

Theory of operation

One of the unique characteristics of a piezoelectric element is that it vibrates at a particular fundamental frequency, depending on its physical size and type of cut. This characteristic can be observed very easily in the laboratory using very simple equipment. The following experiment description should illustrate this point.

If you were to attach the leads of an oscilloscope to a variable-frequency signal generator, you would, of course, be able to observe the variable-frequency output of the radio signal produced by that generator. You could also then observe and measure the amplitude of this signal. Next, place the leads of a piezoelectric crystal element in series with the "hot" signal lead between the generator and the oscilloscope. As you do this you will observe a decrease in signal amplitude on the scope's screen. This amplitude decrease would be caused by the introduction of what at first appears to be a certain amount of electrical impedance created within the crystal itself. This apparent impedance would then naturally produce a voltage drop, causing a subtraction from the signal generator's output voltage, therefore producing a smaller voltage at the scope's input. In reality, what is happening is that the crystal or ceramic element is caused to vibrate mechanically. As it vibrates, it generates a "clone" frequency at its output, giving the appearance of signal current conduction; quartz and piezoelectric ceramic are not electrically conductive. As you continue varying the signal generator's frequency output over a broad range of frequencies, you may find that for a particular frequency, the amplitude of the displayed signal increases suddenly and then decreases once again as you continue to change the generator's frequency. Going back to that frequency, you discover that the crystal appears to "conduct" that frequency much better than all the others. In other words, for a particular frequency a relatively "impedance-free" condition seems to exist. However, in reality, the resonant frequency of the piezoelectric element causes the element to "excite" itself into a vibrating mode, allowing a more vigorous voltage generation that duplicates the input frequency. If a spectrum analyzer were available for this experiment, the peak frequency characteristic would become very obvious on its display screen, as depicted in Figure 5-9.

How can this phenomenon be applied to radio-frequency (RF) oscillators? To begin with, let's review the purpose of an RF oscillator. An RF oscillator is an electrical circuit that has been designed to oscillate at one particular desired frequency. This frequency of oscillation is used in numerous electronic applications for timing purposes, or for electronic communications, as is the case in radio and television transmissions. Precise frequency maintenance is mandatory in these circuits. The piezoelectric crystal element is often used to maintain this required precision. Figure 5-10 shows a typical radio-frequency circuit in which a crystal is used for frequency stabilization, the circuit itself being a radio-frequency oscillator used in a communications transmitter. The frequency of oscillation is determined by the "tank" circuit seen in the transistor's collector circuit (i.e., L_1 and C_1). The setting of the variable capacitor, C_1, will determine the approximate output fre-

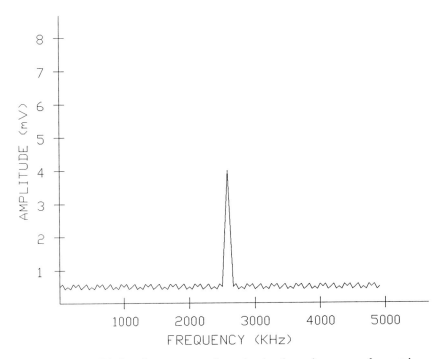

Figure 5-9 Display of a spectrum analyzer showing the peak response of a crystal.

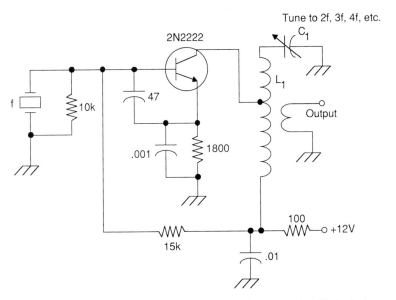

Figure 5-10 Typical radio-frequency circuit whose frequency of oscillation is determined by a crystal. (From *Radio Amateur's Handbook*, 60th ed., American Radio Relay League, Newington, CT.)

quency of this oscillator. As a matter of fact, one of the interesting features of a resonating crystal is that it can be made to oscillate not only at its designed fundamental frequency, but can also be encouraged to oscillate at multiples or harmonics of its fundamental frequency. (This characteristic will be discussed in further detail in the topic that follows.) This is why C_1 is shown as having the capability of being set at twice the frequency, f, three times f, and so on. Typical of many oscillating circuits, if you were to remove the crystal while supplying power to this oscillator circuit, the circuit might well continue oscillating. However, the frequency of oscillation would probably become erratic. However, with the crystal installed, the frequency becomes very stable and predictable. (In reality, if the crystal were removed, the oscillations might quit altogether. This is due to the fact that the crystal does contribute a certain amount of capacitance and inductance to the transistor's base circuit. Removing these quantities could interrupt the feedback signal coming from the tank circuit, causing the oscillator to shut down completely.) Figure 5-11 shows a schematic of a communications transmitter and receiver whose operating frequencies are dependent on crystals. Not only is the transmitter's frequency controlled by a switchable bank of crystals, but its companion receiver's reception frequency is controlled by a similar crystal bank.

On a somewhat related topic, crystals are often used as filters to discriminate against unwanted radio and audio frequencies. To understand how this is done, refer back to our lab experiment with the oscilloscope, signal generator, and crystal. Let's assume that we are transmitting a particular desired frequency within a wired circuit and this frequency contains information, such as a voice transmission or perhaps encoded data. Let's further assume that this transmission is taking place over a considerable distance such that an appreciable amount of undesirable noise has been introduced into the transmission circuit. As is typical with any undesirable random noise, this noise is comprised of many different frequencies all mixed with the single desirable frequency. Taking advantage of the fact that crystals can be produced for a very wide range of frequencies, we can have one manufactured to oscillate at precisely the desired frequency for the signal we want. Because of this characteristic, the crystal acts as a filter and selectively removes all surrounding undesirable frequencies. Figure 5-12 compares two circuits used for the purpose of receiving communications signals. Part (a) shows a portion of a radio receiver's intermediate-frequency circuit and its output frequency response characteristics. The same circuit is shown in part (b) but with crystals added in both legs of the intermediate frequency (IF) transformer; its frequency response is also shown. Notice that with the crystals added, the circuit's output is much narrower, and therefore much more capable of filtering out or blocking unwanted frequencies. We stand a much better chance of receiving only the desired frequency containing the information we want.

Operating characteristics

As mentioned in the discussion above pertaining to the oscillator in Figure 5-10, a crystal can be made to oscillate on harmonically related frequencies in addition to its fundamental frequency. This characteristic is true for many mechanical struc-

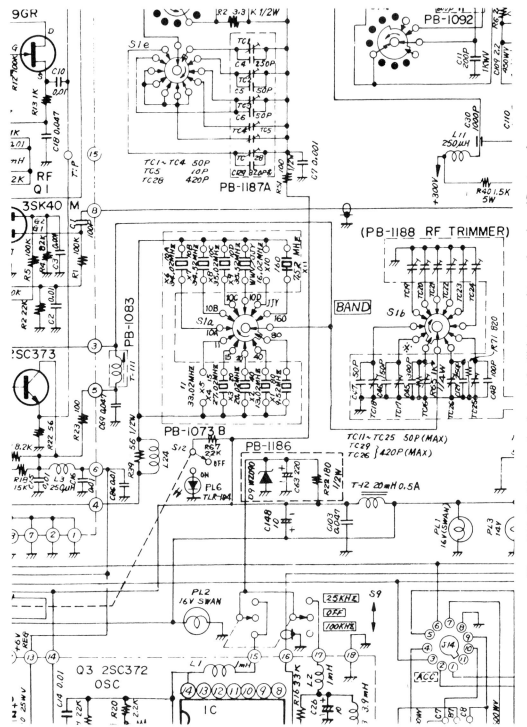

Figure 5-11 Partial schematic of a crystal utilizing transmitter and receiver. (Courtesy of Yaesu USA, Inc., Cerritos, CA.)

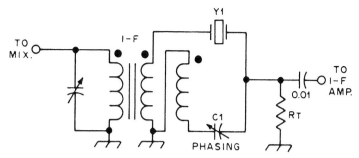

Figure 5-12   How a crystal can be used in communications receivers for the purpose increasing the receiver's selectivity. (From *Radio Amateur's Handbook,* 60th ed., American Radio Relay League, Newington, CT.)

tures, such as bridges, buildings, and suspended wires. All of these structures possess a fundamental resonant frequency and have the capacity of demonstrating resonance at related multiple frequencies. However, one not so obvious characteristic of certain crystal oscillators is that some may be made to oscillate very efficiently at their third or fifth harmonic frequency, depending on the design of the crystal's oscillator circuit. The frequency of this particular crystal is referred to as the third or fifth *overtone frequency* instead of the third or fifth harmonic. A crystal must be specially cut and mounted to allow it to oscillate at these odd-multiple frequencies. The fundamental frequency is not normally used in these crystals. This means that a third overtone crystal made to oscillate at 48 MHz is, in reality, a 16-MHz-cut crystal that is actually being used. Crystals made to oscillate above 10 MHz are usually of the third overtone type. It is not unusual to find crystals of these types that have been made to oscillate above 100 MHz.

A typical third- and fifth-overtone oscillator is shown in Figure 5-13. It is

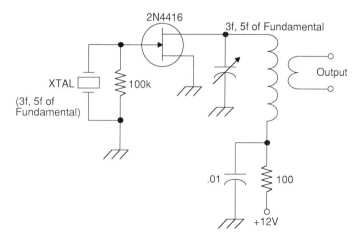

Figure 5-13   Third and fifth overtone oscillator circuit. (From *Radio Amateur's Handbook,* 60th ed., American Radio Relay League, Newington, CT.)

interesting to note that the third and fifth overtone values are only *approximately* three and five times the crystal's fundamental frequency. This is done for stability purposes. Crystals that have been cut to oscillate in this mode are very stable and reliable.

One of the perils of piezoelectric crystal operation in an oscillator is that the crystal tends to become warm. Crystals must be able to withstand heat, because this is the leading cause of crystal failure. If a crystal is part of an oscillating circuit that contains large amounts of circulating ac current, the crystal could easily fracture under the heat stress generated. This is especially true for higher-frequency crystals built from thinner slabs of piezoelectric material. Thinner slabs do not have the heat-dissipating qualities of the lower-frequency units.

Heat generated within a crystal has another detrimental effect: It produces frequency drifting. The heating causes the crystal to expand, therefore causing its mechanical resonance characteristics to change. Even though the crystal circuit's tank circuit is relatively fixed and stable, it does have an appreciable "bandpass" associated with it, meaning that the crystal's oscillating frequency could change and still be within the tuned frequency range of the tank circuit to sustain oscillations.

A typical temperature versus frequency drift for a crystal oscillator is shown in Figure 5-14. Once the drifting characteristics and oscillating frequency have been determined for an oscillator, the crystal oscillator itself can become an excellent calibrated thermometer for temperature measurement. It should be emphasized here that the type of crystal cut has a direct bearing on the temperature versus frequency drift curve shape. In Figure 5-14 the crystal cut is of the DT, BT, or CT style. Also, the operating temperature range for the thermometer is to either side of the zero-temperature-coefficient plateau associated with these types of cuts, the plateau occurring at 20°C in this case. Looking at this curve, we see a classic parabolic shape, with the vertex of the curve occurring at 20°C in this case. By proper crystal shaping, this vertex point can be shifted to virtually any desirable location. The vertex point is a point of zero frequency shift. Moving the temperature slightly to either side of this point does not produce any appreciable frequency shift. Consequently, we would want to avoid crossing this point in our calibrated thermometer application. The desired operating point would be somewhat along the arms of the parabola.

For the sake of maintaining good frequency stability in an oscillator, we would want to use a crystal that has no associated frequency drift. This is especially important for critical timing and communications applications. The Federal Communications Commission has very strict guidelines concerning commercial transmitter operations. In these cases the AT-cut crystal has vastly superior stability qualities compared to the other cuts. A typical frequency–temperature drift curve for an AT-cut crystal is shown in Figure 5-15. Because of the somewhat sinusoidal appearance of this curve, there are two areas, called *turnover points*, where zero-temperature coefficients exist. These are shown as plateaus *a* and *b* on the curve in example 1. Again, by utilizing special crystal-shaping techniques on the crystal slab itself, it is possible to reduce the "amplitude" of the sinusoidal waveform so that the

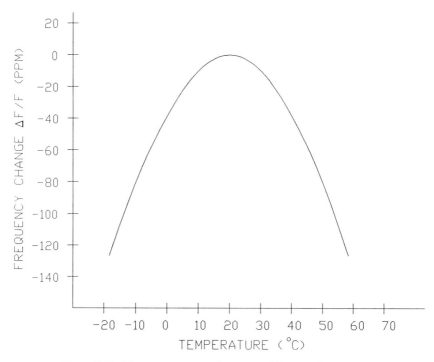

Figure 5-14   Temperature versus frequency drift curve for a crystal.

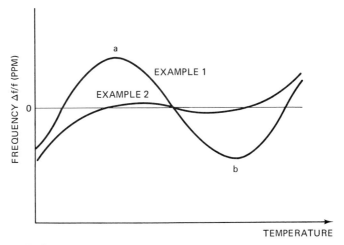

Figure 5-15   Temperature versus frequency drift curve for an AT-cut crystal.

middle portion of the curve, where the slope reverses itself, is quite flat. As a result, one could have a crystal whose frequency-change coefficient is zero over a very broad range of temperatures. This is demonstrated in example 2 in the figure.

Another characteristic associated with crystal-type oscillators is that the frequency of oscillation is affected by the *drive level* of the crystal. In other words, crystals must be supplied with some amount of power, called drive, that is supplied at timed intervals to keep the crystal oscillating. This driving power usually comes from some sort of feedback circuit within the oscillator circuit itself. It is this drive that can produce shifts in oscillation frequency. The rule of thumb is to use as little drive as possible to sustain oscillations, to keep frequency drift to a minimum. Generally, crystals require about 1 or 2 mW for their operation. This can account for a change of frequency of 1 or 2 ppm. Ideally, to minimize drift, driving amounts should be held down in the microwatt range, if at all possible.

Another factor that affects crystal stability is aging. As time passes, a new crystal undergoes a stress-relieving occurrence between the crystal slab and its mechanical holder—electrodes inside the main crystal-holding unit. This, in turn, causes a shift in the crystal's overall vibrating frequency. Also, there is a small movement or shift in the crystal's mass over its surface which, again, affects its vibrating frequency. These two happenings can take place over a relatively short period of time, measured in days, to as long as a year or two before the crystal finally settles down. This settling-down period produces a gradually diminishing aging rate, which produces a slowly reducing frequency shift as time goes on. Figure 5-16 shows a typical aging curve for a crystal mounted in an HC-6 or HC-18 holder (see Figure 5-22 for a description of these holders).

Physical configurations

The crystal oscillator is usually a circuit external to the crystal itself. The crystal is housed inside a small container having either protruding pins for plugging into a socket, or wire leads for soldering onto a circuit board. One of the older crystal containers, or holders as they are sometimes called, was shown in Figure 5-2. The crystal is shown removed from the holder, showing the mounting style and the electrode plates used. This particular unit is called an FT-243 holder, a holder that was very popular during and immediately following World War II. Many of these crystal holders are being used to this day, although very few are being manufactured. This type of crystal was generally operated at its fundamental frequency and was therefore seldom used at the fundamental frequency, above 10 or 12 MHz.

Figure 5-17 shows the more modern crystal holders popular today. These take up only a fraction of the space of the older FT-243 holders and require much smaller sockets or none at all. Many of these crystal types are designed to operate at their third or fifth overtone frequencies. The crystals themselves are mounted inside soldered hermetically sealed tin canisters. Either the air has been removed by means of establishing a vacuum, or a dried inert gas such as nitrogen is sealed in with the crystal to prevent corrosion and moisture from developing.

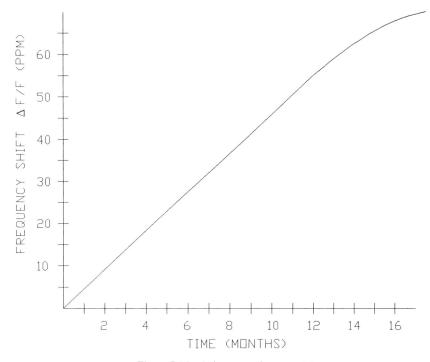

Figure 5-16 Aging curve for a crystal.

Figure 5-17 Popular crystal holders in use today.

Because of the temperature-sensitive nature of crystals, it is not unusual to find crystals being operated in temperature-controlled environments in an effort to minimize drift. One way to create such an environment is to use a *crystal oven*, as shown in Figure 5-18. The oven is simply an electrical heating element surrounding the crystal holder, whose temperature is maintained by some sort of thermostatic circuit.

5-3.2 Piezoelectric Microphone

As mentioned at the beginning of the chapter, we have already discussed the piezoelectric microphone in Chapter 3 under the subject of sonic-type sensors. Refer specifically to Section 3-8. Also, refer to our discussion of the use of piezoelectric sensors in sonar detection devices in Section 3-10.

5-3.3 Measuring Shock and Vibration

Theory of operation

The concept of acceleration was presented in Section 1-5.9 in our discussion of force. Recall from this discussion that we stated that an unbalanced force acting on a mass will produce an acceleration in the mass, all of which is related by the well-known relationship force = mass × acceleration. What we want to be able to do is to measure this acceleration using some sort of acceleration-sensitive transducer. How-

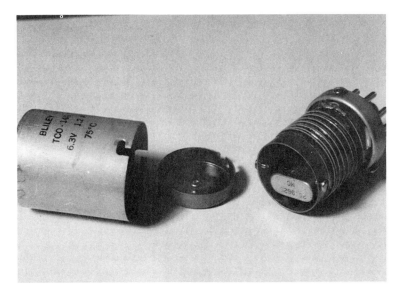

Figure 5-18 Typical crystal oven for maintaining a constant operating temperature environment for the crystal.

ever, as it turns out, devices referred to as "piezoelectric accelerometers" are really not used to measure acceleration. The name *piezoelectric accelerometer* is, in fact, somewhat of a contradiction in terms, as will soon be explained. Piezoelectric accelerometers are used to measure shock (or impact) and vibration, as explained below.

A piezoelectric substance is a mechanical sensor, but responds by producing an electrical output only if there is a change in the applied force. In other words, a constant force applied to a piezoelectric crystal will not produce electrical output. The force's magnitude *must change* to produce any measurable output at all. Additionally, the crystal's output will vary inversely with the magnitude of the time over which the change in force is in effect. Expressed as an equation, we have

$$v_o = K \frac{dF}{dt} \tag{5-5}$$

where v_o = output voltage (V)

K = proportionality constant (V-s/N)

dF = change in applied force (N, lb)

dt = change in time over which the change in force took place (s)

Since $F = ma$,

$$v_o = K \frac{d(ma)}{dt} \tag{5-6}$$

where v_o = output voltage (V)

K = proportionality constant (V-s/N)

$d(ma)$ = incremental change in mass times acceleration (kg-m/s^2).

m = mass (kg, slugs)

a = acceleration (m/s^2, ft/s^2)

dt = incremental change in time (s)

Perhaps now you can see why a piezoelectric accelerometer would never work strictly as an accelerometer. According to eq. (5-3), as long as $d(ma)$ is zero, v_o will remain zero. Even if $d(ma)$ were to assume some value *over an infinitely long period of time* [i.e., $d(t) = \infty$], there would still be no value for v_o. Values for v_o occur only when there is a *change* in the acceleration brought on by a *change* in the applied force over a measurable period of time.

In the case of both shock and vibration, we are talking about changes in acceleration direction. In the case of shock, an object may encounter a sudden change in its velocity, such as in the case of collision with an immovable object. This is where the moving object's velocity is suddenly reduced to zero. A very large deceleration value (i.e., an acceleration value in the direction opposite to its original velocity direction) is produced as an outcome of this event.

In the case of a vibrating object, we are describing what usually consists of rapid and cyclic changes in the object's velocity *direction*. That is, the object first heads in one direction, then travels in the opposite direction. These reversals in velocity direction imply that for an instant, the velocity of the object drops to zero. The object then reassumes a velocity value, but in the opposite direction. These velocity reversals represent changes in velocity that produce acceleration.

To distinguish between acceleration of the type just described and the acceleration produced by a change in velocity while traveling in one direction only (probably the more familiar type of acceleration), we will refer to the latter as *linear* acceleration.

Figure 5-19 represents a simplified view of how a piezoelectric accelerometer operates. The force shown in this illustration can be a "created" force. That is, it can be created by a mass, called a *seismic mass*, that reacts to acceleration. If that acceleration changes [i.e., $d(ma)$ assumes a value over a change in time, $d(t)$], the crystal produces an output response. Just such an arrangement is depicted in Figure 5-20, where we see a seismic mass attached to a piezoelectric crystal inside a cylinder containment. In one example we see the seismic mass riding on the crystal slab, causing compression. In the other example the seismic mass is attached to the crystal so as to produce shear. In either case, any change in the acceleration of the mass will produce an output at the crystal's terminals. Keep in mind, though, that what is being said here concerning acceleration has to do with the acceleration produced by vibration or shock, not with linear acceleration.

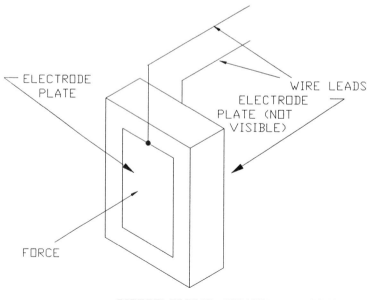

Figure 5-19 How a simple crystal accelerometer functions.

ge: "−40 to +100°C." The temperature range is the range of
within which the transducer has been designed to operate with-
unanticipated results.

rsus temperature effect: "less than 0.1%/°C." Since temperature
the behavior of most transducer devices, this figure indicates to
ee the transducer's output/input is affected.

quirements: "12 to 25 V dc at 4 mA, max." This transducer requires
wer source that can supply dc voltage anywhere between 12 and 25 V.
ower source, should it be able to supply 25 V, must also be able to have
rent capacity of at least 4 mA.

sical configurations

5-21 shows a typical configuration for an accelerometer. The transducer
ng is usually circular in shape and is often constructed of stainless steel. The
ainer itself is usually hermetically sealed to prevent contaminants from entering.
stainless construction makes the transducer virtually corrosion-proof. Physical
tline data are also given. The hermetically sealed casing allows the transducer to
e installed in certain hazardous locations. These locations are usually given in the
specifications in the form of hazardous class types, groups, divisions, and so on. The
various categories are usually defined by a national standardization organization.

The accelerometer casing is usually supplied with a connector containing pipe
threads for attachment to metal conduit. This is an electrical safety requirement

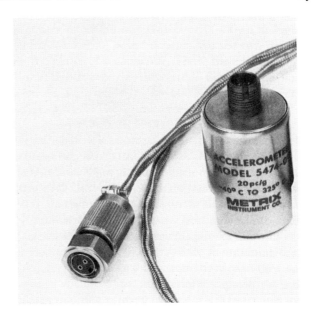

Figure 5-21   Typical accelerometers. (Courtesy of Metrix Instrument Co., Houston, TX.)

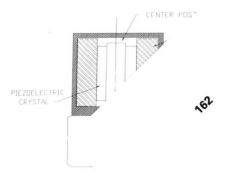

CENTER POST

PIEZOELECTRIC
CRYSTAL

Temperature ra
temperatures
out producin
Sensitivity v
does affec
what deg
Power r
a dc po
This
a cu

Ph

Figure
housi
con
Th
o

Operating characteristic

When analyzing a piezoelectric
several operating characteristics c
back to Section 1-9 for definitions o1
used by sensor/transducer manufacture

Typical characteristics and expected
the behavior of piezoelectric accelerometers
These figures are sample values only and do no.
catalog specification sheets. The reference given in
the definition of that term in this book.

Type: "Piezoelectric." This description refers to 1.
being used.

Sensitivity (Section 1-9.2): "100 mV/g, ±5% at 25°C," wh.
tional constant, or 9.8 m/s². In this case, for every g of accelera
be a 100-mV signal generated, ±5%, assuming that the operatii.
was 25°C. Any acceleration amounts greater than or less than.
produce outputs proportionally greater or lesser. Any other ten.
would produce a different sensitivity figure (see "Sensitivity versus Te.
ture Effect" below).

Cross-axis sensitivity: "5%." The transducer is built to respond to acceleratio.
along one axis only. Any response appearing at right angles to that axis is the
cross-axis sensitivity. In other words, only 5% of the "sensitized" axis signal
will appear at a right-angle axis in this case.

Dynamic range (Section 1-9.7): "0.005 to 50 g's." The term *dynamic* refers to
the transducer's sensing range capability under actual "dynamic" usage. The
dynamic range is given in gravitational constant g units.

Frequency range: "5 Hz to 8 kHz." This is the range of frequencies of the
measurand to which the transducer is capable of responding adequately; in this
case it is the range of vibration frequencies. It should be noted here that the
stated range *includes* the −3-dB down points of the transducer's frequency
response. (Refer to Section 1-9.1 for a discussion of frequency response.)

to prevent exposed wires from occurring in industrial applications. The wires are for supplying power to the transducer and for conducting the output signal from the transducer to an instrument or processing center. It is often possible to specify some other connector style if the transducer is needed for other than industrial applications.

Practical applications

A common problem in servicing industrial equipment is trying to find the source of excessive machine vibration and determining its cause. Often, knowing the frequency of vibration will give a clue as to the vibration's source, as depicted in Figure 5-22. Using the equipment example in this figure, the following observations might very well have been made:

1. If the detected frequency is approximately 60 Hz, the probable sources are electrical motors or vibrating transformers, the vibrations due to 60-Hz magnetic fields interacting with nearby metals.
2. If the detected frequency is approximately 1 to 3 Hz, the probable cause is the part ejector.
3. If the detected frequency is 1600 to 1800 Hz, the probable source is the main drive motor.

The rotational shaft speeds of machines are a frequent target of application for accelerometers sensitive to shock or vibration detection. These accelerometers take advantage of the fact that no rotating shaft is exactly balanced. As a result, the slightest imbalance can usually be detected by these transducers (Figure 5-23).

Another frequent application of accelerometers is in the detection and measurement of shock or impact. This is a one-time-only event as opposed to the repetitive nature of vibration. However, the piezoelectric detector responds just as well to this type of event as it does to the other. Figure 5-24 shows a typical crash test being conducted on one of its products by an automotive firm. The purpose of the test is to ascertain how well the car holds up under an impact so that better safeguards can be incorporated into the vehicle's design. To do this, the car has to be instrumented with transducers, at least one of which is an accelerometer, to measure the deceleration resulting from the car's impact into a solid wall. The signals generated by the accelerometer, along with the signals from the other transducers, are sent through an attached flexible cable to a nearby processing site, where the data are analyzed. Many of these signals may also be transmitted by radio rather than through the attached cable.

Another example of measuring impact is found in the packaging industry. Most manufactured products have to be packaged for shipping. The design of the shipping container dictates that the contained product must be able to withstand a certain impact, usually measured in *g*'s. A shipping simulator is used for testing these containers and can be as simple as the one illustrated in Figure 5-25. In this particular

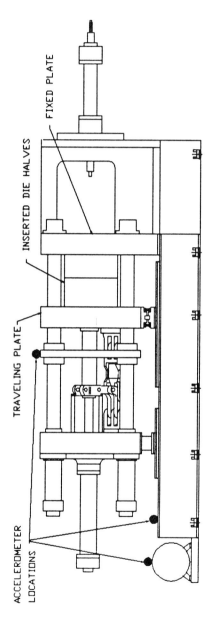

FIXED PLATE

INSERTED DIE HALVES

TRAVELING PLATE

ACCELEROMETER
LOCATIONS

DIE CASTING MACHINE

Figure 5-22 Production machine outfitted with an accelerometer for troubleshooting. (Adapted from Wayne Alofs and James R. Carstens, *Mechanical Maintenance and Evaluation of Die Casting Machines*, copyright 1987. Reprinted by permission of the North American Die Casting Association, River Grove, IL.)

Figure 5-23   Using accelerometer probe to detect vibrations in an electric motor.

Figure 5-24   Crash testing a vehicle equipped with accelerometers. (Courtesy of Chrysler Corp., Detroit, MI.)

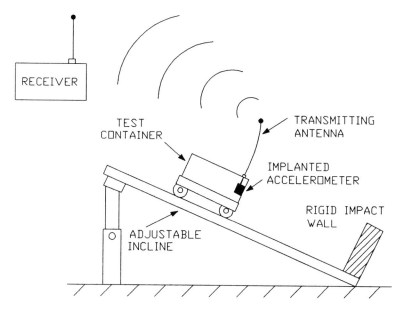

Figure 5-25 Shipping simulator used for the testing of product shipping methods.

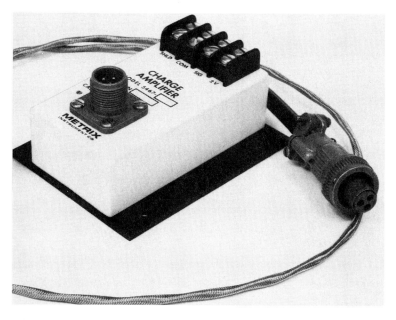

Figure 5-26 Charge amplifier used in conjunction with piezoelectric accelerometers. (Courtesy of Metrix Instrument Co., Houston, TX.)

setup we see nothing more than an inclined plane whose incline is adjustable. The container design is allowed to roll down the incline and smash against a rigid wall at its bottom. Implanted in the container is an accelerometer that continuously radios information to a nearby receiver, where the data are recorded and analyzed. The only data generated and recorded, however, will be the pulse generated at the moment of impact at the incline's bottom.

A piece of hardware related to a piezoelectric accelerometer is the charge amplifier. Because piezoelectric crystals generate voltage at a very high impedance, resulting in very little current generation, the crystal is often thought of as a charge generator rather than a voltage generator, even though either is correct and acceptable. Amplification of this signal requires special circuitry, not the usual amplifier circuits. A typical charge amplifier for an accelerometer is shown in Figure 5-26. This type of amplifier must be able to amplify the charge generated and turn it into a proportional voltage with current drive capabilities. This type of output implies that we now have a signal that is much more readily compatible with other electronic circuitry because of its much lower impedance, typically 50 to 75 Ω or so.

REVIEW QUESTIONS

5-1. Explain what is meant by the term *mode* in relation to a crystal.

5-2. Explain the difference between the motor mode of a crystal and its generator mode.

5-3. Explain the difference between a series-stacked and a parallel-stacked multielement crystal. What are the characteristics of each?

5-4. What is the Curie point of a crystal? Why must this factor be considered in using a piezoelectric transducer in a particular environment? How is the crystal's frequency of oscillation affected?

5-5. Explain in detail how a piezoelectric substance is used in conjunction with a seismic mass to produce an accelerometer.

5-6. If quartz and piezoelectric ceramic are not electrically conductive, explain how it appears that a current is "conducted through" these substances at their resonant frequencies.

5-7. Explain how a piezoelectric crystal could be used as a temperature transducer when used in an RF oscillator.

5-8. Why are higher-frequency crystals more prone to fracturing than lower-frequency crystals?

5-9. Why are AT-cut crystals so superior in their frequency-drift characteristics to other styles of crystal cuts? Explain an AT-cut crystal's turnover points.

5-10. Explain a piezoelectric accelerometer's cross-axis sensitivity.

PROBLEMS

5-11. Determine the voltage produced by a cantilevered series-connected crystal having the following crystal dimensions: length 7 mm, width 15 mm, thickness 2.5 mm, and applied force 2.32 lb. Assume a voltage constant, g_{33}, of -11.1×10^{-3} V-m/N.

5-12. How much force must be applied to a parallel-connected cantilevered crystal to create a 9.5-V signal? Crystal dimensions are: length 7.5 mm, width 12.3 mm, and thickness 1.92 mm. Assume a g_{33} value of -11.1×10^{-3} V-m/N.

5-13. Assuming that the force calculated in Problem 5-12 was applied over a time of 35 μs, calculate the crystal's proportionality constant, K. Be sure to indicate the proper units associated with this constant.

5-14. It was found through measurement that the frequency response of a vibration transducer had the range 24 to 1260 Hz. Prove mathematically that this transducer has a -3 dB reduction in output when its voltage output is reduced by 29.3%.

5-15. A piezoelectric transducer has an output voltage of 4.32 V ac at 3.51 μA. What is its output impedance?

5-16. A shipping tester's oscillograph of a "handled" package is shown in Figure 5-27. This output represents the output of a piezoelectric transducer. If the vertical scale represents a force of 5 N per division (not to be confused with the 1 V/div reading for the same scale) as calibrated from previous tests, what is the crystal's proportionality constant?

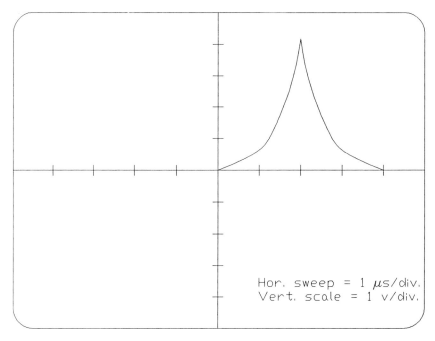

Hor. sweep = 1 μs/div.
Vert. scale = 1 v/div.

Figure 5-27   Figure for Problem 5-16.

5-17. If, in Problem 5-16, the seismic mass contained within the piezoelectric transducer had a mass of 820 g, what acceleration was experienced when the force was 4 N?

REFERENCES

FINK, DONALD J., *Electronics Engineers' Handbook*. New York: McGraw-Hill, 1975.

NORTON, HARRY N., *Sensor and Analyzer Handbook*. Englewood Cliffs, NJ: Prentice Hall, 1982.

The Radio Amateur's Handbook. Newington, CT: American Radio Relay League, 1983.

SANDS, LEO G., and DONALD R. MACKENROTH, *Encyclopedia of Electronic Circuits*. West Nyack, NY: Parker, 1975.

6

Voltage Sensors

CHAPTER OBJECTIVES

1. To review the concept of electrical potential.
2. To understand the purpose of the voltage sensor.
3. To study the operating characteristics and practical applications of the voltage sensor.

6-1 INTRODUCTION

Up to this point we have discussed sensing devices that have been considered by industry to be somewhat classical or conventional in nature. That is, these devices are known to have a specific function and have been proven and accepted by industry for some time. On the other hand, potential difference or voltage-sensing devices are a relatively new category. They are not usually mentioned when electrical transducers are discussed. However, I feel that recent events and developments in the electronics industry justify their being discussed here. Identifying and keeping track of potential differences that may occur between objects or components within a system is becoming more and more important as we become more sophisticated in our system design efforts. For instance, it has only been within the past few years that electrostatic detection, which is a form of voltage detection, has received much interest. This has been due primarily to research efforts to determine the effects of electrostatic charges on the environment and on human behavior. On the consumer level, it has only been within the last 10 or 15 years that adequate ground-fault detection in home and commercial building construction has become a real concern and that devices have been constructed to detect such problems. Because of the newness of these issues and because few sensing systems have been devised for general use for electrostatic and electromagnetic voltage detection, this area appears to be somewhat fertile at present as far as the market is concerned.

6-2 REVIEW OF ELECTRICAL POTENTIAL THEORY

Earlier it was suggested that the term *voltage* was synonymous with the term *pressure*. Both can be created by a resistance to flow, both must be measured relative to a reference point or location, and if their magnitudes become large enough, both can damage or destroy the conductor containing them. To understand voltage, let us first investigate the concept of the *charged particle*.

Most particles in nature can be considered electrically neutral under normal circumstances. That is, they are neither positively nor negatively charged with respect to adjacent particles. Figure 6-1 depicts a typical particle and its subatomic particle makeup. Each of these naturally occurring particles contain just as many electrons (negative subatomic particles) as they do protons (positive subatomic particles). If a particle is not neutral, it is said to be *polarized*. Its polarization will be either positive or negative compared to a reference point or reference charge with known polarization. The particle's polarity cannot be determined simply by looking at it since it is not practical to look inside the particle to determine its subatomic makeup. It is much easier to compare it to a reference particle with known polarity.

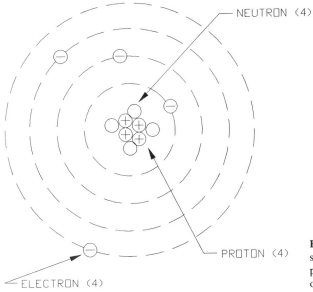

Figure 6-1 One atom of beryllium showing the nucleus comprised of four protons and four neutrons with its four orbiting electrons.

Quite often, just holding a particle of unknown polarity near the body of the reference particle will give away the unknown particle's polarity. Because experimentation has taught us that like charges repel each other while unlike charges attract, we can utilize this fact to determine the unknown particle's polarity simply by observing its motion near the known particle. Although this may be an oversimplification of the polarity-determining process, this is basically how it is done.

Determining the quantity or degree of polarity is quite another problem, however. For this particular exercise, a calibrated voltmeter is required. One of the meter's two probes is connected to the surface of the unknown particle while the other probe is connected to the surface of the reference particle, as demonstrated in Figure 6-2. The voltmeter will then indicate the potential difference that exists between the two. The meter will also indicate the polarity of the unknown particle relative to that of the reference.

The next question may logically be: What particle or object is usually used for making reference voltage measurements? The logical answer is: The earth, in most cases. The earth is considered to be a reservoir, much like the reservoir used in hydraulic systems for supplying and receiving hydraulic fluids circulating inside the hydraulic circuitry. However, in the earth's case, its bulk acts as a supply and return reservoir for electrons. Like the hydraulic reservoir, "pumps" (i.e., batteries, electrical power supplies, etc.) are needed to circulate the electrons through the circuitry to cause the desired work or action to be done and then to return them to the earth's ground, as shown in Figure 6-3. As a result of all of this, a good many of the voltages measured are measured with respect to earth's ground.

Often, it is necessary to measure voltages relative to a point within an electrical circuit other than earth ground. For instance, it may be necessary to measure a voltage drop across a component or several components. This is similar to measuring a pressure drop using a pressure gage or manometer across a hydraulic component, as shown in Figure 6-4. In this instance, a potential difference, or voltage drop, is

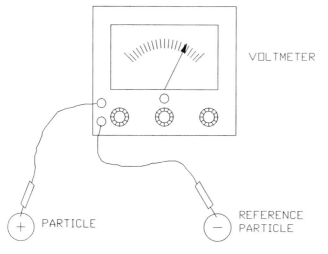

VOLTMETER

PARTICLE

REFERENCE
PARTICLE

Figure 6-2 Using the voltmeter with its negative lead as reference.

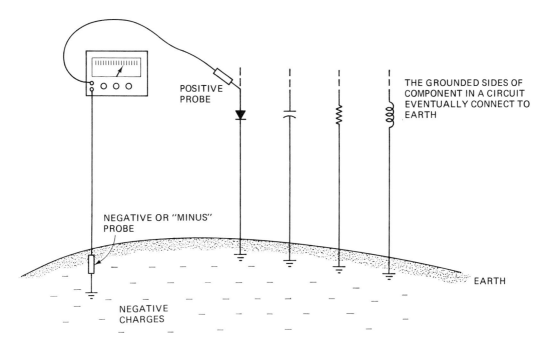

Figure 6-3   Earth acting as a "reservoir" or ground reference.

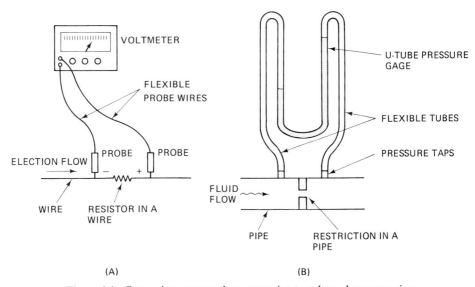

(A)   (B)

Figure 6-4   Comparing pressure drop measuring to voltage drop measuring.

being measured relative to another part of the circuit. If you state what magnitude that voltage is, you have to be specific and also state that the voltage was measured with respect to the indicated reference point where the voltmeter's other probe was placed.

It should be emphasized at this point that the direction of electron flow within a circuit has a profound effect on the polarity of the voltage drop occurring across components. That is, electron flow is always from a negative source to a positive source. Conversely, current flow, referred to in some textbooks as *conventional* electron flow, always flows from a positive source to a negative source, just the opposite of electron flow. (The latter current flow convention was necessary to correct a somewhat embarrassing error that was made in the selection of electron flow direction many years ago based on the information that was available at that time.) In this book the expression "electron flow" refers to *true* electron flow; the expression "current" refers to *conventional* electron flow.

Referring again to Figure 6-4, notice the direction of electron flow in part (a). The side of the component on which the electron flow enters will always be negative with respect to the opposite side or end out of which the flow leaves. This observation should make it easy to determine the polarity of any voltage drop.

We must now also try to understand electrostatic voltage and why it appears to be different from conventional or electromagnetic voltage. (The term *electromagnetic voltage* refers to a voltage that has been created by means of relative motion between conductors and magnetic fields such as may be found in electrical generators or in electrical high-frequency oscillators.) As the name implies, *electrostatic* or *static electricity* is comprised of electrical "charges" that, for the most part, are essentially static, or nonmoving. These charges can create a pressure or force much like the pressure or force produced by moving charges, or electrons, as they move through a resistance inside a conductor. As far as static electricity is concerned, however, an electrical conductor or charge movement is not necessary for developing voltage or potential difference. These charges can remain at rest and still create a voltage. It is these stationary charges that give static electricity its distinguishing characteristic compared to so-called conventional electromagnetic electricity. To understand how this can be, we have to study the relationship between the concepts of mechanical work and electrical charge.

We have already established the fact that it is possible for particles to possess a negative or positive charge as measured relative to some reference. This charge is capable of setting up a "force field" (to use a popular science fiction expression) that can easily be sensed. It can be sensed merely by taking another particle having the same polarity as the original's and noting the repelling reaction that takes place. This repulsion is very similar to the repulsion sensation experienced between opposing magnets. Obviously, a measurable interaction takes place in this process. The effort that is required to move the hypothetical particle closer to the other particle is the expended energy. Equation (6-1) allows us to calculate the *potential difference* existing between any two given points inside one charge's electrostatic field assuming that we know the work required to move another charge between these points and the size or amount of charge that it has:

$$V_{1\text{-}2} = \frac{W}{Q} \tag{6-1}$$

where $V_{1\text{-}2}$ = potential difference (V)

$\qquad W$ = energy or work expended in attempting to move one particle from point 1 to point 2 inside the electrostatic field of another (J)

$\qquad Q$ = amount of charge on the particle that has moved relative to the other (C)

As you study eq. (6-1), think of the similarity between this situation and the situation of potential energy. Figure 6-5 shows a stone of known mass. It has been lifted, against earth's gravitational force field, from point 1 to point 2. In the process we have expended work to create a potential energy (PE) equal to

$$\text{PE}_{1\text{-}2} = mgh \tag{6-2}$$

where $\text{PE}_{1\text{-}2}$ = potential energy capability existing between points 1 and 2 (J)

$\qquad m$ = mass of object (kg)

$\qquad g$ = gravitational constant (m/s^2)

$\qquad h$ = elevation above "ground" (m)

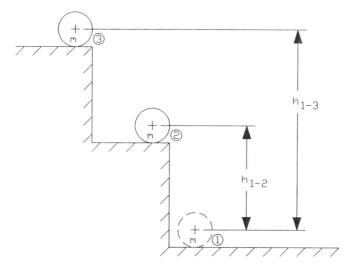

Figure 6-5   Comparing the determining of potential energy to determinig potential difference.

If we had lifted the stone in Figure 6-5 to point 3 rather than to point 2 on the cliff, the stone would, according to eq. (6-2), have a greater potential to expend more energy as it fell to point 1 than it would falling from the lower point, 2.

6-3 TYPES OF TRANSDUCERS PRESENTLY AVAILABLE

Sensing the presence of voltage has been accomplished for many years using an instrument called a *voltmeter*. This is device that requires the usage of two sensing probes, a voltage-sensing network with scaling facilities and some sort of visual readout-indicating device, such as an indicating needle and scale or digital readout. Voltmeters and similar sensing and measuring equipment devices are not usually thought of as being transducers. One reason for this is that transducers are usually thought of as being used for remote sensing and measurement, and can often operate without supervision.

It is possible, however, to construct a device that treats voltage as a measurand and acts as a true sensing device or transducer in the process. One such device is the voltage-controlled oscillator (VCO). This is a circuit that generates a variable output audio-frequency signal and whose frequency varies proportionally with the input dc voltage level. A typical circuit using discrete components (as opposed to using integrated circuits) is shown in Figure 6-6. In this particular circuit, the first operational amplifier encountered by the input voltage whose value is being determined acts as a low-pass filter. This is to filter out unwanted ac signals but allows only dc to be processed. The second op amp is a conditioning stage of amplification allowing for range adjustments and calibration. The third op amp and its associated FET transistor are the actual active voltage-controlled oscillator components in this

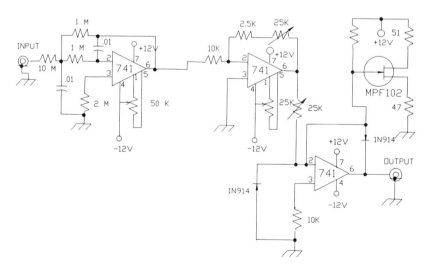

Figure 6-6 Circuit for a VCO (voltage-controlled oscillator).

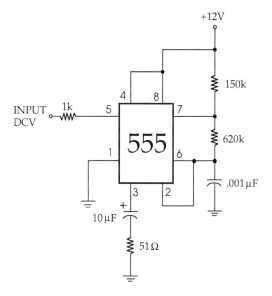

Figure 6-7 Circuit for a VCO using an integrated circuit. (Circuit adapted from H. M. Berlin, *The 555 Timer Applications Sourcebook,* copyright 1976, p. 111. Reprinted by permission of the author.)

circuit. The third op amp operates as an oscillator whose frequency of oscillation is determined by the input voltage at pin 2. The 1N914 diode located in the op amp's feedback circuit (pins 2 to 6) allows for positive feedback with the op amp, creating the needed oscillations. For an input voltage range of 0 to 50 V ac, the output frequency range is 0 to 5000 Hz.

Figure 6-7 shows a very compact design for a VCO circuit. It uses an integrated circuit, thereby reducing the number of external components needed for its construction. Only a half-dozen or so components are needed in this case. This entire circuit can easily be built within a 1-in.³-volume enclosure and attached to the measurand sensing area. Temperature stability for this circuit is to some degree dependent on the temperature stabilities of the external resistor and capacitor components used in the VCO's construction. However, it is not unreasonable to expect an overall stability or thermal drift of 0.005% per °C. If one were to construct a typical voltage versus frequency calibration chart using the circuit in Figure 6-7, it would look something like the curve illustrated in Figure 6-8. A characteristic of this particular

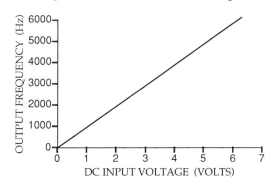

Figure 6-8 Typical output response from circuit in Fig. 6-7 showing its linear output capability.

circuit is that the output response is quite linear. This is highly desirable for transducer operation since data interpretation is made much easier.

Another integrated circuit VCO design is shown in Figure 6-9. A block diagram representation of the integrated circuit (IC) is shown only because of the complexity of the overall internal circuitry. This particular IC design uses more external components because of its more complex design and versatility. This circuit is typically less affected than other circuits by variations in temperature. This is due to the use of temperature-compensated components (resistors and capacitors) in fabrication of the IC, whose values self-adjust to varying temperature conditions. The desired voltage range that produces the output frequency range is selected by a binary input control. This selection is usually made with an external DIP switch (shown in inset in Figure 6-9). The proper combination of switch settings at the DIP switch allows for the selection of 16 different voltage ranges. Selecting one of these ranges causes a particular set of timing components within the timer unit to be connected to the VCO oscillator. This, in turn, causes the VCO unit to oscillate within the designed oscillator range. A typical frequency output for this IC would be something like 0 to 1000 Hz, or perhaps 100 to 10,000 Hz, where 100 Hz would be equivalent to 0 V dc in the latter case. Notice, too, that the user has a choice of output waveforms. Either a square wave or a triangular wave can be selected. Typically, the square-wave output would be chosen for easy signal processing within a digital system at the receiving end. However, the triangular waveform does have the advantage of producing less interfering harmonic radiation during transmissions. This can be a serious

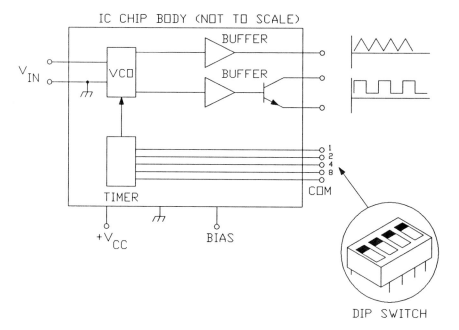

Figure 6-9 More complex design for a VCO having temperature independency.

problem if proper shielding techniques are not utilized in square-wave data transmissions.

Another type of voltage sensor, a sensor used primarily for sensing voltage levels, utilizes an entirely different circuit approach for sensing. Figure 6-10 shows an op amp wired as a differential voltage comparator amplifier. (The circuit shown has been simplified for better understanding and will function poorly in its present configuration.) This is a type of amplifier whose output will depend on the input voltage exceeding a particular preset value or reference voltage "dialed in" (at the variable resistor labeled V_{REF}) by the user. Only when this reference voltage is exceeded will a signal appear at the op amp's output. When this happens the output voltage may drive a transistor relay amplifier to produce a relay switch closure or opening, as shown in the figure. In Figure 6-10 the input voltage signal is shown superimposed over the output voltage to emphasize the voltage switching times of the output.

6-4 OPERATING CHARACTERISTICS

Some of the performance characteristics for voltage sensors have already been mentioned. They are noted again here together with additional characteristics that you should look for in voltage sensing.

Operating temperature range: Example: "−20 to +80°C." This is the range of temperatures in which the transducer can safely operate without causing significant deviations from the other performance characteristics stated.

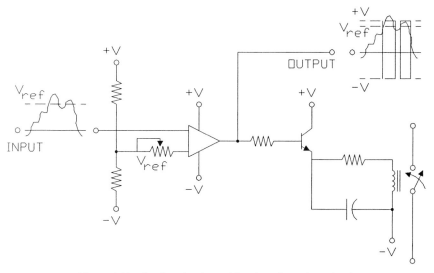

Figure 6-10   Sensing circuit used for detecting voltage levels.

Input voltage response range: Example: "0.01 to 100 V dc." This is the range of voltages that the transducer has been specifically designed to detect within the accuracy range specified.

Percent accuracy (Section 1-9.4): Example: "±0.02% for 0.01 to 1 V dc range; ±0.005% for 1 to 25 V dc range; ±0.01% for 25 to 100 V dc range."

Responsiveness (Section 1-9.5): Example: "0.02% at 50 V dc input." A sample calculation follows.

Example 6-1

A change of 0.01 V dc at 15 V dc is needed to produce a measurable change in the transducer's output (the "indicator" being the output frequency in this case). The 0.01 V dc change, in other words, just managed to produce a change of 1 Hz in the transducer's output. Find the responsiveness.

Solution:

$$\text{Responsiveness} = \frac{0.01 \text{ V dc}}{15 \text{ V dc}} \times 100$$

$$= 0.067\% \text{ (at 15 V dc input}$$

Linearity (Section 1-9.9: The following sample specs apply only to a VCO sensor: "+2 Hz, 0.01 to 10 V dc; −3 Hz, 10 to 25 Hz; −2 Hz, 25 to 100 V dc."

Sensitivity (Section 1-9.2): Example: "100 Hz/V."

Type of output signal: For a VCO sensor: Example: "FM carrier, 6.3 V ac p-p (peak to peak)." See Chapter 2 for other possible output signals. For a voltage comparator sensor: Example (if an electromechanical relay is used): "contacts, N.O. (normally open)"; (if solid-state relay): "can supply (or 'source') 10 V dc at 10 mA max." or "can handle (or 'sink') 10 V dc at 10 mA max."

Output frequency range: Example: "10 to 5000 Hz."

Number of output bits: Example: "10-bit word output." This designation is for a VCO having a binary bit output rather than an analog voltage output. Basically, this sensor would then be a voltage analog-to-digital converter whose output would be a parallel output. Its capability would have a maximum voltage reading of 2^{10} parts, that is, 1024 discrete readings per input voltage range (see the discussion of resolution in Section 1-9.11). Assuming the input range to be 10 to 100 V, or a span of 90 V, resolution would then be [using eq. (1-34)]

$$\text{resolution} = \frac{90 \text{ V}}{1024 \text{ bits}}$$

$$= 0.0879 \text{ V/bit}$$

The transducer would be able to read to the nearest 0.0879 V.

Power supply requirements: Example: "6 to 15 V dc at 8.5 mA max." The maximum current requirement is 8.5 mA while the supply voltage can be within the range 6 to 15 V dc. For the purpose of calculating the power consumed, it is usually assumed that the maximum current may very well occur at 15 V dc; consequently, the maximum consumed power is determined using these two values. Power = IE, or (0.0085 A)(15 V), or 0.1275 W.

Hysteresis (Section 1-9.10): Used for voltage-level sensors. Example: "Hysteresis, variable." Can be user-selected. Adjustable for a maximum of 30% of selected increasing switching voltage. Example: If the selected switching voltage were 1000 V dc, the diminishing voltage for the deactivating switch could be as much as 30% less, or 700 V dc.

6-5 SOME PRACTICAL APPLICATIONS

One of the most frequently used applications for the voltage sensor is in the remote monitoring of supply voltages in circuits whose proper operation is critical. Numerous examples can be found onboard space satellites and space research vehicles, the most notable being the NASA space shuttles. Their complex control circuitry must be supervised continuously from ground-level monitoring stations for the purpose of interpreting their performance. Circuitry malfunctions can be analyzed and corrective command signals radioed back to the vehicle to prevent further problems from developing.

Space exploration satellites are another example of using voltage-monitoring sensing circuits. The satellite's power supply voltages are continuously monitored (there most likely being more than one voltage per supply) and the signal-converted information sent to the satellite's transmitter for transmission back to earth. All of the voltage sensors being used onboard the spacecraft are similar to the VCO sensors described earlier.

Figure 6-11 shows another example of an application of a voltage sensor. This is an example that can be used in the electrical power industry for the monitoring of line voltages along a power-line system. The line voltage is read by a VCO voltage sensor in which the incoming line voltage is inductively coupled through a transformer and then converted to a dc voltage by means of rectification. The converter then changes the voltage to audio oscillations for remote processing and interpretation. These monitoring stations can be located at periodic intervals along power-line routes for troubleshooting line conditions. The voltage sensors' audio output signals are transmitted by small low-powered transmitters located at each tower to passing service vehicles for "in-motion" line-voltage supervision from these vehicles.

A common problem associated with commercial radio and television transmission towers is the following. During electrical storms very high voltages can often develop at the antenna. These voltages are either static voltages created during the passage of overhead, highly charged storm clouds, or high-voltage spikes induced into the antenna as a result of nearby lightning strikes. In either case, voltage-level sensors can be installed on the antenna so that when a certain high voltage value is attained, the outgoing radio transmission is, automatically, interrupted momentarily. The antenna is shorted to ground until the antenna's electrical charge is harm-

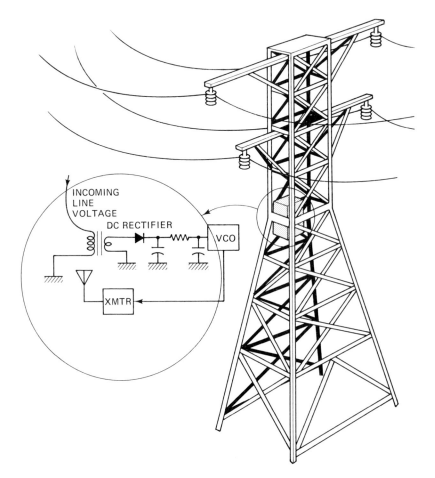

Figure 6-11 Voltage-sensing transmitter for monitoring line voltages on a high-voltage transmission line.

lessly bled off. Figure 6-12 shows such an installation. When the voltage value at the antenna falls below the switching value, radio transmissions are continued. This type of voltage-level switch performs well when a certain amount of hysteresis is used. That is, the lowering voltage value that reactivates the switch can be at a lower value than the original rising voltage-level switching value. This prevents switch "chattering" from taking place in case the voltage on the antenna varies slightly on either side of the voltage sensor's preset switching voltage value.

These are but a very few examples of voltage-sensing applications presently being used. Voltage sensing, especially static voltage sensing, is a relatively new sensing application, and more and more applications are being developed by industry.

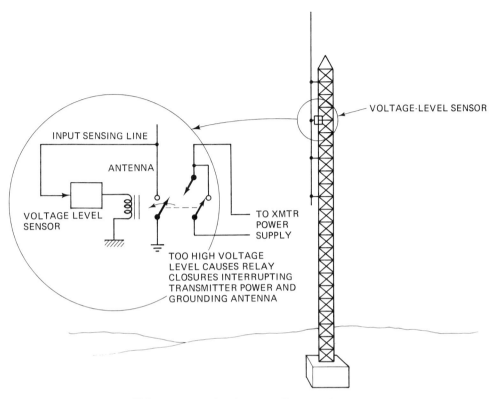

Figure 6-12 Voltage sensor placed on a radio tower for detecting high static voltages.

REVIEW QUESTIONS

6-1. Explain the difference between electrostatic voltage and nonelectrostatic voltage. How do you think the difference affects the design of the voltage sensors that are used for detecting each?

6-2. Explain how the VCO can be used in a voltage detection circuit.

6-3. Show how an operation amplifier design can be used as a voltage-level detection device. How does this design differ from a VCO?

6-4. Explain the difference between "sinking" an output signal or "sourcing" such a signal. Draw schematics to illustrate your explanation.

6-5. Give a practical example of how a voltage sensor may be used at a remote site or installation to monitor a critical supply voltage.

6-6. Explain how a voltage detection circuit may be modified to monitor current usage in a particular circuit or system.

PROBLEMS

6-7. Calculate the potential difference between the old position and the new position of a charge resulting from moving a charge of 7.8 C that required expending 17 J of work to make the move.

6-8. Determine the energy that was expended in moving a charge of 37 C over a potential difference of 10 V.

6-9. If a change of 0.12 V ac is required to produce measurable change in a voltage-sensing transducer's output, what is the transducer's responsiveness? The voltage change measurement was made at a voltage of 25.0 V ac.

6-10. Determine the resolution for a digital voltage-sensing circuit having a 50-V span whose output is capable of handling 8-bit data words.

6-11. A designer wants to design a voltage sensor that can detect voltages as small as 50 mV out of a range of 0 to 100 V. How many data lines will be necessary to handle the resolution desired?

REFERENCES

BERLIN, HOWARD M., *The 555 Timer Applications Sourcebook*. Indianapolis, IN: Howard W. Sams, 1979.

BUCHSBAUM, WALTER H., *Encyclopedia of Integrated Circuits*. Englewood Cliffs, NJ: Prentice Hall, 1981.

DeMAW, DOUGLAS, *ARRL Electronics Data Book*. Newington, CT: American Radio Relay League, 1976.

SESSIONS, KENDALL W., *Master Handbook of 1001 Practical Electronic Circuits*. Blue Ridge Summit, PA: TAB Books, 1975.

7

VARIABLE-RESISTANCE SENSORS

CHAPTER OBJECTIVES

1. To review the relationship between voltage and resistance in order to understand the characteristics of the potentiometer.
2. To understand the behavior of the linear and rotary displacement potentiometers.
3. To study the strain gage and its characteristics.

7-1 INTRODUCTION

Of all the sensing devices available for transducing, the potentiometer device is perhaps the easiest to understand, the least expensive to construct, and the simplest to install. Its basis of operation lies within the conversion of a resistance change to a change in voltage or current: namely, the application of Ohm's law. However, another type of variable-resistance device, the strain gage, is perhaps not quite as straightforward as the potentiometer in operation, as we will find out in the discussions below. However, to understand either device, an understanding of Ohm's law and the concept of resistivity is certainly helpful. A review of how voltage divider networks operate will also contribute to an understanding of these devices.

7-2 OHM'S LAW, RESISTIVITY, AND THE VOLTAGE DIVIDER RULE

7-2.1 Ohm's Law

The quantities of resistance, voltage, and current are all related by the expression

$$I = \frac{E}{R} \tag{7-1}$$

where I = current (A)

E = voltage (V)

R = resistance (Ω)

This relationship is called *Ohm's law*. The relationship holds regardless of whether you are dealing with direct current or voltage or with alternating current or voltage. Furthermore, if you are dealing with alternating current or voltage, it makes no difference if you are working with their rms values, their peak-to-peak values, or simply with their peak values. Ohm's law works equally well within any of these concepts.

Another important idea that is a key to the understanding of how strain gages operate is the concept of resistivity. *Resistivity* is the electrical resistance measured for any material having a uniform cross-sectional area and is usually stated in terms of the material's length and/or cross-sectional area. In other words, resistivity is resistance that has been stated in terms of a unit length or area. The unit of resistivity in the English system of measurements is the Ω-ft. or the CM-Ω/ft, where the abbreviation CM represents the circular mil. In the SI system of measurements the unit of resistivity is the ohm. The circular mil unit of area is often associated with wire because of its circular cross-sectional area. This unit is discussed in more detail below.

7-2.2 Resistivity

Resistivity is determined by the following four quantities:

1. The type of material (cork, iron, glass, water, etc.)
2. The length of the material
3. The material's temperature
4. The material's cross-sectional area

In our discussion here we are specifically interested in the resistivity of wire. A wire's cross-sectional area is usually circular in shape. Its composition is usually steel, copper, or some other similar highly conductive material. The electrical resistance of wire can be varied either by changing its length, changing its temperature, or changing its cross-sectional area. If the wire's length is changed, its resistance will vary directly as the length is changed; if the cross-sectional area changes, the resistance will vary inversely as the area changes. Mathematically,

$$R \propto \frac{l}{A} \tag{7-2}$$

We can eliminate the proportional symbol in eq. (7-2) by replacing it with a proportionality constant, ρ (rho), and inserting an equals sign. If we do this, eq. (7-2) now becomes

$$R = \rho \frac{l}{A} \tag{7-3}$$

where R = resistance (Ω)

ρ = proportionality constant, called resistivity (Ω-CM/ft, Ω-m)

l = length (ft, m)

A = area [usually expressed in *circular mils* (CM; see the discussion below); also, m^2]

The unit circular mil is often used in stating wire dimensions since wire diameters are often expressed in mils rather than inches. Therefore, since

$$1 \text{ inch} = 1000 \text{ mils}$$

and

$$A = \frac{\pi d^2}{4} \tag{7-4}$$

a wire having a diameter of D_M mils will have for its cross-sectional area,

$$A = \frac{\pi D_M^2}{4} \qquad \text{square mils} \tag{7-5}$$

By definition, 1 circular mil (CM) is equal to $\pi/4$ square mils. That is,

$$1 \text{ CM} \equiv \frac{\pi}{4} \qquad (\equiv \text{ means "by definition")}$$

Therefore, by dividing eq. (7-5) by $\pi/4$, we can determine the number of circular mils contained within the square-mil results that were calculated with eq. (7-5). In other words,

$$A = \frac{\pi/4}{\pi/4}(D_M)^2 \tag{7-6}$$

or

$$A_{CM} = (D_M)^2 \tag{7-7}$$

where A_{CM} = area (CM)

D_M = diameter (mils^2)

Example 7-1

Find the resistance of a 500-ft length of copper wire 0.013 in. in diameter. Assume a temperature of 68°F and $\rho = 10.37$ ohm-CM/ft.

Solution: Converting the diameter to mils, we get 12 mils. Note that this conversion is done simply by moving the decimal point to the right three places. Then, using eq. (7-7), we obtain

$$A_{CM} = D_M^2$$

$$= 13^2$$

$$= 169 \text{ CM}$$

Placing this result into eq. (7-3) to find R yields

$$R = \frac{(10.37 \text{ ohm-cm/ft})(500 \text{ ft})}{169 \text{ CM}}$$

$$= 30.68 \ \Omega$$

7-2.3 Voltage Divider Rule

Figure 7-1 shows a circuit comprised of three resistors, R_1, R_2, and R_3, wired as shown to a voltage source E_s. We would like to find the value of E_{out}. To analyze this circuit we first notice that E_s also occurs across the two resistors, R_1 and R_2. We can neglect R_3 since it does not affect the value of E_s. We must now decide on how E_s is going to divide across R_1 and R_2. The voltage across R_2 is the same voltage, E_2, which is the voltage we are interested in finding.

To solve this problem, we first determine the total current flowing through both R_1 and R_2. Using Ohm's law [eq. (7-1)] we find this total current, I_T, to be

$$I_T = \frac{E_s}{R_T} = \frac{E_s}{R_1 + R_2} \tag{7-8}$$

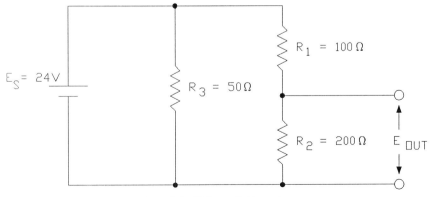

Figure 7-1   Voltage divider rule.

Now knowing the value of I_T, we can find E_{out} by again using Ohm's law:

$$E_{out} = R_2 I_T = R_2 \frac{E_s}{R_1 + R_2} \tag{7-9}$$

Equation (7-9) is the basic form of the voltage divider rule. The only change in the equation is to transpose the E_s and the R_2 to make the resultant equation a little easier to memorize. In other words,

$$E_{out} = E_s \frac{R_2}{R_1 + R_1} \tag{7-10}$$

To find the voltage drop across R_1 in Figure 7-1, eq. (7-10) would be rewritten as

$$E_1 = E_s \frac{R_1}{R_1 + R_2} \tag{7-11}$$

Example 7-2

Calculate the voltage drops across both R_1 and R_2 in Figure 7-1.

Solution: To find the voltage drop across R_1, we use eq. (7-11):

$$E_1 = 24 \text{ V}\left(\frac{100 \ \Omega}{100 \ \Omega + 200 \ \Omega}\right)$$

$$= 8 \text{ V}$$

To find the voltage drop across R_2, we use eq. (7-10):

$$E_{out} = 24 \text{ V}\left(\frac{200 \ \Omega}{100 \ \Omega + 200 \ \Omega}\right)$$

$$= 16 \text{ V}$$

To check our two results we add E_1 and E_{out} to see that they total 24 V, the value of E_s across both resistors.

7-3 POTENTIOMETER

As stated earlier, the potentiometer is a relatively easy device to understand once the concept of the voltage divider is understood. We now investigate this device to learn how it operates and why it is such a popular and versatile sensing device in many industrial applications.

7-3.1 Theory of Operation

A potentiometer is an electromechanical device containing a rotatable wiper arm that makes an electrical contact with a resistive surface and can move across this surface. The wiper is coupled mechanically to a movable member or linkage. The wiper and resistive surface form a voltage divider circuit when voltage is applied across the entire resistance within the potentiometer. A variable voltage can then be produced at the wiper arm relative to one end of the resistor as the wiper is moved. Figure 7-2 shows simplified schematics of two popular configurations used in potentiometer construction. The resistive element in the potentiometer is usually constructed either in a linear fashion as shown in part (a) of the figure, or in a circular fashion, as shown in part (b). In either case the construction behaves as a voltage divider circuit. A voltage source is shown attached across the entire length of the resistive element in both cases. The potentiometer's output may be considered to be across the middle terminal and either of the other two terminals.

7-3.2 Typical Construction and Operating Characteristics

The resistive material used in construction of the potentiometer can be one of five typical kinds of construction: (1) wire wound, (2) carbon film, (3) metal film, (4) conductive plastic film, and (5) cermet (a ceramic–metal combination).

Typical potentiometer constructions are shown in Figure 7-3. Figure 7-4 shows a typical mechanical construction for a linear potentiometer. The wiper assembly, the heart of the potentiometer's operation, determines the success of this sensing device. The wiper is usually made from a copper alloy such as beryllium or from a phosphor bronze material.

Because of the continuous motion of the wiper, the rotating shaft of an angular displacement pot is usually mounted on roller bearings for heavy-duty applications, whereas the more inexpensive bronze sleeve bearings are used for lighter-duty applications. In the case of a linear pot, the wiper is attached to a piston-like rod that slides through a sleeve bearing or packing gland at one end of the pot's enclosure.

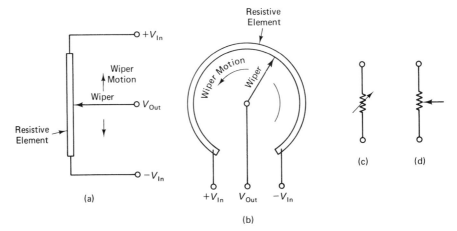

Figure 7-2 How variable voltages are obtained with potentiometers. (From James R. Carstens, *Automatic Control Systems and Components,* copyright 1990, p. 88. Reprinted by permission of Prentice Hall, Englewood Cliffs, NJ.)

Figure 7-3   Typical potentiometer styles.

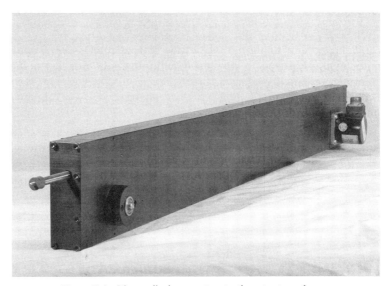

Figure 7-4   Linear-displacement potentiometer transducer.

Wire-wound potentiometers

The wire-wound potentiometer was the earliest type of construction used in the manufacture of potentiometers. In this construction many turns of insulated wire are wound around a form, or *mandrel*, which is used for rigidity. The mandrel can be either straight or circular in shape, depending on the pot's application. The mandrel itself is usually constructed from glass, plastic, or an insulated metal. Each wire winding around this form is wound adjacent to the next winding (see Figure 7-5). This is done in order to obtain as many turns as possible within the confined space of the potentiometer's container or housing, to allow the maximum resolution possible for each increment of the wiper's travel, as we will see in a moment. The windings themselves can be either circular or rectangular; that is, the mandrel's cross-sectional area can be either of these shapes. For a given potentiometer size, the smaller the winding's wire diameter, the higher the total resistance of the potentiometer. A typical resistance range for a wire-wound pot that is to be used for transducing position is a few tens of ohms to 1 MΩ or so.

Using a greater number of turns in a potentiometer's construction increases that pot's resolution. The type of wire used is normally one with high electrical resistance. Usually, platinum or a nickel alloy is used. Their resistance values do not change significantly with age and they have a fairly broad temperature range over which they can operate without affecting these values adversely. This is generally in the range −50 to + 150°C.

In Figure 7-5 we see how the wiper is mounted relative to the potentiometer's numerous windings. Figure 7-6 shows the wiper and winding construction of a wire-wound potentiometer. The insulation on each winding is removed in the vicinity of the wiper so that the wiper can make an electrical contact with that particular winding. The wire itself is made from a specially formulated metal composition to increase its electrical resistance over that of ordinary copper or steel wire.

Wire-wound potentiometers can handle fairly large amounts of power. This is one of their advantages. On the other hand, they have the disadvantage of suffering from discontinuous output, as shown in Figure 7-7. The characteristic stair-step voltage output curve is created by the fact that the wiper's electrical contact leaves one wire contact and makes contact with the next, which results in a discontinuity in the measured output resistance at the wiper during this winding transition. Another disadvantage is that due to their coiled nature, wire windings are inductive. If the windings are around a metal mandrel, the inductance is increased. An ac signal

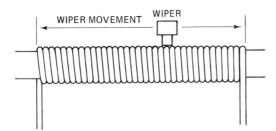

Figure 7-5 Mounted wiper assembly on the mandrel and winding.

Figure 7-6   Wire-wound pot showing wiper and winding assembly.

Figure 7-7   Characteristic stair-step output of a wire-wound pot. (From Richard W. Miller, *SERVOMECHANISMS: Devices and Fundamentals,* copyright 1977, p. 16. Reprinted by permission of Reston Publishing Co., Inc., Reston, VA.)

traveling through this coil will experience an inductive reactance proportional to the signal's frequency. To prevent either situation, one of the other four types of pots described below should be considered instead.

Carbon film potentiometers

The carbon film potentiometer has one principal advantage over the wire-wound potentiometer: The resistive element in the carbon film pot provides a continuous, uninterrupted contact surface for the wiper contact to travel over. This means that the pot's output, whether measured in ohms or volts, is a continuous curve, as demonstrated in Figure 7-8. This is possible due to a continuous band of conductive or partially conductive material that has been bonded in the form of a thin layer or film over a plastic or metal form. This material is comprised of a mixture of carbon and a nonconductive claylike substance whose mixing proportions have been formulated with the carbon to create the desired resistive value within the potentiometer. Since the pot is to be used for transducing position, it is very important to produce a variable resistor whose output is very predictable and stable. In many cases it is desirable to have this output be extremely linear. That is, over any given unit displacement of the wiper, the measured resistance must be the same. However, in a few instances it may be desirable to have a logarithmic output from the pot. The purpose of this would be to compensate for an otherwise nonlinear mechanical linkage for position measurement. The logarithmic output of the pot would then compensate for the nonlinear performance of the linkage, thus producing a combined linear output response. Another example is use of a logarithmic pot's output to drive an otherwise logarithmic amplifier's input to create a linear output.

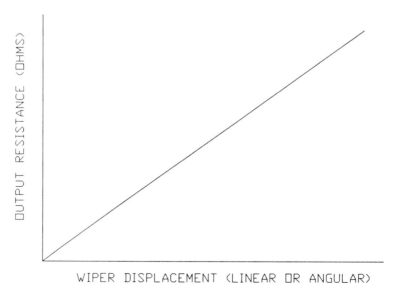

Figure 7-8   Characteristic output of a carbon film pot.

Metal film potentiometers

A metal film potentiometer is constructed very much like the carbon film pot. The basic difference between the two types is in the resistive film used to form the variable resistor. In this particular construction, a partially conductive metal, similar to the metals used for wire-wound pots, is sprayed in its molten state onto the mandrel to form the variable-resistance surface over which the wiper travels.

The operating characteristics for this particular type of potentiometer are very similar to those of the carbon film pot. The metal film pot's output can be manufactured to be quite linear and continuous (as opposed to the characteristic discontinuous output of the wire-wound pot), or it can be constructed to have a nonlinear output. The power-handling capabilities are not quite as high as those of wire-wound pots, however. One distinguishing characteristic is that the metal film pot is considered to be more rugged than the carbon film pot. It can withstand more shock and vibration before the conductive path is fractured.

Continuous plastic film potentiometers

This type of pot is composed of a conductive plastic molded into a film that has been mounted on a stiffened nonconductive backboard or substrate. Otherwise, the behavior of the conductive plastic film pot is very similar to that of the carbon film pot, although it may not be as rugged as the metal film pot.

Cermet potentiometers

Cermet is a resistive material made from a mixture of metal particles and a ceramic material. Typical metals used are chromium, silver, and an oxide of lead. The ceramic clays contain silicon oxides, which when mixed with the metal, form the various desired resistances. When fired at high temperatures, a very hard and durable resistive surface is formed. However, the exact resistances are often difficult to obtain using this method of fabrication, and usually after the firing process is completed, some sort of trimming process must be used on the resistor's body to obtain the precise values needed. From the standpoint of durability, the cermet pot is considered to be fairly rugged because of the ceramics used in its manufacture. Otherwise, its operating characteristics are similar to those of the metal, plastic, and carbon film types.

Up to this point we have assumed that all the pots were constructed using a rotating or linearly traveling wiper that moved across a resistive path. There is, however, one other very popular type of construction that allows the wiper to move on a threaded shaft such that as the shaft is rotated, the wiper will move at a rate dependent on the number of threads machined on the shaft and on the mating threads on the wiper body itself. Figure 7-9 shows this type of construction. The purpose of this construction is to reduce the wiper motion for a given input motion. A common reduction is a "10-to-1" motion, otherwise referred to as a "10-turn pot." That is, for every 10 turns of the pot's input shaft, the wiper will move the equivalent

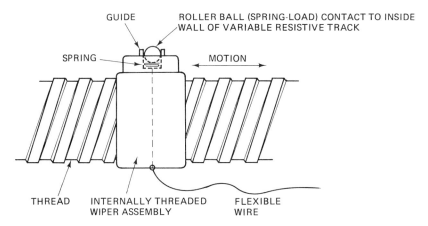

Figure 7-9 Construction of a 10-turn pot.

of what would have been considered one turn in a normal pot. Collectively, these specially designed pots are referred to as *multiturn pots*.

The potentiometers described above have one limitation in common. Assuming that it is being used for rotary-motion detection, the input shaft cannot rotate and produce a continuous resistance reading through a complete 360° line of travel. Figure 7-10 shows the reason. Because a gap must exist between the two ends of the resistance path so that one end will not touch or short out the other end, there is a complete break in continuity of the resistance values within the pot. The wiper must also be able to "jump" this gap mechanically so that its travel is smooth and

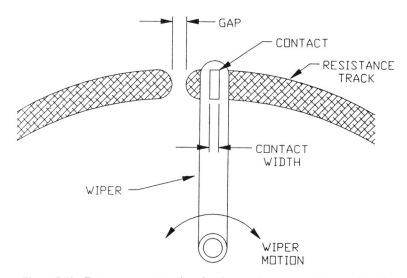

Figure 7-10 Rotary pot construction showing gap between resistive track ends.

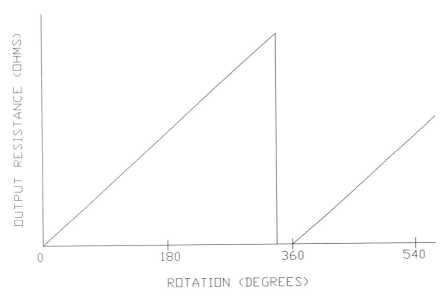

Figure 7-11   Resulting output curve from pot having resistive gap described in Figure 7-10.

uninterrupted. The pot's output resistance would appear something like that shown in Figure 7-11.

It is obvious that the larger the diameter of the pot, the greater the resistance length and therefore the greater the displacement that can be detected. Greater displacements can also be accommodated by using the multiturn pot construction described earlier.

7-3.3 Practical Applications

The potentiometer is most often used for the purpose of sensing position or location. It is also sometimes used for sensing linear or angular velocity. If resolution is a problem in the degree of sensing required in either application, a wire-wound pot should not be used, due to its characteristic stair-step output.

Figure 7-12 demonstrates one practical application of the potentiometer used as a position sensor. In this device a string, similar to the string on a yo-yo, is attached to a moving member. The drum around which the string is wound inside the motion sensor is spring-loaded. The drum, in turn, is mechanically coupled to a potentiometer. When the drum is rotated by pulling on the string, the drum's motion is transferred to the pot, whose motion is translated to a proportional output voltage created by an onboard dc power supply wired to the pot. This type of sensor, appropriately referred to as a "string pot" or "yo-yo pot," is used extensively in industry. Often, the pot being used in the device's design is a 10-turn pot rather than a single-turn pot. This higher turns ratio allows for an increase in the amount of overall travel by the string compared to that of the one-to-one ratio design.

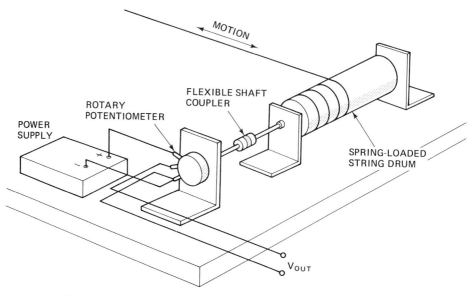

Figure 7-12 Practical application of a rotary pot used as a position sensor.

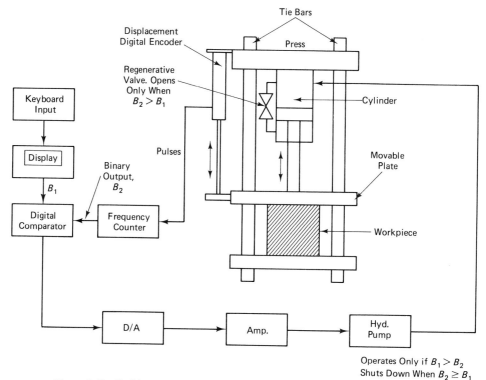

Figure 7-13 Position sensor used on a hydraulic press to detect its moving platen displacement. The sensor is part of an overall digitally controlled process control system. (From James R. Carstens, *Automatic Control Systems and Components*, copyright 1990, p. 296. Reprinted by permission of Prentice Hall, Englewood Cliffs, NJ.)

Another practical application of position sensing is demonstrated in Figure 7-13. A linear motion pot is attached to the moving platen of a hydraulic press so that the platen's motion can be monitored and remotely controlled. In this application the pot's positional voltage information can be fed to an automatic control system so that the hydraulic cylinder controlling the platen's location can be adjusted precisely. What is often done in an application such as this is to measure the change in voltage output from the pot over a small period of time such as 1 second or less. Since the voltage is easily translated to a known displacement, the pot's moving wiper, and therefore the platen's velocity in this case, can easily be calculated.

Figure 7-14 illustrates yet another common application of displacement measurement, this time in a precision laboratory pressure gage. The moving indicator needle of a diaphragm pressure gage is replaced by the wiper of a specially designed pot that has been built into the gage's casing. As the gage responds to a change in pressure, the wiper is caused to move across the face of the resistive element. This, in turn, causes a proportional voltage to appear at the gage's output terminals. In the illustration shown, an increase in pressure creates an increasing voltage output, while a decrease in pressure produces a decrease in voltage output.

7-3.4 Typical Specifications

Figures 7-15 through 7-17 show typical catalog design data for various pot types.

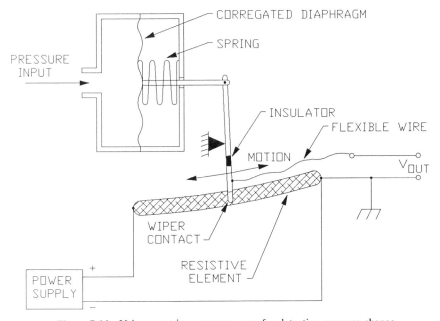

Figure 7-14   Using a pot in a pressure gage for detecting pressure change.

75-M90   75-M91   75-M92   SHOWN ACTUAL SIZE

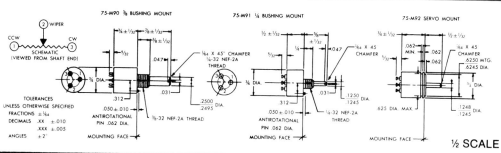

ELECTRICAL AND MECHANICAL CHARACTERISTICS				
	75-M90 STANDARD	75-M91 STANDARD	75-M92 STANDARD	RANGE OF SPECIAL DEVIATIONS
Total Resistance	(See Chart)	(See Chart)	(See Chart)	10 to 200,000 ohms
Total Resistance Tolerance	±5%	±5%	±5%	±20% to ±1%
Theoretical Resolution	(See Chart)	(See Chart)	(See Chart)	To .09% minimum
Independent Linearity	(See Chart)	(See Chart)	(See Chart)	±5% to 0.3%
Terminal Based Linearity	(See Chart)	(See Chart)	(See Chart)	±7% to ±0.6%
Theoretical Electrical Travel	320°	350°	350°	5° to 360°
Theoretical Elect. Travel Tolerance	±3°	±2°	±2°	±5° to ±1°
Total Mechanical Travel	330°	360° cont.	360° cont.	90° to 340° with stops
Total Mechanical Travel Tolerance	±5°	±3°*	±3°*	±10° to ±1°
Temp. Coefficient of Resistance Wire	(See Chart)	(See Chart)	(See Chart)	To .000005 min. (ohm/ohm/°C)
Operating Temperature Range	−55° to +135°C	−55° to +135°C	−55° to +135°C	−200°C to +325°C
Power Rating (at 85°C)	2 watts**	2 watts**	2 watts**	To 1 watt max. at 250°C
Dielectric Strength, Min. (6Q Sec.)	1000 volts RMS	1000 volts RMS	1000 volts RMS	500 V DC to 2000 V RMS
Terminals	3	3	3	To 7 maximum
Taps	0	0	0	To 4 maximum
Tap Tolerance	±2°	±2°	±2°	±5° to ±0.7°
Life Expectancy (at 40 RPM)	25,000 Cycles	25,000 Cycles	1,000,000 Rev.	To 5,000,000 rev. max.
Moment of Inertia (Single Cup)	0.1 gm cm²	0.1 gm cm²	0.1 gm cm²	
Starting Torque	3 to 8 oz.-in.††	0.3 oz.-in., max.†	0.3 oz.-in., max.†	.01 to 12 ounce-inches†
Running Torque	3 to 8 oz.-in.††	0.2 oz.-in., max.†	0.2 oz.-in., max.†	.008 to 12 ounce-inches†
Weight	.6 oz.	.5 oz.	.5 oz.	
Stop Strength, Minimum	4 lb.-in.	4 lb.-in.*	4 lb.-in.*	To 10 pound-inches, maximum
Number of Cups	1	1	1	To 3 maximum, terminals on side
Bearings	Teflon sleeve	Bronze sleeve	Bronze sleeve	Stainless steel ball bearings, Oilite

Figure 7-15 Typical specification sheet for wire-wound rotary potentiometers. (Courtesy of Maurey Instrument Corp., Chicago.)

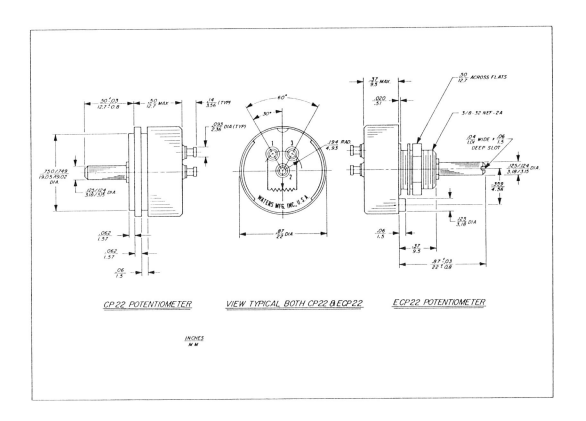

CP 22 POTENTIOMETER

VIEW TYPICAL BOTH CP22 & ECP22

ECP22 POTENTIOMETER

INCHES
MM

MECHANICAL SPECIFICATIONS

	ECP 22 Bushing	CP 22 Servo-Mount
Torque	1.0 in. oz. Nom.	0.2 in. oz. Max.
Mechanical rotation	320°	Continuous
Shaft bearings	N.A.	Precision Polyimide
Shaft T.I.R.	N.A.	.003"
Stop strength	5 in. lbs.	N.A.
Terminals	Gold plated brass, solder type	
Rotational Life	1 x 10⁶ cycles	5 x 10⁶ cycles

ELECTRICAL SPECIFICATIONS

	ECP 22 Bushing	CP 22 Servo-Mount
Resistance values	1K, 2K, 5K, 10K	1K, 2K, 5K, 10K
Resistance tolerance	±20%	±20%
Linearity, standard (ind.)*	1%	1%
Linearity, best (ind.)*	0.5%	0.25%
Output smoothness	.03%	.03%
Resistance-temperature characteristic	5%	5%
Function angle	310°	340°
Temperature range	−20°C to 100°C	
Power rating	2 watts to 80°C 0 watts at 100°C	
Resolution	infinite	
Dielectric strength	100 volts RMS	

*Linearity is measured and specified over 5% to 95% of function angle.

Figure 7-16 Typical specification sheet for carbon rotary potentiometer. (Courtesy of Waters Manufacturing, Inc., Wayland, MA.)

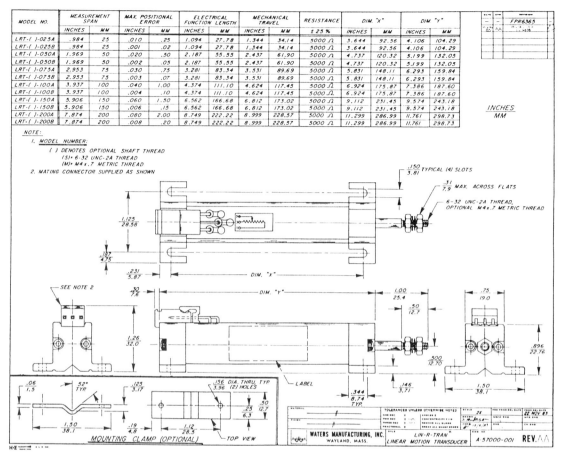

MODEL NO.	MEASUREMENT SPAN		MAX. POSITIONAL ERROR		ELECTRICAL FUNCTION LENGTH		MECHANICAL TRAVEL		RESISTANCE	DIM "X"		DIM "Y"	
	INCHES	MM	INCHES	MM	INCHES	MM	INCHES	MM	± 25%	INCHES	MM	INCHES	MM
LRT-( )-025A	.984	25	.010	.25	1.094	27.78	1.344	34.14	5000 Ω	3.644	92.56	4.106	104.29
LRT-( )-025B	.984	25	.001	.02	1.094	27.78	1.344	34.14	5000 Ω	3.644	92.56	4.106	104.29
LRT-( )-050A	1.969	50	.020	.50	2.187	55.55	2.437	61.90	5000 Ω	4.737	120.32	5.199	132.05
LRT-( )-050B	1.969	50	.002	.05	2.187	55.55	2.437	61.90	5000 Ω	4.737	120.32	5.199	132.05
LRT-( )-075A	2.953	75	.030	.75	3.281	83.34	3.531	89.69	5000 Ω	5.831	148.11	6.293	159.84
LRT-( )-075B	2.953	75	.003	.07	3.281	83.34	3.531	89.69	5000 Ω	5.831	148.11	6.293	159.84
LRT-( )-100A	3.937	100	.040	1.00	4.374	111.10	4.624	117.45	5000 Ω	6.924	175.87	7.386	187.60
LRT-( )-100B	3.937	100	.004	.10	4.374	111.10	4.624	117.45	5000 Ω	6.924	175.87	7.386	187.60
LRT-( )-150A	5.906	150	.060	1.50	6.562	166.68	6.812	173.02	5000 Ω	9.112	231.45	9.574	243.18
LRT-( )-150B	5.906	150	.006	.15	6.562	166.68	6.812	173.02	5000 Ω	9.112	231.45	9.574	243.18
LRT-( )-200A	7.874	200	.080	2.00	8.749	222.22	8.999	228.57	5000 Ω	11.299	286.99	11.761	298.73
LRT-( )-200B	7.874	200	.008	.20	8.749	222.22	8.999	228.57	5000 Ω	11.299	286.99	11.761	298.73

NOTE:
1. MODEL NUMBER:
   ( ) DENOTES OPTIONAL SHAFT THREAD
   (S) = 6-32 UNC-2A THREAD
   (M) = M4 x .7 METRIC THREAD
2. MATING CONNECTOR SUPPLIED AS SHOWN

WATERS MANUFACTURING, INC. WAYLAND, MASS.
LIN-R-TRAN LINEAR MOTION TRANSDUCER A-57000-001 REV. ΛΛ

Specifications:

MECHANICAL:

Stroke length
25-200 mm

Shaft diameter
6 mm (0.236") with threaded end (US 6-32 or M4 × .7)

Hysteresis
< .001 mm (.00004")

Incremental Sensitivity
< .001 mm (.00004") displacement

Repeatability
< .001 mm (.00004")

Life
10×10^6 cycles at 50 mm/sec.

ELECTRICAL:

Resistance Values and Accuracy:

Type	Level	Max. Positional Error
LRT 25	– A	± .25 mm
	– B	± .02 mm
LRT 50	– A	± .50 mm
	– B	± .05 mm
LRT 75	– A	± .75 mm
	– B	± .07 mm
LRT 100	– A	± 1.00 mm
	– B	± .10 mm
LRT 150	– A	± 1.50 mm
	– B	± .15 mm
LRT 200	– A	± 2.00 mm
	– B	± .20 mm

Total Resistance
5.0 kohms

Resistance Tolerance
±25%

Power Rating
0.75 Watts/25mm stroke length

5-87 5000

Figure 7-17 Catalog data for carbon linear potentiometer. (Courtesy of Waters Manufacturing, Inc., Wayland, MA.)

7-4 STRAIN GAGE

The strain gage is a resistive device whose operation is very closely tied to eq. (7-3), explained earlier. The strain gage is used primarily for the detection of stress and strain, although it can readily be adapted to sense other measurands as well. We now investigate how the strain gage works by looking first at its theory of operation.

7-4.1 Theory of Operation

Closely associated with the strain gage sensor is a sensitive electronics circuit used for detecting very small electrical resistance changes. These changes occur in the wirelike conductors that comprise the gage's main construction feature. The resistance changes are produced by the stretching of these wires, caused, in turn, by a force applied to the gage's body. In another form of strain gage construction the wire is replaced by a solid-state semiconducting material whose internal electrical resistance changes with the stress applied. The sensing of stress or strain using this method takes advantage of the phenomenon called *piezoresistivity*. However, we must first define the terms *stress* and *strain*.

Stress is defined as follows:

$$\text{stress} \equiv \frac{\text{force}}{\text{area}} \qquad (7\text{-}12)$$

where stress = units of pressure (i.e., N/m^2, lb/in^2, etc.)

  force = force applied either in tension or in compression on an object, perpendicular to a surface on that object (N, lb)

  area = area of an object that is perpendicular to the force applied (m^2, ft^2)

Strain is defined as follows:

$$\text{strain} \equiv \frac{\text{change in length of object due to stress applied}}{\text{original length of object}} \qquad (7\text{-}13)$$

where strain is measured in the units m/m, in./in., and the like.

Figure 7-18 illustrates how the piezoresistivity process takes place. A force, F, is applied to both ends of loops of conducting material such as metallic wires. As this happens, the conductors become stretched, causing their overall length to increase. This, in turn, causes the conductor's cross-sectional area to decrease, which causes an increase in the conductor's total resistance according to eq. (7-3). The particular gage illustrated in Figure 7-18 is called an *unbonded strain gage*. This gage type will be discussed in more detail later in the chapter.

Wheatstone bridge

The change in electrical resistance described above is quite small, as pointed out earlier. As a result, a circuit especially designed to detect very small resistance changes must be used along with the strain gage to detect the gage's very weak

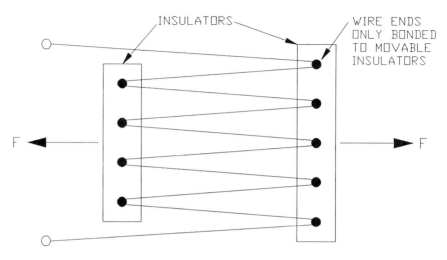

Figure 7-18 Typical unbonded strain gage construction.

responses to stress. This circuit, shown in simplified form in Figure 7-19, is called a *Wheatstone bridge*. The circuit is named after Sir Charles Wheatstone, an English physicist, who developed this circuit in the mid-1800s. The circuit works like this: Four resistors are wired in the manner shown, with a galvanometer as a very sensitive current detector. A voltage source, E, produces two opposing current flows in the legs of the resistance bridge as shown. If the resistance values are adjusted properly so that I_1 equals I_2, it is possible to cause a net current of zero to occur through the galvanometer. This happens only if the following resistor ratios are maintained:

$$\frac{R_1}{R_2} = \frac{R_3}{R_4} \tag{7-14}$$

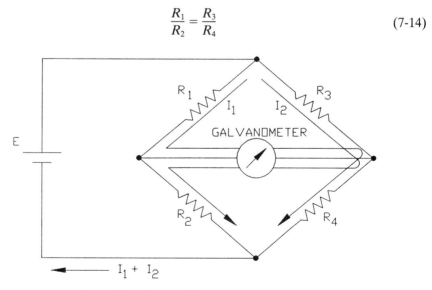

Figure 7-19 How the Wheatstone bridge works.

As you can see, there are virtually an infinite number of resistor combinations that will satisfy the conditions above. In reality, there are certain desirable combinations of resistors that work better than others for strain gage applications, but that is beyond the scope of our discussion.

If one of the resistors in eq. (7-4) is made variable, say R_1, and one of the other resistors, say R_4, is replaced by the dc resistance of our strain gage, it is possible to adjust R_1 so that the ratio formed by R_3/R_4 is precisely equal to the ratio formed by the fraction R_1/R_2. When this occurs, there will be no current flow through the galvanometer; the galvanometer's indicator needle will read zero in a straight-upward position. A mismatch of these two ratios will produce an imbalance of currents in each of the two current paths of the bridge, causing the galvanometer's needle to deflect to one side or the other of center zero. Once the balanced or nulled condition has been reached, any movement of the galvanometer's needle to the right or left of center zero will be a certain indication of a change in the strain gage's resistance brought on by a stressed condition to the gage.

7-4.2 Operating Characteristics

In theory, what was just described is certainly true. However, in reality, another more subtle factor can produce significant changes in the strain gage's internal resistance. That factor is temperature. As we will discover in later chapters dealing with temperature-sensing devices, certain metals in the form of wire make excellent temperature indicators. Since the strain gage is composed primarily of lengthy wire strands or metallic paths made from a metallic foil sometimes bonded (see Section 7-4.3) to a base or substrate, they respond no less differently to temperature. This is one of the more notable and undesirable characteristics of a strain gage.

So how can we nullify the effects of temperature? The technique is surprisingly simple. Figure 7-20 illustrates how it is done. Using the example above, where R_1 was the variable balancing resistor for the Wheatstone bridge and R_4 was the strain

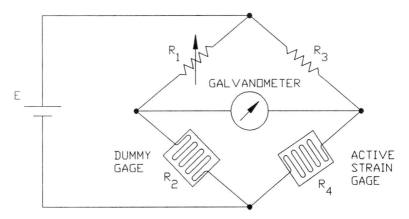

Figure 7-20   Nullifying the effects of temperature in a strain gage circuit.

gage, we could insert a duplicate gage into the bridge circuit that was not subject to the measured strain. This duplicate gage would be subjected to the same temperature as that experienced by the stressed gage. The duplicate gage, or *dummy gage* as it is often called, can be inserted in place of R_2 to offset the temperature-created resistance in the stressed strain gage, R_3.

Another circuit scheme is shown in Figure 7-21. Here we see all four of the Wheatstone bridge resistors replaced with strain gages. Balancing resistors R_{b1} and R_{b2} are applied in series with two adjacent arms for equalizing purposes, one compensating for thermal drift, the other for bridge balancing. Any temperature variations external to the bridge will cause all four gages to respond equally (assuming that all four gages are equally matched), causing any temperature-induced resistance variations to be canceled.

Another notable characteristic of the strain gage has to do with its output signal when subjected to stress. Associated with each strain gage is its *gage factor*. The gage factor for any strain gage compares that gage's output, expressed as a ratio of resistance change to the gage's original resistance, to its input, expressed as a strain. In other words, this comparison is the gage's sensitivity. In this particular case, the figure calculated has no units:

$$\text{gage factor (GF)} = \frac{\Delta R/R}{\epsilon} \qquad (7\text{-}15)$$

where ΔR = change in resistance due to stress (Ω)

$\quad R$ = original resistance of strain gage (Ω)

$\quad \epsilon$ = strain (in./in., m/m, etc.)

Since the modulus of elasticity of any material is defined as stress/strain, we can say that

$$\epsilon = \frac{p}{E} \qquad (7\text{-}16)$$

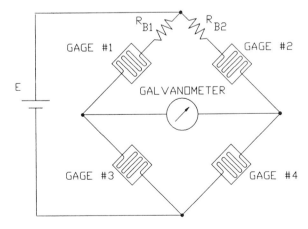

Figure 7-21 Another method for reducing temperature effects in a Wheatstone bridge, strain gage circuit.

where ϵ = resultant strain (in./in., m/m)

   p = applied stress (lb/in^2, N/m^2)

   E = modulus of elasticity of the gage's wire (lb/in^2, N/m^2)

Equation (7-17) may now be rewritten as

$$GF = \frac{(\Delta R/R)E}{p} \tag{7-17}$$

As can be seen from eq. (7-16) or (7-17), the higher the change in resistance for a given applied stress to the gage, the higher the gage factor. Or conversely, the higher the gage factor figure associated with each gage, the higher its sensitivity to the stress applied.

The gage factor equations above are usually associated with wire strain gages. However, as pointed out in Section 7-4.1, strain gages are also constructed from certain semiconductor materials having internal strain-resistance behavior similar to that of wire. Often, for these materials the gage factor is defined somewhat differently:

$$GF = 1 + 2\mu + \gamma E \tag{7-18}$$

where GF = gage factor

   μ = Poisson's ratio for semiconductor material

   γ = longitudinal piezoelectric coefficient

However, eq. (7-16) or (7-17) can be used for either type of strain gage.

Note: Poisson's ratio is defined as the ratio obtained by dividing the stressed length of a material by its width. The longitudinal piezoresistive coefficient is a figure not often found published but can be obtained from the strain gage's manufacturer.

Semiconductor gages display some interesting characteristics compared to metal gages. Gage factors for semiconductor gages tend to run much higher, often by factors of 20 or greater. These gages are usually smaller, due to their higher gage factors, and they display a much wider resistance range for a given applied stress. This, of course, accounts for the higher gage factors. Semiconductor gages tend to have less hysteresis than do metal gages. Also, semiconductor gages tend to be more rugged than their metal counterparts. Unfortunately, the semiconductor strain gage suffers from the same temperature-sensitivity problem as does the wire strain gage.

Typical gage factors for wire and foil strain gages range between 2 and 5. Gages made from semiconductor material typically range from around 40 to as high as 180. All strain gages have a most sensitive response direction or axis relative to the applied stress. This is illustrated in Figure 7-22. Notice that the most sensitive axis is parallel to the lengths of conductors in the grid.

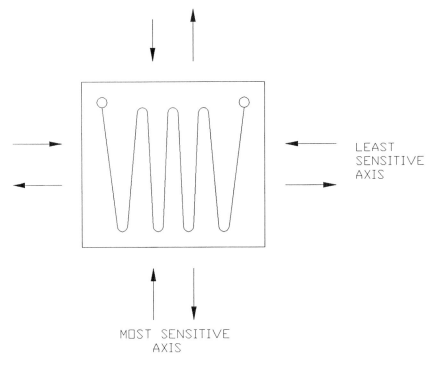

Figure 7-22   Strain gage showing sensitivity axes.

7-4.3 Typical Construction

As already mentioned, strain gages are manufactured from either wire or foil strain
detectors, or from semiconductor materials (primarily silicon, and germanium). As
for all other sensing transducers, strain gages come in a variety of sizes and shapes.
Looking first at their construction, there are two major types, *bonded* and *unbonded*
construction. In bonded construction the strain-sensing material forming the grid
(see Figure 7-23) is bonded or fixed rigidly to a base called the substrate or backing.
The backing is constructed of a moisture- and shock-resistant material usually made
from an epoxy or plastic material. The strain grid itself is bonded to the backing by
means of an epoxy, or the grid may be chemically etched onto the substrate. When
the bonded gage is subjected to a stress, the stress is applied to the entire bonded
structure.

   In an unbonded strain gage, the grid is allowed to have independent motion
relative to its backing so that the applied stress is applied only to the grid and not
to the backing, as in the case of the bonded gage. (Refer to Figure 7-18 to see how
this is done.)

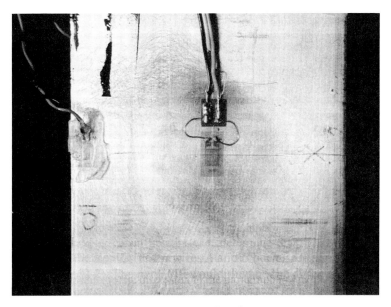

Figure 7-23   Bonded strain gage.

7-4.4 Practical Applications

The following examples are practical applications of strain gages showing how they are installed and how the output data are handled.

Measuring stress on a rigid beam member

This is perhaps the most common application of a strain gage—measuring stress on a mechanical member. The strain gage in this case is a singular unit acting as a sensor. All the signal-conditioning circuitry is located at a distance from the gage installation. Figure 7-24 shows a gage installed on an I-beam to detect compressive stress due to an applied load. All beams are designed to withstand a maximum load before collapsing. The strain gage in this particular application determines the existing load so that the design load is not exceeded. For this application the gage is oriented so that its sensitive axis is aligned vertically with the beam in order to respond to the vertical compressive stress created by the massive concrete load from above. The typical strain gage indicator needed to read out the strain indication directly in pounds or tons is shown in Figure 7-25.

Load cell

Strain gages are often used in industrial situations where it is necessary to determine the forces or pressures created between the platens of presses or casting machines. Strain gages are permanently fastened to the inside of a large steel drum designed

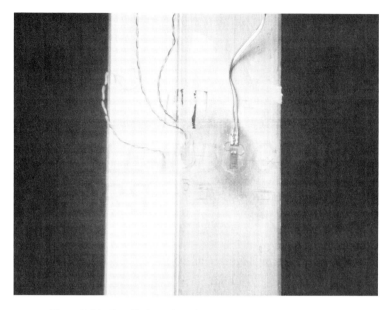

Figure 7-24   Installation of strain gage on an I-beam installation.

Figure 7-25   Strain gage readout indicator.

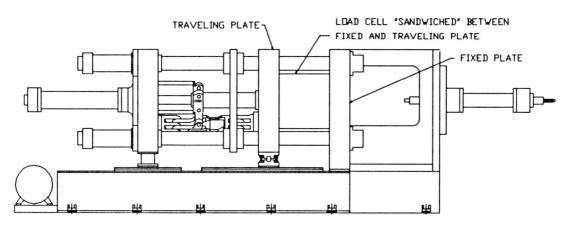

TRAVELING PLATE

LOAD CELL 'SANDWICHED' BETWEEN FIXED AND TRAVELING PLATE

FIXED PLATE

Figure 7-26 Installation of a load cell in a die-casting machine.

to withstand the measured pressures. The gages are distributed systematically around the drum's interior in a manner to produce the best averaged reading of the applied pressure or force. These drum devices, which with their assembled strain gages are called *load cells*, are used extensively in industry for testing and research.

When in use, a load cell is inserted between the pressure-creating plates of a press or die casting machine and the machine is closed, squeezing the cell in the process. An attached strain indicator placed in a safe location near the test site can then read out and record the tonnage produced by the machine. Figure 7-26 shows the actual installation of a load cell, and Figure 7-27 shows another typical type of load cell. Load cells may also be used in applications where weighing is done, such as at truck scales on interstate highways or at granaries (Figure 7-28).

Measuring fluid pressures

Strain gages are used extensively for measuring pressures of liquids and gases in vessels and pipes. They are gradually replacing many of the older mechanical pressure gages now in use. Because they are much smaller than mechanical gages, they can be installed in much tighter and more remote locations.

The electrical circuitry in a pressure strain gage is usually quite condensed compared to the ordinary strain gage. Also, the user is not required to balance the device on a Wheatstone bridge; the gage circuitry is designed so that the null reading is zero pressure, and any off-center null readings are interpreted proportionally to

Figure 7-27 Typical load cell. (Courtesy of Sensotec, Inc., Columbus, OH.)

Figure 7-28 Load cell application used for the weighing of trucks.

the actual pressure readings. Nulling is done at the factory before the sensor is shipped. Figure 7-29 shows the circuitry for a typical pressure gage. All circuit adjustments are done internally and are usually not made accessible to the user. Figure 7-30 illustrates typical installation of a pressure gage.

Measuring displacement

Strain gages are sometimes used to indicate or measure displacement. The way this is done is illustrated in Figure 7-31. A strain gage is attached to a metallic strip one of whose ends is attached to a fixed location. The other end is allowed to move so that the entire mechanism forms a cantilever. Since the amount of the strip's deflection to an applied force varies by a known amount, it is a relatively simple matter to measure the stress within the strip and correlate that stress to the amount

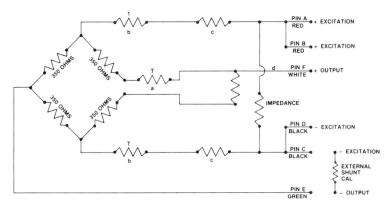

Figure 7-29 Circuitry for a typical strain gage pressure gage. (Courtesy of Sensotec, Inc., Columbus, OH.)

Figure 7-30   Installation of a strain gage pressure gage.

of deflection. Systems like this are used in industry for measuring extremely small deflections where accuracy is very important.

7-4.5 Typical Specifications

Typical specifications for a bonded metallic strain gage are shown in Figure 7-32. Specifications for a strain-type pressure gage are shown in Figure 7-33.

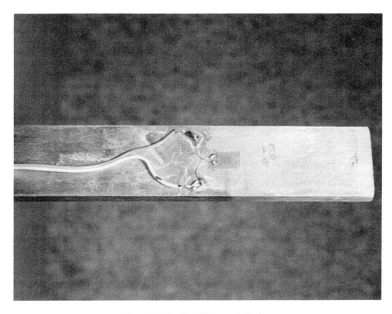

Figure 7-31   Cantilevered strain gage.

Strain gage design		Foil strain gage with embedded measuring grid
Measuring grid		
Material		Constantan foil
Thickness	μm	3 or 5, depending on gage type
Carrier		
Material		Phenolic resin with glass fiber reinforcement
Substrate thickness	μm	35 $\pm$10
Cover thickness	μm	25 $\pm$ 8
Connections		Nickel plated Cu ribbons about 30 mm long
Nominal resistance	Ω	120, or 350 depending on type
Resistance tolerance per package	%	$\pm$0.2**
Gage factor		Approx. 2
Nominal value of gage factor		Given on package
Gage factor tolerance	%	$\pm$1
Reference temperature	°C [°F]	23 [73]
Service temperature range		
for static measurements	°C [°F]	$-70\ldots+200$ [$-90\ldots+390$]
for dynamic measurements	°C [°F]	$-200\ldots+260$ [$-330\ldots+500$]
Temperature characteristic		Given on package
Temperature characteristic matched to thermal expansion coefficient		
α for ferritic steel	1/K [1/°F]	$11\cdot10^{-6}$ [6.1 ppm] (all gages)
α for aluminum	1/K [1/°F]	$23\cdot10^{-6}$ [12.7 ppm] (see table)
α for plastic	1/K [1/°F]	$65\cdot10^{-6}$ [36.1 ppm] (on request)
α for austenitic steel	1/K [1/°F]	$16\cdot10^{-6}$ [8.9 ppm] (on request)
α for titanium	1/K [1/°F]	$9\cdot10^{-6}$ [5 ppm] (on request)
α for molybdenum	1/K [1/°F]	$5.4\cdot10^{-6}$ [3 ppm] (on request)
α for quartz	1/K [1/°F]	0 (on request)
Tolerance of temperature characteristic	1/K [1/°F]	$\pm1\cdot10^{-6}$ [$\pm$0.5 ppm]
Temperature range of matching	°C [°F]	$+10\ldots+120$ [$+50\ldots+250$]
Mechanical hysteresis*		
at reference temperature and at strain $\epsilon = \pm1000\ \mu$m/m		
at strain gage type 6/120 LG 11		
for 1st load cycle and adhesive EP 310	μm/m	0.5
for 3rd load cycle and adhesive EP 310	μm/m	0.5
for 1st load cycle and adhesive X 60	μm/m	3
for 3rd load cycle and adhesive X 60	μm/m	1.5
Maximum elongation*		
at reference temperature, adhesive Z 70 used		
at strain gage type 6/120 LG 11		
strain amount ϵ in positive sense	μm/m	20,000 (= 2%)
strain amount ϵ in negative sense	μm/m	50,000 (= 5%)
Fatigue properties*		
at reference temperature, adhesive Z 70		
and oscillating strain $\epsilon_w = \pm1,000\ \mu$m/m used at strain gage		
type 6/120 LG 11		
No. of load cycles and		
zero point change $\Delta\epsilon_m \leq 300\ \mu$m/m		$7\cdot10^5$
$\Delta\epsilon_m \leq 30\ \mu$m/m		$1\cdot10^4$
Smallest bending radius, at reference temperature	mm	3
Adhesives used		
cold curing adhesives		Z 70; X 60
hot curing adhesives		EP 310

*Data depend on various application parameters, therefore only given for representative examples.
**At 0.6 mm grid length the nominal resistance may deviate by $\pm$1%.

Figure 7-32 Specifications for a bonded metallic strain gage. (REPRODUCED WITH THE PERMISSION OF OMEGA ENGINEERING, INC.)

Strain gage design		Foil strain gage with embedded measuring grid
Measuring grid		
Material		Constantan foil
Thickness	μm	3.8 or 5, depending on gage type
Carrier		
Material		Polyimid
Substrate thickness	μm	40 $\pm$5
Cover thickness	μm	12 $\pm$2
Connections		Nickel plated Cu ribbons about 30 mm
except for LY 41/43, LY 61/63, RY 31/33, RY 71/73, RY 81/83		integral terminals about 1.5 mm long and 1.6 . . .2.2 mm wide
Nominal resistance	Ω	120, 350 or 700 depending on type**
Resistance tolerance per package	%	$\pm$.3 without; $\pm$.35 with leads
except for KY types, per chain	%	$\pm$0.5
Gage factor		approx. 2
Nominal value of gage factor		given on package
Gage factor tolerance	%	$\pm$1 ($\pm$1.5 .6mm and 1.5 mm grid length)
Reference temperature	°C [°F]	23 [73]
Service temperature range		
for static measurements	°C [°F]	-70 . . . $+200$ [-90 . . . $+390$]
for dynamic measurements	°C [°F]	-200 . . . $+200$ [-330 . . . $+390$]
Temperature characteristic		given in package
Temperature characteristic matched to thermal expansion coefficient		
α for ferritic steel, i.e. 430 SS	1/K [1/°F]	$10 \cdot 8 \times 10^{-6}$ [6.0 ppm] (all gages)
α for aluminum	1/K [1/°F]	$23 \cdot 10^{-6}$ [12.7 ppm] (see table)
α for plastic	1/K [1/°F]	$65 \cdot 10^{-6}$ [36.1 ppm] (see table)
α for austenitic steel	1/K [1/°F]	$16 \cdot 10^{-6}$ [8.9 ppm] (on request)
α for titanium	1/K [1/°F]	$9 \cdot 10^{-6}$ [5 ppm] (on request)
α for molybdenum	1/K [1/°F]	$5.4 \cdot 10^{-6}$ [3 ppm] (on request)
α for quartz	1/K [1/°F]	$0.5 \cdot 10^{-6}$ [.3 ppm](on request)
Tolerance of temperature characteristic	1/K [1/°F]	$\pm 1 \cdot 10^{-6}$ [$\pm$0.5 ppm]
Temperature range of matching	°C [°F]	-10 . . . $+120$ [$+14$ . . .250]
Mechanical hysteresis*		
at reference temperature and at strain $\epsilon = \pm 1000 \mu$m/m		
gage type 6/120 LY 11		
for 1st load cycle and adhesive Z 70	μm/m	1
for 3rd load cycle and adhesive Z 70	μm/m	0.5
for 1st load cycle and adhesive X 60	μm/m	2.5
for 3rd load cycle and adhesive X 60	μm/m	1
Maximum elongation*		
at reference temperature, adhesive Z 70 used		
at strain gage type 6/120 LY 11		
strain amount ϵ in positive sense	μm/m	50000 (5%)
strain amount ϵ in negative sense	μm/m	50000 (5%)
Fatigue properties*		
at reference temperature adhesive X 60 used		
at strain gage type 6/120 LY 61		
No. of load cycles achieved at		
oscillating strain $\epsilon_w = \pm 1000 \mu$m/m and zero point change		
$\Delta\epsilon_m \leq 300 \mu$m/m		$>> 10^7$ (test stopped after 10^7)
$\Delta\epsilon_m \leq 30 \mu$m/m		$> 10^7$ (test stopped after 10^7)
Smallest bending radius, longitudinal and lateral, at reference temperature		
for strain gages with ribbons	mm	0.3
for strain gages with integral terminals		
near the connections	mm	0.3
near the terminals	mm	2
Adhesives used		
cold curing adhesives		Z 70; X 60
hot curing adhesives		EP 310
Transverse sensitivity		
at reference temperature, using adhesive Z 70		
on strain gage type LY 11 6/120	%	-0.1

*Data depend on various application parameters, therefore only given for representative examples.
**At 0.6 mm grid length, the nominal resistance may deviate by $\pm$1%.

General Purpose Gage/Absolute Pressure Transducers

Models Z and A-5

Many options

Stainless steel

0.5 to 30,000 psi

0.5 to 1500 PSI

2000 to 10000 PSI

The SENSOTEC Models Z and A-5 are designed as general industrial pressure transducers with a wide variety of available options to meet specific application requirements. Pressure ranges span from 0.5 to 30,000 psi. All models are constructed of stainless steel and utilize complete four arm 350 ohm strain gage bridges. Models A-5 and Z transducers use a standard gage design. SENSOTEC's proprietary "True Gage" design is available as an option on the Model Z. The absolute models have an internal sealed 0 psia reference.

DIMENSIONS

Model A-5 Gage
(Order Code AP141)

Available Ranges	D″	L″
0.5; 1; 2; 5 psig	2.25	2.45
10; 15; 25 psig	1.50	2.37
50; 75; 100; 150; 200; 300 psig	1.50	2.37
500; 750; 1000; 1500 psig	1.50	2.37
2000; 3000; 5000; 7500; 10,000 psig	1.50	1.90
15,000; 20,000; 30,000 psig	1.50	1.90

Models Z and A-5 Absolute
(Order Codes AP132/AP142)

Available Ranges*	D″	L″
1, 2, 5 psia	2.25	2.54
10; **15; 25; 50;** 75; **100;** 150 psia	1.50	2.37
200; 300; **500;** 750; **1,000;** 1500 psia	1.50	2.37
2000; **3000; 5000;** 7500; **10,000** psia	1.50	1.90
15,000; 20,000; 30,000 psia	1.50	1.90

Model Z Gage
(Order Code AP131)

Available Ranges	D″	L″
0.5; 1; 2; 5 psig	2.25	1.93
10; **15** psig	1.75	2.00
25; 50; 100; 150 psig	1.50	2.00
200; 300; **500** psig	1.50	2.00
750; **1000;** 1500 psig	1.50	2.35
2000; **3000; 5000;** 7500; 10,000 psig	1.50	1.90
15,000; 20,000; 30,000 psig	1.50	1.90

*Stocked Ranges (semi-complete only) for Model Z are in bold face print for 3-6 week delivery. Model A-5 is not stocked.

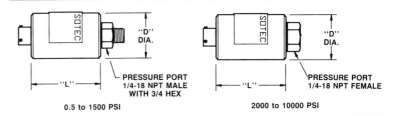

0.5 to 1500 PSI

2000 to 10000 PSI

OPTIONS
Temperature compensated 1b, 1c, 1d, 1e, 1f; Internal amps 2a, 2j; Amp enhancements 3d; Pressure ports 5a, 5b, 5c, 5d; Electrical terminations 6e, 6f, 6g, 6h, 6i; Int. shunt cal 8a; Special calibration 9a (≥5 psi gage only) (See P. 154) (For True Gage design on Model Z call SENSOTEC)

PREMIUM OPTIONS
1g (≥15 psi only), 1i (≥15 psi only); 2c, 2q; 3a, 3b, 3c; 6b, 6c, 6j; 9b (Gage only, ≥5 psi); 10a; 12a, 12b (See P. 154)

ACCESSORIES
Mating connectors and connector/cable assemblies; Pressure port adapters. (See P. 138)

Figure 7-33 Specifications for a strain gage pressure gage. (Data courtesy of Sensotec, Inc., Columbus, OH.))

		Model Z	Model A-5
		True Gage: Order Code AP131 Gage: Absolute: Order Code AP132	Gage: Order Code AP141 Absolute: Order Code AP142
PERFORMANCE	Pressure ranges	0.5 to 30,000 psi	0.5 to 30,000 psi
	Accuracy (min)	+/-0.25% F.S.	+/-0.5% F.S.
	Non-Linearity.	+/-0.15% F.S.	+/-0.25% F.S.
	Hysteresis	+/-0.10% F.S.	+/-0.13% F.S.
	Non-Repeatability	+/-0.05% F.S.	+/-0.07% F.S.
	Output (standard)	3mv/v*	3mv/v*
	Resolution	Infinite	Infinite
ENVIRONMENTAL	Temperature, Operating	-100° F to 325° F	-65° F to 250° F
	Temperature, Compensated . . .	60° F to 160° F	60° F to 160° F
	Temperature Effect		
	- Zero (max).	0.005% F.S./° F	0.0075% F.S./° F
	- Span (max)	0.005% Rdg./° F	0.01% Rdg./° F
ELECTRICAL	Strain Gage Type	Bonded foil	Bonded foil
	Excitation (calibration)	10VDC	10VDC
	Excitation (acceptable)	Up to 12VDC or AC	Up to 12VDC or AC
	Insulation Resistance	5000 megohm @ 50VDC	5000 megohm @ 50VDC
	Bridge Resistance	350 ohm	350 ohm
	Shunt Calibration Data	Included	Included
	Wiring code (std)	#2 (See P. 143)	#2 (See P. 143)
	Electrical Termination (std)	PT1H-10-6P or equiv. (Hermetic stainless)	PT1H-10-6P or equiv.
	Mating Connector (not incl.) . . .	PT06A-10-6S or equiv.	PT06A-10-6S or equiv.
MECHANICAL	Media	Gas, Liquid	Gas, Liquid
	Overload-Safe	50% over capacity	50% over capacity
	Overload-Burst		
	0.5 to 5,000 psi	300% over capacity	300% over capacity
	7500 to 10,000 psi	200% over capacity	200% over capacity
	15,000 to 30,000 psi	70% over capacity	70% over capacity
	Pressure Port		
	0.5 to 1500 psi	1/4-18NPT male	1/4-18NPT male
	2000 to 10,000 psi	1/4-18NPT female	1/4-18NPT female
	15,000 to 30,000 psi	Autoclave AE F250-C	Autoclave AE F250-C
	Dead Volume		
	0.5 to 5 psi	0.32 cu.in.	0.32 cu.in.
	10 to 15 psi.	0.25 cu.in.	0.17 cu. in.
	25 to 1500 psi	0.17 cu.in.	0.17 cu.in.
	2000 to 30,000 psi	0.12 cu.in.	0.12 cu.in.
	Wetted Parts Material	17-4 PH Stainless	17-4 PH Stainless
	Type (Gage, Abs.)	True Gage or Absolute	Gage or Absolute
	Weight	10 oz. (50 psi)	10 oz. (50 psi)
	Case Material	Stainless steel	Stainless steel
INTERNALLY AMPLIFIED UNITS (Optional)	Outputs Available	0-5VDC, 4-20ma	0-5VDC, 4-20ma
	Additional Length	1.12″	1.12″

NOTES *Output for 0.5, 1 psi units is 1-2mv/v
*0.5 psi is not available in absolute pressure
*Gage pressure units greater than 500 psi are sealed at atmospheric pressure.

GENERAL INFORMATION   How to order (see page 153)
Gage/Absolute pressure selection flow chart (see page 8)
Wiring Codes (see page 142)

REVIEW QUESTIONS

7-1. Explain what is meant by the term *circular mil*.

7-2. What is a major disadvantage of using a rotary resistive potentiometer?

7-3. What is an advantage of using a carbon composition resistance element versus using a wire-wound element in the construction of a linear potentiometer?

7-4. What is the major advantage of using a resistance-type position sensor for measuring position compared with other sensing mechanisms?

7-5. List the five most commonly used potentiometer constructions and list a characteristic of each.

7-6. What is a string pot? Describe how it works and give an example of an application or installation.

7-7. Describe how you would use a variable-resistance potentiometer to measure flow rate in a liquid-filled pipe.

7-8. Explain the function of a dummy gage in a strain gage setup.

7-9. Explain the difference between a bonded strain gage and an unbonded strain gage.

7-10. List an advantage and a disadvantage in using strain gage sensors for measuring stress.

PROBLEMS

7-11. A rotary potentiometer is to be used to monitor the position of a reciprocating arm on a piece of machinery. Make a sketch showing how this may be done.

7-12. A linear potentiometer has a total resistance of 1200 Ω. The supply voltage to this pot is 13 V ac. If the pot's wiper arm has a total travel distance of 6.50 in., what is the pot's output voltage when the wiper has traveled 2.78 in.? What would be its output resistance?

7-13. The output of a rotary potentiometer was found to have a value of 8.92 V dc for a particular angular rotation. Calculate this angle assuming that the pot has an output of 0 V at 0° rotation and a maximum output of 14.5 V dc at 357.2°. Assume a linear response output for the pot.

7-14. Find the dc resistance of 450 ft of No. 36 AWG copper wire used in manufacturing a linear wire-wound potentiometer. Assume a temperature of 68°F.

7-15. A thin-film carbon linear potentiometer is to be fabricated having a total resistance of 356 Ω. The cross-sectional area of the conducting surface measures 1.2 × 2.3 mm. Find the total conducting length needed to create this potentiometer. (ρ = 21000 CM-Ω/ft)

7-16. Calculate the total change in length for a strain indicator wire in a strain gage having a GF of 3.7. The wire's original resistance was 0.713 Ω and its final "strained" resistance was 0.803 Ω. The wire's total prestrained length was 57.8 mm.

7-17. In designing a simple Wheatstone bridge to be used as part of a strain gage indicating circuit, determine the desired unstrained resistance value needed for a gage in order to produce a balanced bridge having the following resistance values in each of its legs: R_1 = 100.00 Ω, R_2 = 150.00 Ω, and R_3 = 149.3 Ω.

REFERENCES

ALLEN, RICHARD L., and ROBERT R. HUNTER, *Transducers*. Albany, NY: Delmar, 1972.

BOYLESTAD, ROBERT L., *Introductory Circuit Analysis*. Columbus, OH: Charles E. Merrill, 1982.

HORDESKI, MICHAEL F., *Microprocessor Sensor and Control Systems*. Reston, VA: Reston, 1985.

MILLER, RICHARD W., *Servomechanisms: Devices and Fundamentals*. Reston, VA: Reston, 1977.

NORTON, HARRY N., *Sensor and Analyzer Handbook*. Englewood Cliffs, NJ: Prentice Hall, 1982.

The Pressure Strain and Force Handbook, 1989–90 ed. Omega Engineering (P.O. Box 4047, Stamford, CT 06907).

8

VARIABLE-CAPACITANCE SENSORS

CHAPTER OBJECTIVES

1. To review the concept of capacitance.
2. To study the linear motion capacitor.
3. To understand proximity sensing and gaging using the capacitive sensor.
4. To study the rotary capacitor and its multiple-plate designs.

8-1 INTRODUCTION

The capacitance sensor is one of the more recent developments in sensor and transducer technology. Its function depends on the characteristics and behavior of the variable capacitor. To understand how the variable capacitance sensor works, we must first review the theory of the capacitor itself.

8-2 REVIEW OF CAPACITANCE

A capacitor is comprised of two parallel plates of conducting material separated by an electrical insulating material called a dielectric (Figure 8-1). The plates, along with the "sandwiched" dielectric, may be either flattened (i.e., parallel with each other) or rolled into some other convenient shape. Electrodes are attached to each capacitive plate for the purpose of making electrical connections within a circuit.

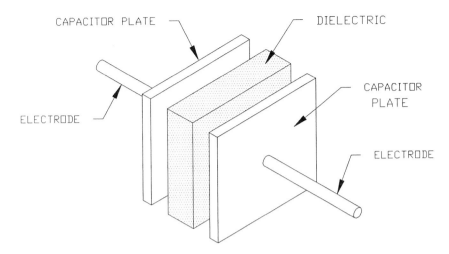

Figure 8-1 Simplified construction of a capacitor.

The purpose of the dielectric is to help the two parallel plates maintain their stored electrical charges. In other words, the dielectric tries to discourage any negative electrical charges stored on one plate from traveling over to the other plate and becoming neutralized. If this segregation of charges can be maintained within the capacitor, the capacitor becomes a very effective device for storing charges until they are needed, much like the behavior of a battery. (However, in a battery, the segregated charges are maintained through an active chemical process rather than through a somewhat passive one in the case of the capacitor.) Equation (8-1) shows the relationship between capacitance, size of capacitor plate, amount of plate separation, and the dielectric, often referred to as the *permittivity*.

$$C = \frac{\epsilon A}{d} \tag{8-1}$$

where C = capacitance (F)

ϵ = permittivity (F/m)

A = area of capacitor plates (m^2)

d = separation distance of plates (m)

Quite often in discussions of capacitors the term *relative permittivity* is used, because the permittivities of materials are often compared to the permittivity of a vacuum, the vacuum's value being 8.85×10^{-12} F/m. Table 8-1 lists the relative permittivity values of some of the more common materials used in the manufacture of capacitors.

It is important here to emphasize that the amount of capacitance can be varied by altering any one or all of the three variables shown in eq. (8-1). The only item necessary to measure or record a change in capacitance is, of course, a capacitance-

TABLE 8-1 RELATIVE PERMITTIVITY VALUES FOR COMMON
MATERIALS IN CAPACITORS

Material	Relative permittivity
Vacuum	1.0
Air	1.0006
Paper	2.5
Mica	5.0
Glass	7.5
Ceramic	7500.0

Source: R. L. Boylestad, *Introductory Circuit Analysis*, 4th ed.
(Columbus, OH: Merrill, 1983).

measuring circuit of some sort. However, what is often done is not to measure
capacitance or capacitance change directly, but rather to sense some electrical
characteristic that is closely associated with this change in capacitance. This concept
is explained in more detail in the next section.

The "active" area of a capacitor is the area that is meshed; this is the area that
participates in the determination of the capacitance of that capacitor. To determine
this area, one must take into account the surface areas directly opposite the two
parallel plates making up the capacitor, as Figure 8-2 illustrates. Assume that you
have a capacitor made up of two plates that rotate relative to each other on the same
shaft, as shown in the figure. The area that actually determines the capacitance of
this capacitor is the active area, A. The remaining area does not affect the capacitance except for a minor interaction referred to as *fringing*. Fringing is discussed later
in the chapter. For the present, however, we will neglect this characteristic and

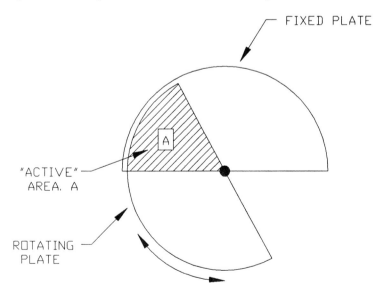

Figure 8-2 Rotary capacitor.

concentrate on the active areas of a capacitor. The capacitor described in Figure 8-2 is put to good use in at least one displacement-measuring sensor, which we discuss in Section 8-5.

To be sure that the theory of capacitance is fully understood, we will now calculate the distance of separation between two parallel plates using the device in Figure 8-2 as an example.

Example 8-1

The capacitor shown in Figure 8-2 has a capacitance of 38 pF when the two plates are 100% meshed. These plates have a diameter of 20 cm and are rotated so that they are 30% meshed. The dielectric is air. Calculate the amount of separation between the two plates.

Solution: We first calculate the area of each plate in our problem. Since area is found by using the equation area $= \pi d^2/4$, and since Figure 8-2 shows each plate to be semicircular in shape, we will assume each plate's area to be one-half of the calculated value. Therefore,

$$\text{area of capacitor plate} = \frac{\pi(d^2)}{4} \times 0.5$$

$$= \frac{\pi(0.2 \text{ m})^2}{4} \times 0.5$$

$$= 0.0157 \text{ m}^2$$

Because the capacitor is 30% meshed, we use 30% of 0.0157 m^2, or 0.00471 m^2, for the active plate area in our capacitor. We must also note that because of the 30% mesh, we have also reduced the initially stated capacitive value of 38 pF to only 30% of this value, or 11.4 pF.

Having been told that air has a relative permittivity of 1.0006 (Table 8-1) and recalling that the permittivity value for a vacuum is 8.85×10^{-12} F/m, the permittivity of air is then the product of these two figures, or 8.855×10^{-12}. We can now use eq. (8-1) to calculate the plates' spacing by solving this equation for d:

$$d = \frac{\epsilon A}{c}$$

$$= \frac{(8.855 \times 10^{-12})(0.00471 \text{ m}^2)}{11.4 \times 10^{-12} \text{ m}}$$

$$= 0.00366 \text{ m}$$

$$= 3.66 \text{ mm}$$

8-3 COMMONLY SENSED MEASURANDS

One of the most frequently sensed measurands using capacity-sensing devices is displacement or some variation of displacement. Also, a typical variation of displacement sensing would be the measurement of force or pressure, in which either would cause movement of a variable plate capacitor (Figure 8-3).

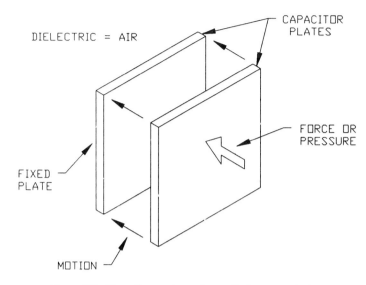

Figure 8-3 Capacitor used as a linear displacement device.

Temperature is another measurand that depends on displacement for its detection. Figure 8-4 shows such an application. In this application we see a bimetallic coil whose expansion and contraction with temperature are converted to a capacitive displacement. In this application the capacitive element has nothing to do with the sensing of temperature. It is used simply as a translation device to convert the motion of the actual sensing device, the bimetallic strip, into an electrical signal.

In the sections that follow we study the detection methods used and obtain a better idea of how capacitive sensors work.

8-4 LINEAR MOTION CAPACITOR

The linear motion capacitor is a variable capacitor whose movable plate moves in a straight path toward and away from the fixed plate. At the same time, the faces of both plates are kept parallel to each other, maintaining an evenly spaced gap to allow for a dielectric to be present. The term *linear* refers to the type of straight-line motion of the movable capacitor plate and does not refer to its output response. We will now describe several of these configurations and discuss how they operate.

8-4.1 Parallel-Plate Capacitive Sensor

Detecting displacement

Detecting displacement may be done with a measuring device employing the concept shown in Figure 8-3. The one movable plate of the capacitor is allowed to move the same displacement amount as that being measured or detected. A variation of this

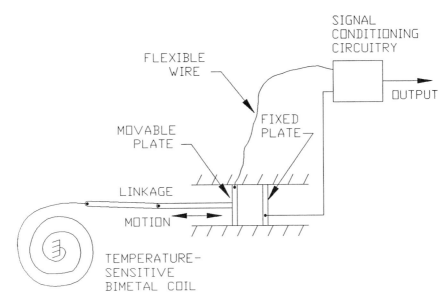

Figure 8-4 Using a linear displacement capacitor to indicate temperature with a bimetal coil.

application is similar to the one just described except that one plate is allowed to move over the other stationary plate. The overlapping areas then represent an analog of the displacement being measured.

Detecting force or pressure

Now that we have described how displacement is detected using a linear motion capacitor, another often detected measurand is force or pressure. Actually, measuring force or pressure is nothing more than modifying the displacement sensor so that the applied force or pressure, when either is applied to the movable plate of a capacitive sensor, produces a plate displacement instead. Once again, the resonant frequency of a circuit in which the sensor has been installed can be monitored and the readout modified by scaling circuits to produce a direct readout in pounds, newtons, psi, and so on.

Detecting acceleration

Acceleration is often detected by using a slight variation of the system just described. Figure 8-5 shows an accelerometer that uses a *seismic mass*. A seismic mass is a mass attached to the movable capacitor plate which causes the plate to move away or toward the fixed plate when the entire structure is accelerated or decelerated. The resultant output of the capacitor is processed just like the other examples we studied using a resonant circuit.

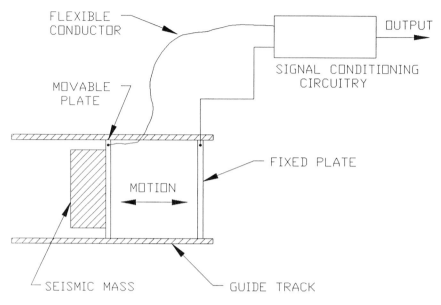

Figure 8-5 Using a linear motion capacitor with a seismic mass for detecting acceleration.

Measuring liquid depths

Another application of a linear motion capacitor is shown in Figure 8-6. This is a method often used for measuring liquid depths. We see two parallel metal bars or plates representing the plates of a capacitor. These have been submerged in a tank of liquid whose liquid depth is to be measured. Because the liquid represents a variable dielectric as its depth changes by covering and uncovering more of the capacitor's plate area, with the air above its surface representing another variable dielectric as it varies with the liquid's depth, it becomes a matter of calibrating the attached output display to show the correct depth. So in essence, we have a capacitive detector whose capacitance varies linearly with the change in its dielectric.

Another depth-measuring system is depicted in Figure 8-7. Instead of having the depth-measuring capacitor inside the main tank, it is located over an adjoining sight glass. Inside the sight glass a float rides on the liquid level. Attached to the float is a connecting rod attached to a rod "plate" that moves into and out of a second stationary plate. The second plate is a machined cylinder allowing the movable inside rod to move like a piston inside its cylinder while maintaining the desired spacing for the dielectric (i.e., air) to exist. In this system the area of the two capacitor plates is varied, recalling once again that for any variable capacitor it is the *opposite area faces* on the two plates that determine a capacitor's actual capacity (a concept that was reviewed in Example 8-1).

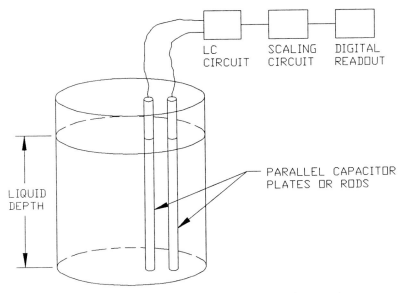

Figure 8-6   Measuring liquid depth using a linear motion capacitor.

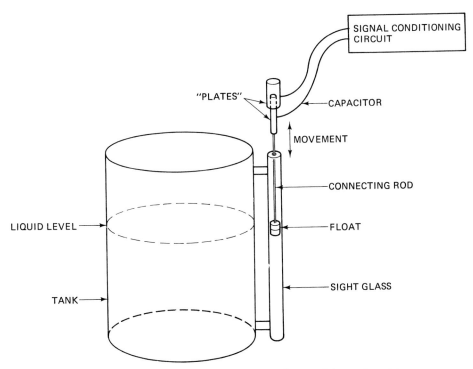

Figure 8-7   Another depth-measuring system using parallel capacitor plates.

8-4.2 Multiple-Plate Linear Displacement Design Configurations

All of the capacitive sensors that we have discussed up to this point are of the *single-plate-pair* design. That is, only one pair of plates—one plate fixed, the other plate movable—is involved in the capacitor's design. However, multiple-plate designs are also common and have one outstanding advantage over the single-plate design: For a given linear or rotary displacement, the change in capacitance is much larger than in single-plate-pair design. Multiple plates are comprised of individual plate pairs that form a parallel capacitive circuit with each other. They are coupled mechanically so that all plate pairs move in unison. Because of this parallel circuit, large capacitive values are formed, resulting in easier displacement detection. Figure 8-8 summarizes some of the more typical designs, each of which is discussed below.

In Figure 8-8(a) the two movable plates are coupled together mechanically to cause movement in opposite directions due to the same motion-creating displacement. This, in effect, causes a doubling in the increase or decrease in capacitance since the two movable plates are wired in parallel with each other. Figure 8-8(b)

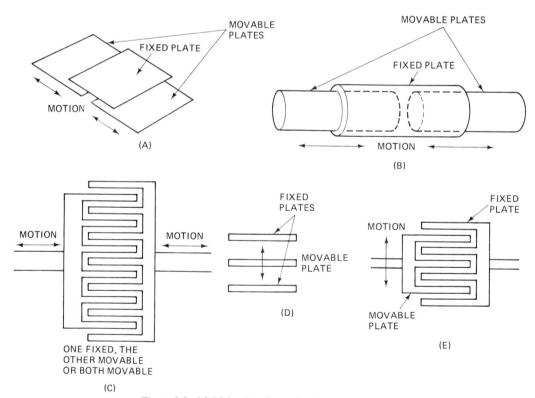

Figure 8-8   Multiple plate linear displacement capacitor designs.

works on the same principle as part (a). For a given displacement or force, both movable plates move in opposite directions to produce a doubling in the capacitance variations.

Figure 8-8(c) shows a multiple-plate capacitor in which either or both sets of plates move relative to the other, producing a change in effective capacitive area. If both plates move due to the same applied motion, the amount of capacitance change varies twice as rapidly as having only one of the sectioned plates move.

Capacitive bridge circuit

Figure 8-8(d) and (e) are similar in principle. As the movable plate moves, the distance between its plate and the fixed plates vary. In addition, a nulling-type circuit may be used with these configurations to produce a center-zero condition; part (a) may also utilize this circuit. A typical nulling circuit based on the operation of the Wheatstone bridge (refer to Section 7-4.1 for a review of how the Wheatstone bridge functions) is shown in Figure 8-9. However, instead of using resistors for the circuit-balancing elements, the capacitive reactances produced by the signal generator and the bridge capacitors are used instead. Recalling that $R_1/R_2 = R_3/R_4$ in a Wheatstone bridge (refer to Figure 7-21), this equation can be modified to read

$$\frac{X_{C1}}{X_{C2}} = \frac{X_{C3}}{X_{C4}} \tag{8-2}$$

In general, for a capacitor,

$$X_C = \frac{1}{2\pi f C} \tag{8-3}$$

where X_C = capacitive reactance (Ω)

f = frequency (Hz)

C = capacitance (F)

Then substituting eq. (8-3) into eq. (8-2), we get

$$\frac{C_1}{C_2} = \frac{C_3}{C_4} \tag{8-4}$$

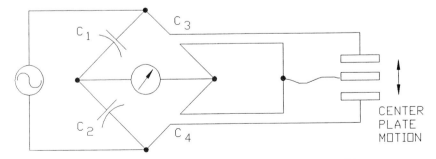

Figure 8-9 Nulling circuit to produce a "center-zero" condition.

Equation (8-4) shows the relationship between the four capacitances in each of the four legs of the Wheatstone bridge. Assuming that a constant generator frequency is maintained in the bridge circuit, the reactance within each leg will determine the amount of current flow within the bridge since, in general,

$$I = \frac{V}{X_c} \tag{8-5}$$

where I = current (A)

V = voltage (V)

X_c = reactance (Ω)

8-5 CONFIGURATIONS AND PRACTICAL APPLICATIONS

Capacitive sensors and transducers are extremely popular in industry for proximity detection and gaging. The size and shape of the sensor are determined by its designed application. Next we investigate a few of these applications and look at some typical application specifications.

8-5.1 Proximity Sensing

Sensing the proximity of nearby objects using a capacitive sensor is a common application. However, there are two primary considerations in selecting this type of sensor: (1) how large or small the target material is, and (2) the material's conductivity. The size of target determines the ability of the sensor to sense the target. A larger target, that is, a target having a larger surface area, certainly becomes easier to sense than a smaller target. If a sensor is unable to sense a small target area, using a sensor having a larger capacitive sensing surface area will usually cure the problem.

If the material being sensed is conductive, the material becomes one of the two plates of the sensor's capacitor, as shown in Figure 8-10. In an application like this the conducting properties of the target material will have a profound effect on the sensor's performance. A highly conductive material will increase the sensitivity of the sensor, whereas a poorly conducting material will have just the opposite effect.

In applications where the target material is a poor conductor, that material does not become the other capacitive plate. Instead, the material becomes part of the dielectric near the sensor's two existing plates. To understand how this can be, we must first look at one of the interesting peculiarities of capacitors. Figure 8-11(a) shows a parallel single-plate capacitor whose dielectric is a material having a very high permittivity. The ends of the capacitor, however, are exposed to air. Since the high-permittivity material has a greatly reduced electric field flowing through it (resulting, by the way, in a high-value capacitor) the ends of this capacitor will contain electric field patterns comprised of electric field lines trying to "sneak"

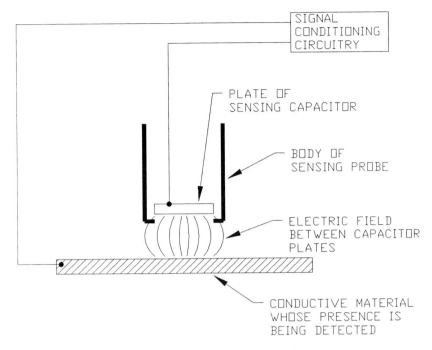

Figure 8-10   Using a capacitor for proximity sensing.

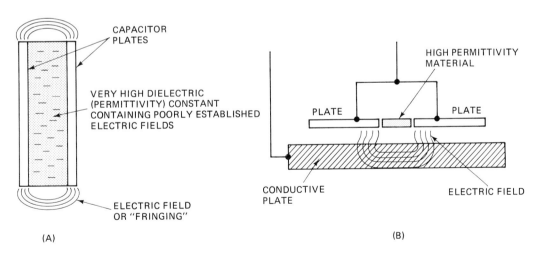

Figure 8-11   Fringing characteristics of a capacitor.

around this dielectric through the weaker dielectric, air. These lines are called *fringing fields*. If the capacitive plates are constructed side by side as shown in part (b), the fringing fields are more pronounced and are more subject to interaction with nearby materials. Any distortion of the fringing patterns created by nearby material will affect the sensor by changing its capacitance and therefore its output.

8-5.2 Gaging

Classically, the term *gaging* refers to the use of certain instruments (usually mechanical or electromechanical in their operation) for the purpose of making quantitative measurements. We are interested in seeing how capacitive sensors are used for this application.

One application of capacitive sensing is found in the measurement of surface features made on relatively flat surfaces. A probe is placed on the material's surface (Figure 8-12) to determine the degree of roughness of a metal's machined surface. The mean surface is indicated to emphasize the surface's variations. To obtain a degree of roughness comparison, a comparison is then made between the capacitance generated by this surface and by that of a perfectly flat surface.

In another application the capacitance probe is used to measure depth or depression features in a material's surface (Figure 8-13). This application is fairly similar to the surface roughness application in that properly shaped surface feature measurements may be compared to production samples, or actual measures can be made on the production samples. Either method works well and both are used extensively.

A drawback to using a capacitive sensor is that the air (the sensor's dielectric

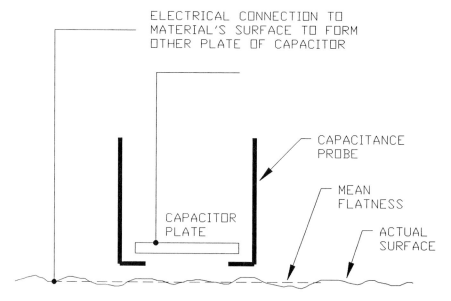

Figure 8-12   Capacitor gaging for detecting surface roughness.

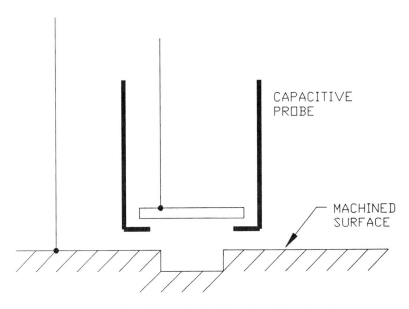

Figure 8-13   Capacitor gaging for detecting surface depressions.

in these cases) must be clean, that is, free of all suspended particulates. Also, no moisture can be present in the air. Otherwise, the results may be erroneous.

8-6 ROTARY CAPACITOR

As the name implies, the rotary capacitor is used for applications involving rotary motion for displacement measurements. An example of this style of capacitor was shown in Figure 8-2. In our discussions below, we refer to this type of capacitor as a *flat rotary-plate capacitor*. Other construction styles are also possible, as shown in Figure 8-14. We will now analyze all four of these rotary styles to obtain a better understanding of how they work.

8-6.1 Flat Rotary-Plate Capacitor

Figure 8-15 illustrates the flat rotary (rotating)-plate capacitor, showing its critical dimensions. Assuming that all segments comprising the two plates' construction are identical, and neglecting area loss due to the shaft attached to the rotatable plate, the capacitance of the entire system is calculated as follows (note that we are calculating the area of the meshed plates only, which is the shaded area in Figure 8-15). The shaded meshed area, A, is equal to $(\phi - \theta)\pi a^2/360$, and since $C = \epsilon A/d$ [eq. (8-1)],

$$C_{\text{frp}} = \frac{\epsilon(\phi - \theta)a^2 \pi}{360d} \tag{8-6}$$

where C_{frp} = capacitance of flat rotary-plate capacitor (F)

ϵ = permittivity (F/m)

ϕ = subtended angle of plate segment (deg)

θ = angle of rotation of movable plate (deg)

a = radius of movable plate segment (m)

d = plate separation (m)

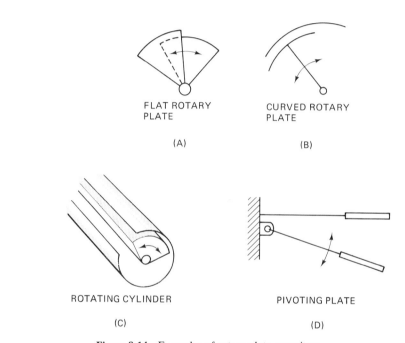

Figure 8-14   Examples of rotary-plate capacitors.

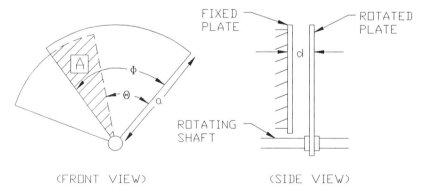

Figure 8-15   Detailed dimensions for style (a) in Figure 8-14.

If the permittivity of air is substituted for ϵ in eq. (8-6) along with the value of π, eq. (8-6) may be written as

$$C_{\text{frp}} = \frac{7.723 \times 10^{-14}(\phi - \theta)a^2}{d} \qquad (8\text{-}7)$$

Example 8-2

A flat rotary-plate capacitor is constructed from a 72° segmented disk having a radius of 1.7 in. The plate separation for this capacitor is 0.075 in. The movable (rotatable) plate has been attached to the shaft of a bimetallic temperature sensor that rotates 0.13 degree per °C, measured from a fully meshed position of the capacitor's two plates. If a temperature change of 36° occurs, what is the resultant capacitance? What is the change in capacitance assuming that the capacitor's plates are initially fully closed? Assume the dielectric to be dry air.

Solution: First, we must convert any dimensions expressed in inches to meters in order to use either eq. (8-6) or (8-7). Therefore, for our a value, 1.7 in. × 0.0254 m/in. = 0.0432 m. Also, for our plate separation value, d, 0.0075 in. × 0.0254 m/in. = 0.000191 m.

The segment angle is given as $\phi = 72°$. We now have to calculate θ. This can be done by noting that the angle of rotation is given as 1.13°/°C and we were told that a temperature change of 36°C occurred. Therefore, $\theta = 1.13°/°C \times 36°C = 40.68°$. Using eq. (8-10) to calculate the final resulting capacitance, we get

$$C_{\text{frp}} = \frac{7.723 \times 10^{-14}(72.00° - 40.86°)(0.0432 \text{ m})^2}{0.000191 \text{ m}}$$

$$= 23.5 \times 10^{-12} \text{ F}$$

$$= 23.5 \text{ pF (final value)}$$

To find the change in capacitance, we must next calculate the capacitance with the plates fully closed (i.e., $\theta = 0°$). Therefore,

$$C_{\text{frp}} = \frac{7.723 \times 10^{-14}(72.00° - 0°)(0.0432 \text{ m})^2}{0.000191 \text{ m}}$$

$$= 5.43 \times 10^{-11} \text{ F}$$

$$= 54.3 \text{ pF (initial value)}$$

The change in capacitance as a result of the 36°C change in temperature is 54.3 pF − 23.5 pF, or 30.8 pF. Another, perhaps easier method for determining the change in capacitance would be merely to use the value of θ, or 40.68°, to represent the value of $(\phi - \theta)$ in eq. (8-10), since this is the amount of capacitance change that was experienced during the 36°C temperature change.

8-6.2 Curved Rotary-Plate Capacitor

The curved rotary-plate capacitor is similar to the flat rotary-plate capacitor. The only difference is in the cross-sectional shapes of the capacitor's two plates. Both the fixed and movable plates are curved to form sections of a cylinder's surface. The advantage of this style of capacitor is that additional plate surface area is obtained for a given capacitor size, resulting in a higher overall capacitance. This type of construction is shown in Figure 8-16. The angle ϕ, not shown in this figure, is the angle subtended by the radii at each end of the plate's arc. Angle θ is the angle of rotation of the movable plate, similar to that used in Example 8.2 for the flat rotary plate. Dimension a is the length of each arc. Dimension d is the spacing between the two plates.

To find the capacitance of this system, we first find the area of the plates. We assume that the two plates are identical in size. To begin with, the area of a cylinder is found by the equation

$$A_{cyl} = 2\pi r a \qquad (8\text{-}8)$$

where A_{cyl} = area of cylinder (m², ft², etc.)

r = radius of cylinder (m, ft, etc.)

a = length of cylinder (m, ft, etc.)

Since we are dealing with only a section of this cylinder, the plates' areas are equal to

$$A_{crp} = \frac{\pi r a \phi}{180} \qquad (8\text{-}9)$$

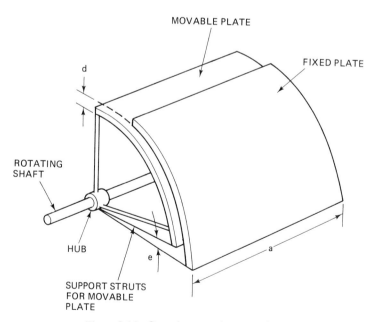

Figure 8-16   Curved rotary-plate capacitor.

where A_{crp} = area of curved rotating plate (m², ft², etc.)

r = radius of cylinder (m, ft, etc.)

a = length of cylinder (m, ft, etc.)

ϕ = angle subtended by plate's curve (deg)

Substituting eq. (8-9) into eq. (8-1), the general capacitance equation, we get

$$C_{crp} = \frac{\epsilon \pi a \phi}{180d} \qquad (8\text{-}10)$$

But because of partial rotation of the movable plate, the actual capacity for any given rotation amount will be

$$C_{crp} = \frac{\epsilon \pi a (\phi - \theta)}{180d} \qquad (8\text{-}11)$$

Combining all constants, assuming that ϵ for air = 8.85×10^{-12} F/m, we get

$$C_{crp} = \frac{1.545 \times 10^{-13} ra(\phi - \theta)}{d} \qquad (8\text{-}12)$$

Remember that to use eq. (8-12) properly, r, a, and d must be expressed in meters. This will result in C_{crp} being expressed in farads.

8-6.3 Pivoting Rotary-Plate Capacitor

The pivoting rotary-plate capacitor, shown in Figure 8-14(d), is very simple in construction. One of its two plates is allowed to swivel or pivot away from its fixed plate so that a somewhat modified "linear"motion is created for very small movements. However, because of the rotary component imposed on the movable plate, this system suffers from a lack of linearity. However, this effect can be reduced somewhat by increasing the length of the pivoting arm on which the movable plate is mounted and mounting the plate as far away from the pivot as possible. This also decreases the change in capacitance per degree of rotation to some extent, if this is a concern.

8-6.4 Cylindrical Rotary-Plate Capacitor

The cylindrical rotary-plate capacitor behaves similarly to the curved rotary-plate capacitor described in Section 8-6.2. The only distinguishing difference between these two systems is in their construction. The cylindrical capacitor [Figure 8-14(c)] is considered to be far more rugged but more costly to construct than the curved plate type [Figure 8-14(b)]. The curved cylinder is usually a machined part having better bearing construction and support, whereas the cylindrical style is made from individual stamped-metal components and has a lighter bearing construction for the moving plate. However, because of the more massive construction, the cylindrical style requires greater torque to cause shaft rotation than in the curved plate design. This could be a severe limitation in some very low torque rotational applications.

The inner moving plate is totally insulated from the outer fixed plate through insulated bearings at both ends of its mounting to the fixed-plate ends. The electrical capacity of this unit is calculated just like the curved rotary-plate capacitor using eq. (8-15).

8-6.5 Multiple Rotary-Plate Designs

As we found for the multiple-linear-motion-plate designs (Section 8-4.3), there are also numerous multiple-rotary-plate designs. Figure 8-17 shows three multiple-curved-plate designs, two of which [(a) and (c)] are differential designs. These two designs, similar to the differential designs discussed in Section 8-4.3, are generally used with Wheatstone bridge conditioning circuits. The capacitor shown in part (b) is very similar to those found in the tuned circuits of electronic communications equipment. The only difference between the multiple rotary capacitors used there and the ones used in sensing circuits has to do with the bearing design. The sensing circuit designs usually incorporate a more rugged bearing, to facilitate the higher cycle rates encountered. Otherwise, the applications are very similar.

Multiple-plate capacitors have the advantage over single-plate designs that for a given displacement of the rotary shaft, a much larger change in capacitance can be created. This, in turn, produces greater sensitivity and increases the resolution in detecting the measurand. Furthermore, there is a reduction in the amount and precision of circuitry needed to detect the smaller capacitive changes in the single-plate-pair capacitive systems.

A disadvantage of multiple-plate designs is that because of the greater mass attached to the rotary shaft, the multiple-plate capacitor requires a greater torque input to the shaft to cause rotation. This could create problems in those areas where very small torques are available from the measurand.

Figure 8-18 illustrates installation of a differential cylindrical rotary-plate capacitor being used to sense fluid flow in a process control system. A flow-balancing

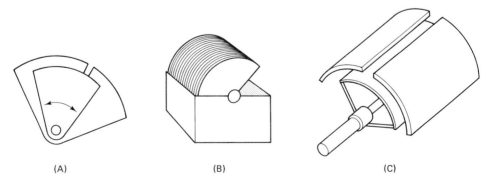

(A)   (B)   (C)

Figure 8-17   Multiple-rotary-plate designs.

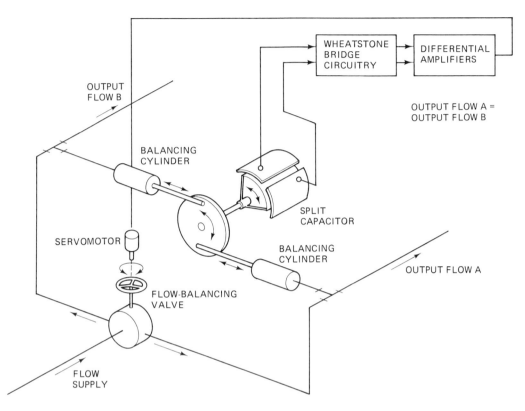

Figure 8-18 Differential cylindrical rotary-plate capacitor being used to maintain a flow balance in two fluid-carrying pipes.

valve is used to maintain a balanced flow in each of the two supply pipes shown. The valve is operated by a servomotor whose operating voltages are received from an amplifier and Wheatstone bridge circuit. If an imbalance is sensed by the bridge due to uneven displacements of the pistons inside their pressure-sensing cylinders, the out-of-balance voltage produced by the bridge is sent to the servomotor to cause the balancing valve to compensate for this imbalance. This general balancing scheme is used in many process control systems to create a self-supervisory system.

REVIEW QUESTIONS

8-1. Explain what is meant by *fringing* in a capacitive sensor.

8-2. Why cannot capacitive sensors be used to measure displacement over very large distances?

8-3. Design a circuit showing how a rotary capacitive sensor can be used to measure angular rotation with the sensor as part of an RF oscillator.

8-4. How would a capacitive sensor be used to measure temperature? Make a sketch of your device and furnish a written explanation to accompany your drawing.

8-5. Why is a bending-plate capacitive sensor not used much for the detection of displacement?

8-6. Why cannot every liquid depth be sensed by a capacitive depth sensor?

8-7. Describe at least two types of mechanical construction techniques used in making rotary capacitive sensors.

8-8. When using a capacitive sensor, what is the difference between proximity sensing and analog sensing?

8-9. Explain how gaging is done when using a capacitive sensor. What effect does the material being gaged have on the gaging system's sensitivity?

8-10. What advantage is there is using multiple-plate capacitive sensors rather than single-plate sensors?

PROBLEMS

8-11. Determine the capacitance existing between two plates of aluminum separated by a distance of 1.7 mm, each plate having a dimension of 2.8 × 2.5 cm. Assume that one plate is directly opposite the other and that the dielectric between the plates is glass.

8-12. Two rods with diameters of 0.25 in. form a capacitor in a tank containing material of relative permittivity 2.75. The rods are separated by a distance of 1 cm. If the rods extend the full depth of the tank and the tank is two-thirds full of this material, the remaining third being air, calculate the total capacitance between the rods. Tank depth is 15 feet.

8-13. A transducer design consists of a variable single-plate-pair capacitor having a maximum capacity of 57 pF. The capacitor's plates are semicircular in shape. If the plates are to be separated by 0.50 mm, determine the plates' diameter.

8-14. To produce a radio-frequency signal of 2.85 MHz for a telemetry application while using a 35-μH inductor in the radio-frequency generating circuit, what size capacitor is required?

8-15. A capacitance bridge is to be constructed from four capacitors: $C_1 = 0.02$ mF, $C_2 = 0.03$ mF, $C_3 = 0.05$ mF, and $C_4 = ?$. What must the value of C_4 be to produce a balanced bridge? If the voltage supplied is 6 V ac at a frequency of 35 kHz, what capacitive reactances are experienced at each of the four capacitors?

8-16. If a displacement sensor is produced from a capacitor having one fixed and one movable plate, how would you prove that its capacitive output is a linear function of the displacement? If the capacitive output of this sensor were made part of a resonant circuit, what sort of frequency output relationship would you expect compared to the displacement? Support your proof mathematically.

8-17. A flat rotating-plate capacitor is constructed from a 52° segmented disk of radius 0.852 in. Assuming a plate separation of 0.036 in. and a 37% mesh, calculate the capacitance assuming air to be the dielectric.

REFERENCES

ALLOCCA, JOHN A., and ALLEN STUART, *Transducers: Theory and Applications*. Reston, VA: Reston, 1984.

BOYLESTAD, ROBERT L., *Introductory Circuit Analysis*. Columbus, OH: Charles E. Merrill, 1983.

DeMAW, DOUGLAS, *ARRL Electronics Data Book*. Newington, CT: American Radio Relay League, 1976.

FINK, DONALD G., *Electronics Engineers' Handbook*. New York: McGraw-Hill, 1975.

9

VARIABLE-INDUCTANCE SENSORS

CHAPTER OBJECTIVES

1. To review the concept of inductance.
2. To study the linear and rotary motion inductor.
3. To understand the advantages and disadvantages of the inductive sensor.

9-1 INTRODUCTION

Sensors that vary the inductance of a looped or coiled conductor to perform their measurand sensing are called variable-inductance sensors. In the section that follows, we first discuss the theory of inductance and then look at the various forms of variable-inductance sensors that fall under this category. Unfortunately, from an intuitive approach, inductance is not as easily understood as capacitance. Hopefully, though, the following review will help to clarify this situation.

9-2 REVIEW OF INDUCTANCE

An inductor is a coiled loop made of a conducting material whose purpose is either to "choke" back alternating current to prevent it from flowing in a circuit, or to encourage oscillations at a particular frequency when used either in series or in parallel with a capacitor. More exactly, inductance is an indication of the ability of a coil to *oppose any change in current flowing through that coil*. Sometimes the term *self-inductance* is used to describe that ability, but often the term *self* is dropped. It should be pointed out here that inductance must not be confused with inductive reactance even though the two terms are related. The amount of inductance for a given coil may be calculated using the following equation:

$$L = \frac{\mu N^2 A}{l} \qquad (9\text{-}1)$$

where L = inductance [henry (H)]

   μ = permeability "constant" [see discussion below; units are webers/ampere turns-meter (Wb/At-m)]

   N = number of loops or turns

   A = area of the inside core of loop (m^2)

   l = length of core (m)

When we talk about inductance, we also have to talk about permeability. *Permeability* refers to a material's ability to support the existence of magnetic lines of flux that pass through the core of an inductor whenever the inductor becomes energized with a current flow. Strictly speaking, the permeability constant, μ, referred to in eq. (9-1) is not really a constant since its value depends on the flux density, B, and the magnetizing force, H. Both are present inside the coil's core when current is flowing through the coiled conductor. Because $\mu = B/H$, it can be seen that μ is dependent on the amount of current flowing through the coil at the time. Also, eq. (9-1) is valid only for a single-layered coil winding. Multiple-layered coils require a considerably modified equation. Note, too, that the unit of inductance in eq. (9-1) is the henry. Equation (9-1) is considered a "basic" equation that is derived from the basic definition of induction. The use of this equation is demonstrated in the following example.

Example 9-1

Calculate the inductance of the coil in Figure 9-1. Assume air as its core material. The permeability of air is 12.57×10^{-7} Wb/At-m.

Solution: We first calculate the area of the core having a diameter of 2.54 cm according to Figure 9-1. Since

$$A = \frac{\pi d^2}{4}$$

then

$$A = \frac{\pi (0.0254 \text{ m})^2}{4}$$

$$= 5.067 \times 10^{-4} \text{ m}^2$$

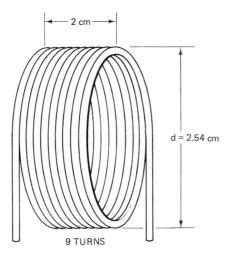

Figure 9-1 Illustration for Example 9-1.

Note that in the calculations above, we converted all lengths to meters, the basic unit of length in the SI system. This is because we are working with an equation containing no conversion factors. We now use eq. (9-1) to get

$$L = \frac{\mu N^2 A}{l}$$

$$= \frac{(12.57 \times 10^{-7} \text{ Wb/At-m})(9 \text{ turns})^2 (5.067 \times 10^{-4} \text{ m}^2)}{0.02 \text{ m}}$$

$$= 2.58 \times 10^{-6} \text{ H} \quad \text{or} \quad 2.58 \text{ } \mu\text{H}$$

Table 9-1 lists the *relative permeability* of several common materials. (The relative permeability is merely a relative comparison or ratio of the permeability of a substance to that of a vacuum, the permeability of a vacuum being 12.57×10^{-7} Wb/At-m.) In reality, all nonferromagnetic materials have a relative permeability approximating 1.0.

TABLE 9-1 APPROXIMATE RELATIVE PERMEABILITIES OF SOME COMMON MATERIALS

Nickel	50
Cast iron	90
Machine steel	450
Transformer iron	5500
Air	1.0006
Aluminum	1.00000
Silver	0.99999
Wood	0.99999

Source: Adapted from W. H. Buchsbaum, *Practical Electronic Reference Data*, 2nd ed. (Englewood Cliffs, NJ: Prentice Hall, 1980).

Notice in Example 9-1 that no core was involved in construction of the coil. The coil, by itself, possessed an amount of inductance simply because air was, in effect, the core. If we had inserted a core of some different material, say brass or copper, we would have introduced a different variable permeability factor that would have depended on how far the core was inserted into the coil. In other words, we would have had a permeable material comprised of a mixture of brass or copper, and air. The problem's solution would have been somewhat complicated. Note the following problem and its solution.

Example 9-2

Using the coil in Example 9-1, we now insert an iron core one-third of the way into it. Determine its total inductance.

Solution: We break this problem into two solutions, each solution resulting from treating that portion of the coil having its own particular core material as a separate coil. The final answer will be the sum of these two solutions. The first solution results from using a coil having $\frac{1}{3}$ of 9 turns, or $N = 3$; $\frac{1}{3}$ of l, or 0.667 cm; and using iron as the core material. The second solution results from using the same coil but with using $\frac{2}{3}$ of 9 turns, or $N = 6$; $\frac{2}{3}$ of l, or 1.33 cm; and using air as the core material.

Solution for coil portion having iron core: From Table 9-1 we see that iron as a relative permeability of 5500. Therefore,

$$L = \frac{(12.57 \times 10^{-7} \text{ Wb/At-m})(5500)(3 \text{ turns})^2 (5.067 \times 10^{-4} \text{ m}^2)}{0.00667 \text{ m}}$$

$$= 0.00473 \text{ H}$$

$$= 4.73 \text{ mH}$$

Solution for coil portion having air core:

$$L = \frac{(12.57 \times 10^{-7} \text{ Wb/At-m})(6 \text{ turns})^2 (5.067 \times 10^{-4} \text{ m}^2)}{0.0133 \text{ m}}$$

$$= 0.00000172 \text{ H}$$

$$= 1.72 \text{ μH}$$

We can see that the air-core portion of the coil contributes an insignificant amount of inductance to that of the iron core; consequently, we can ignore the 1.72 μH value and state that the total inductance of the coil is 4.73 mH.

9-3 COMMONLY SENSED MEASURANDS

The inductive sensor must be designed in such a manner that the measurand alters the self-inductance of the inductive sensor's coil to produce either a change in current through the coil or a change in the voltage across the coil. One of the most common ways of producing this situation is to design a movable core that moves back and forth inside a coil (Figure 9-2) so that as the core is displaced, a variable permeability is

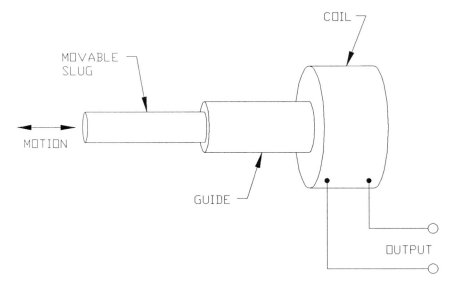

Figure 9-2   Coil with movable slug for varying its inductance.

created within the coil. The core is usually made of a fairly highly permeable material such as ferrite or iron.

Using this scheme it is possible to sense a wide range of measurands. As a matter of fact, the measurands that are sensed in this manner are those that were mentioned in Section 8-3 for the capacitor. In that section it is noted that for the capacitor, a common variation of displacement sensing is the detection of force or pressure. In the case of the induction sensor, either can cause a core to change its position inside the sensor's coil.

Another frequent measurement is liquid-level sensing. However, virtually any other measurand can be detected also, provided that it can be transformed into motion.

9-4 TYPES OF INDUCTIVE SENSING

Both linear motion and rotary motion detectors are used in industry. There is also a third category of inductive sensing, which falls under the heading of proximity sensing. This is the sensing of unknown materials that are generally hidden from view. We cover the theories of operation of each of these categories in the next three sections.

9-4.1 Linear Motion Inductor

The linear motion inductor is an inductor having a movable slug that slides into and out of the coil's central core. As in the case of the linear capacitor, the term *linear* refers to the straight-line mechanical motion of the movable slug and not necessarily to the output characteristic of the inductor's signal.

The construction of a typical linear motion inductor is illustrated in Figure 9-3, a cutaway view of the inductor's interior. The core is usually made from a powdered ferrite material that has been pressed into a core's desired shape. Often, elements such as nickel, zinc, and cobalt are added to the ferrite to obtain certain desirable characteristics. Ferrite itself is an iron material mixed with certain binders and other elements to create a material having very high permeability characteristics for certain current and frequency amounts.

Attached to the core's one end is a guide shaft that maintains alignment between the core and the coil's central core area. The shaft rides inside a bearing surface, usually a sleeved bearing made from a lubricant-impregnated brass- or bronze-like material, to permit smooth, relatively friction-free motion.

One of the characteristics of the linear motion inductance sensor is its relatively linear output. As the inductor's core is moved into and out of its coil, the value of μ is varied in linear fashion. As the inductance, L, is varied linearly, the reactance for any given frequency of current passing through the coil is also produced in a linear fashion:

$$X_L = 2\pi f L \tag{9-2}$$

where X_L = inductive reactance (Ω)

f = frequency (Hz)

L = inductance (H)

Since inductive reactance by itself is an imaginary mathematical expression denoted either by the imaginary symbol $+jX_L$ or its polar equivalent, $+X_L\underline{/90°}$, the impedance created by the inductor's coil is equal to the vector sum of its dc resistance, R, and its imaginary component. In other words,

$$Z_L = \sqrt{R^2 + jX_L^2} \tag{9-3}$$

where Z_L = impedance across inductor (Ω)

R = coil's dc resistance (Ω)

jX_L = inductive reactance (Ω)

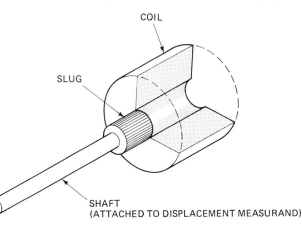

COIL

SLUG

SHAFT
(ATTACHED TO DISPLACEMENT MEASURAND)

Figure 9-3 Internal construction of a linear-motion inductor.

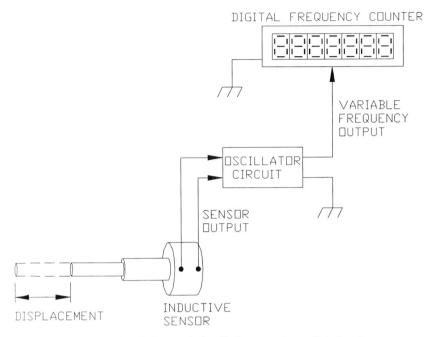

Figure 9-4   Circuit for translating displacement into a digital readout.

Then the voltage drop across the inductor will vary because

$$V_L = I_L Z_L \tag{9-4}$$

where V_L = voltage drop across coil (V)

I_L = current through coil (A)

Z_L = coil's impedance (Ω)

Equation (9-4) tells us that by monitoring the coil's impedance created by the changing inductance, it is possible to develop a relationship between the resultant impedance and the displacement of the coil's core.

There is, however, a somewhat easier method of correlating the changes in inductance with the changes in displacement. This method was also used in capacitive sensing. For our inductor, we can make it part of an *LC* resonant circuit where the capacitance, C, is fixed, and the inductance, L, is now made variable. [Refer to eq. (8-2) in Section 8-4.1 to review the relationships among frequency, inductance, and capacitance in an oscillating circuit.] With this system it now becomes a matter of monitoring the oscillation frequency, which is made variable by our inductive sensor. (See Figure 9-4. Also, refer to Review Question 9-6 at the end of the chapter.)

9-4.2 Rotary Motion Inductor

A linear motion inductive sensor can be modified to sense rotary motion. All that one has to do is to modify the linear shaft's sliding bearing surface by including machined threads so that the surface becomes a screw-type mechanism. By rotating the shaft, the core is now made to move in and out of the coil according to the direction of shaft's rotation. A variation on this concept is to install a ball-screw mechanism, which may help reduce the rotational friction common to threaded surfaces. An advantage of the threaded or ball-screw system is that the pitch of thread used will determine the rotational resolution of the system. That is, it may be possible to have several 360° turns of the shaft to cause a desired incremental change in core displacement as opposed to having only one shaft revolution to produce the same incremental change.

Another rotary motion type of construction is illustrated in Figure 9-5. This particular design requires that the inductive-sensing coil is built surrounding a circular shaft containing the high-permeability core. There are two obvious disadvantages to this system: (1) the coil must be wound on a form having the same radius of curvature as the rotating circular shaft, and (2) there is no possibility of increasing the rotational resolution of the system as there was with the threaded or ball-screw systems described earlier.

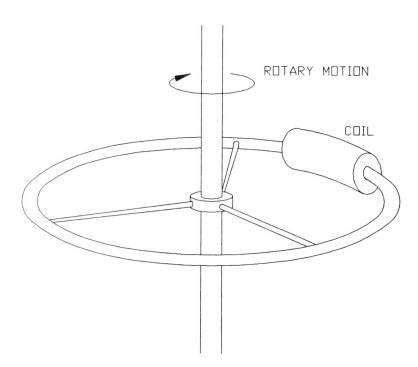

Figure 9-5   Another rotary-motion inductor design.

9-4.3 Inductive Proximity Sensing

Inductive proximity sensing is used primarily for the detection or location of metals. This particular activity has become rather popular in recent years among hobbiests trying to locate buried minerals or metallic treasures. In industry this type of sensing is used for the detection of minerals in the earth.

The inductive proximity sensor is usually comprised of a large sensing coil which is part of an *LC* RF circuit (Figure 9-6). The circuit's frequency of oscillation is typically in the hundreds or thousands of kilohertz. When a concentration of metal is brought near the sensing coil the frequency of oscillation is changed because of the altering of the oscillator circuit's inductive value. This shift in frequency causes a voltage imbalance to take place in an otherwise balanced voltage differential circuit. This imbalance either sounds an alarm or varies an indicating circuit.

9-5 ADVANTAGES AND DISADVANTAGES OF USING THE INDUCTIVE SENSOR

The inductive sensor is a relatively inexpensive sensor, which explains its popularity with users. Also, it is considered to be capable of producing fairly linear outputs. However, the inductive sensor has one major disadvantage. This has to do with its limited motion capability. The length of coil used in the inductor's construction determines the overall length of displacement sensing. As long as the displacements are relatively small, that is, the distances are only a few centimeters or inches in length, the inductive sensor will function very well.

Figure 9-6 Pickup head in a metal detector unit for detecting the location of metallic substances.

REVIEW QUESTIONS

9-1. Why is inductive sensing considered to be such a versatile method of sensing compared with, say, a piezoelectric sensor?

9-2. Explain how you would set up an inductive sensor for the purpose of separating copper metal from its ore rock.

9-3. How would you design a mechanism for the purpose of measuring hydraulic pressure in a pressure line using a variable inductive circuit?

9-4. How would you characterize the input impedances of most inductive sensors? Would you say the impedance would probably be very high or rather low? Justify your answer with an explanation.

9-5. Explain the difference between the terms *self-inductance* and *inductive reactance*.

9-6. In Section 9-4.1 the statement was made that a resonant circuit made from a fixed capacitor and a variable inductor could be used to monitor a measurand by monitoring this circuit's output frequency. Assuming a linear relationship between the coil's inductance and the measurand, what would be a major disadvantage of this system? Support your answer mathematically.

PROBLEMS

9-7. Calculate the inductance for a coil to be used in a displacement transducer that is to be made from 250 turns of wire in a single-layered winding wound on a 10-mm-diameter form. The coil will be 70 mm in length and have an air core.

9-8. Find the inductance of a coil having 85 turns of a single-layered wire winding that is 0.45 in. in diameter and is 2.33 in. long. The core is made from steel that is inserted 60% into the winding's center.

9-9. For the coil described in Problem 9-7, calculate the impedance resulting from supplying the coil with a 230-kHz signal at 6.3 V ac. The coil's dc resistance is 3.9 Ω. How much current does the coil require?

9-10. How many windings are necessary to create a 30-mm-long coil 6 mm in diameter that will use a movable iron core coupled to the measurand? The coil will be supplied with a 3.5-V ac 3.25-MHz signal. Assume no dc resistance in this coil and a current draw of 4 mA.

9-11. Make a sketch showing how a variable-inductance device may be used to sense air pressure. Show how this device may be part of a remote telemetry system that transmits the sensed pressure values by radio signals. Use block diagramming rather than discrete components to show the circuitry involved.

REFERENCES

BLATT, FRANK J., *Principles of Physics*. Boston: Allyn and Bacon, 1986.

BUCHSBAUM, WALTER H., *Practical Electronic Reference Data*. Englewood Cliffs, NJ: Prentice Hall, 1980.

HORDESKI, MICHAEL F., *Microprocessor Sensor and Control Systems*. Reston, VA: Reston, 1985.

MILLER, RICHARD W., *Servomechanisms: Devices and Fundamentals*. Reston, VA: Reston, 1977.

10

VARIABLE-RELUCTANCE SENSORS

CHAPTER OBJECTIVES

1. To review the concept of reluctance.
2. To study and understand the LVDT and RVDT.
3. To understand the behavior of the synchromechanism.
4. To understand the microsyn, the resolver, and the inductance bridge.

10-1 INTRODUCTION

Unfortunately, the terms *inductance* and *reluctance* sound alike and, as a result, are easily confused. As discussed in Chapter 9, inductance (or self-inductance) refers to the inducing of voltages into adjacent windings of a coil to oppose the existing direction of current flow. Reluctance, on the other hand, has an entirely different meaning, as we will soon discover.

10-2 REVIEW OF RELUCTANCE

The term *reluctance* means unwillingness or hesitancy. In electronics the term refers to the unwillingness of certain materials to support magnetic lines of flux. Therefore, reluctance is the "opposite" of permeability. If we let $\mathcal{R}$ represent reluctance, then mathematically,

$$\mathcal{R} = \frac{l}{\mu A} \qquad (10\text{-}1)$$

where $\mathcal{R}$ = reluctance (At/Wb)

l = length of magnetic path (m)

μ = permeability (Wb/A-m)

A = cross-sectional area through which magnetic flux is flowing (m^2)

According to eq. (10-1), reluctance varies inversely with μ.

Now that we have established what reluctance is mathematically, let's see what reluctance can do, practically speaking, in a sensor or transducer. Figure 10-1 shows two coils that are mounted side by side with air as the coils' cores. One of the coils is supplied with an ac voltage from a signal generator, as shown in the diagram. The magnetic flux lines from coil 1 are "coupled," although somewhat weakly, to coil 2 through the existing air core. The lines are weak since air has a fairly high reluctance to magnetic flux. However, by inserting an iron core through both coils or near both coils, the air core becomes mostly or at least partially replaced with an iron core. Because iron has a much lower reluctance (i.e., a much higher permeability) than air, the magnetic lines of flux emanating from coil 1 are now much stronger and more efficiently coupled to coil 2. The result is a much stronger induced signal from coil 1 into coil 2. Coil 2 will now experience a larger current flow through it. In other words, by inserting the iron core near the two coils, the reluctance of the path between the two coils was lowered significantly, causing a greater signal to be induced in the second coil. Also, by varying the amount of iron inserted into the path, a variable amount of current is introduced into coil 2.

By general agreement, when you speak of reluctance transducers, you refer to a transducer whose measurand is converted to an ac voltage resulting from a change in the reluctance path between two or more coils. It could be argued that the term *variable permeability* could just as easily have been used to describe the system's behavior in Figure 10-1. As it turns out though, *permeability* is reserved for single

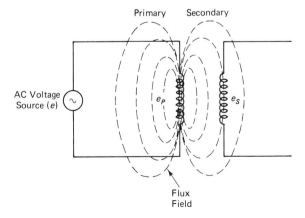

Figure 10-1 Field coupling between two side-by-side coils. (From James R. Carstens, *Automatic Control Systems and Components,* Copyright 1990, p. 35. Reprinted by permission of Prentice Hall, Englewood Cliffs, NJ.)

coils exhibiting self-inductance, whereas *reluctance* is used for multiple-coil flux-coupled systems.

10-3 COMMONLY SENSED MEASURANDS

The measurands normally sensed by variable-reluctance transducers are displacement (both linear and rotary) and pressure–force. This list is similar to the measurands detected by the variable-inductance and variable-capacitance transducers discussed in previous chapters. Other measurands can certainly be sensed, but the ones listed are the principal ones. In the sections that follow we discuss the variable-reluctance transducers currently being used in industry.

10-4 TYPES OF TRANSDUCERS PRESENTLY AVAILABLE

The types of transducers presently being used fall into the following categories: (1) linear variable differential transformer, (2) rotary variable differential transformer, (3) synchromechanism, (4) microsyn, (5) resolver, and (6) inductance bridge. We will discuss each of these types in detail and become acquainted with their characteristics and applications. We begin with the linear variable differential transformer.

10-4.1 Linear Variable Differential Transformer

Figure 10-2 is a circuit diagram of a *linear variable differential transformer* (LVDT). This device is comprised of three separate coils, one of which is supplied with a high-frequency ac signal, typically around 10 to 20 V p-p (peak to peak) at a frequency of 1 kHz to 10 MHz. A movable core of high-permeability material is allowed to move between it and two twin coils located on either side of the signal coil. As the core is moved, varying amounts of the signal are induced into either or both of the twin receiving coils, depending on the position of the core. It is important to note at this point how the two receiving coils are wired. It is also assumed that both coils have their windings wound in the same direction. Notice, however, that while it may appear that the two coils are in series with each other, one coil is actually

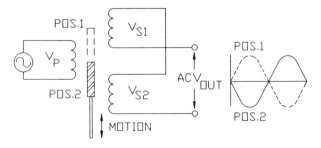

Figure 10-2 Circuit of a linear variable differential transducer.

"doubled back" so that what was once this coil's end farthest away from the approaching core now becomes the closer end. In effect, any voltage signals induced in the upper coil will now subtract from any signals induced in the lower coil (or vice versa). In other words, voltage V_p is in phase with only one of the two secondary voltages, V_{s1} and V_{s2}. The other secondary voltage is out of phase with V_p.

Next let's analyze what happens for various positions of the movable core. When the core is centered exactly midway between the two receiver coils, both coils receive exactly the same magnitude of induced voltage from the primary or signal coil. However, because of the subtracting nature of the receiving coils, zero volts appears at the coils' output. If the core is positioned first to one side of the center position and then to the other side by an equal displacement amount, the induced voltage is the same into the receiver coils, except that the phases of these voltages are reversed. The phase is therefore an indication as to which side the core is located relative to center; the voltage magnitude that results is an indication of the amount of the core's displacement from this center.

Typical signal-conditioning method and how it works

Figure 10-3 is a schematic of a typical signal-conditioning circuit used in conjunction with the LVDT. Its operation is very simple. Basically, this circuit is nothing more than two half-wave rectifiers so wired that depending on which secondary receiving coil receives the more induced voltage from the primary or signal coil, either a positive or a negative dc voltage is produced at the input terminals of the dc amplifier. This type of rectification circuit is sometimes referred to as a *differential rectifier* circuit. To state the operation of this circuit in another way, ac voltages of one phase are rectified by one of diodes while ac voltages of the opposite phase are rectified by the other diode. The capacitor across the output of this circuit is a filtering capacitor that becomes charged either negatively or positively, depending on the predominant charging polarity at the time. These ac phase transformations are depicted in Figure 10-4. The crossover point of the dc voltage line in the lower curve is an indication that the LVDT's moving core has passed through the center point of the two receiving coils.

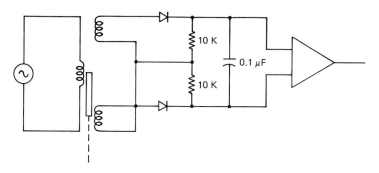

Figure 10-3   Typical signal-conditioning circuit used with an LVDT.

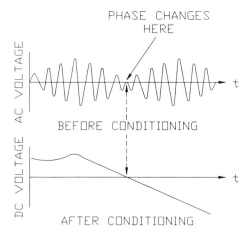

Figure 10-4   Phase changing of output signal occurring in an LVDT.

Practical applications

Figure 10-5 shows a typical application of the LVDT. We see the LVDT device being used to measure tension in a wire spool-winding operation and in gaging the thickness of sheet metal.

Another application of the LVDT (Figure 10-6) is in automatic control systems. The system shown in Figure 10-6 is used to control liquid flow rates. It works like this: A liquid U-tube manometer with an LVDT is installed downstream from a servo-controlled flow valve in a liquid-carrying pipe. The LVDT is part of the manometer system and is mounted so that its movable core rides on a float that is built into one leg of the manometer itself. Any pressure differential sensed by the

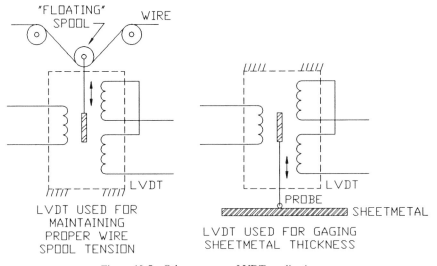

Figure 10-5   Other common LVDT applications.

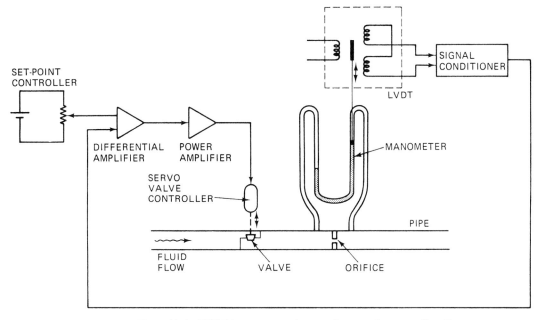

Figure 10-6 LVDT in an automatic control system for controlling flow.

manometer causes the core to move either up or down, depending on the differential's magnitude and the liquid's direction in the pipe. The differential is created by an orifice permanently mounted inside the pipe. A set-point controller is used initially to position the flow valve to the desired setting for the flow rate desired. Assuming zero initial flow in the pipe, the full voltage from the controller is sent to the valve, since the differential amplifier has no other canceling voltage at its other input from the LVDT. As soon as the manometer and LVDT react to the increase in flow, the output voltage from the LVDT begins canceling the set-point controller voltage until full cancellation takes place (i.e., when the LVDT's output voltage equals that of the set-point controller) and the flow control valve stops opening. If for any reason the flow inside the pipe increases or decreases at this point, the valve will either close or open automatically to compensate for this change. This compensating action attempts to maintain the desired set-point value that was set when the system was started up.

A rather interesting variation of the LVDT is seen in this next design. A load cell (see also Section 7-4) is manufactured with a primary and a secondary coil. However, its core contains a series of laminated materials that are designed to flex under the stress of various applied loads to the cell. The applications of these stresses cause the core's permeability to vary, which, in turn, causes variable coupling to take place between the primary and secondary coils. The variably induced voltages are then rectified to dc voltage signals for additional processing, as shown in Figure 10-7. Notice, though, that a differential transformer system cannot be used here because of the symmetry of the deflecting column, or "core." However, one could be used

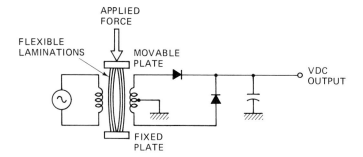

Figure 10-7 LVDT used in a load cell application.

if the usual moving-core system is used, as illustrated in Figure 10-8. Notice the mechanical provision made for zeroing the system. This is one advantage in using a differential transformer circuit.

Typical design specifications

Figure 10-9 shows typical design specifications for a LVDT.

10-4.2 Rotary Variable Differential Transformer

Just as the resistive, capacitive, and inductive linear sensors had their rotary counterparts, so does the LVDT. The construction of the rotary variable differential transformer (RVDT) is somewhat more complex, however. The core must be allowed to rotate between the three coils; this design we will refer to as the *rotating core* design. In another design, however, the primary coil is allowed to rotate in the vicinity of the two secondary coils. There is no moving core. We will call this the *rotating coil* design (sometimes referred to in industry as a *rotary induction potentiometer*).

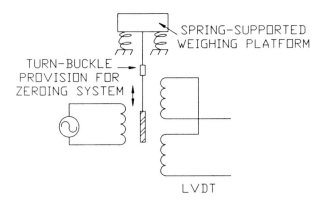

Figure 10-8 LVDT used for weighing.

HR SERIES—GENERAL APPLICATIONS

- OPTIMUM PERFORMANCE FOR THE MAJORITY OF APPLICATIONS

- LARGE CORE-TO-BORE CLEARANCE — 1/16 INCH (1.6 mm) RADIAL

The HR high reliability series of LVDT's is suitable for most general applications. The HR series features large core-to-bore clearance, high output voltage over a broad range of excitation frequencies, and a magnetic stainless steel case for electromagnetic and electrostatic shielding.

GENERAL SPECIFICATIONS

Input Voltage . . . . . 3 V rms (nominal)
Frequency Range . . . 400 Hz to 10 kHz
Temperature Range . . -65°F to +300°F
(-55°C to +150°C)
Null Voltage . . . . . Less than 0.5% full scale output
Shock Survival . . . . 1000 g for 11 milliseconds

Vibration Tolerance . 20 g up to 2 kHz
Coil Form Material . . High density, glass-filled polymer
Housing Material . . . AISI 400 series stainless steel
Lead Wires . . . . . . 28 AWG, stranded copper, Teflon-insulated, 12 inches (300 mm) long (nominal)

PERFORMANCE SPECIFICATIONS AND DIMENSIONS (2.5 kHz) *METRIC DIMENSIONS IN BLUE*

LVDT MODEL NUMBER	NOMINAL LINEAR RANGE		LINEARITY ±PERCENT FULL RANGE				SENSITIVITY mV Out/ Volt In Per		IMPEDANCE Ohms		PHASE SHIFT	WEIGHT Grams		DIMENSIONS			
														A (Body)		B (Core)	
	Inches	mm	50	100	125	150	.001 In.	mm	Pri.	Sec.	Degrees	Body	Core	Inches	mm	Inches	mm
050 HR	±0.050	±1.25	0.10	0.25	0.25	0.50	6.3	250	430	4000	−1	32	4	1.13	28.5	0.80	20
100 HR	±0.100	±2.5	0.10	0.25	0.25	0.50	4.5	180	1070	5000	−5	48	6	1.81	46	1.30	33
200 HR	±0.200	±5.0	0.10	0.25	0.25	0.50	2.5	100	1150	4000	−4	60	8	2.50	63.5	1.65	42
300 HR	±0.300	±7.5	0.10	0.25	0.35	0.50	1.4	56	1100	2700	−11	77	10	3.22	82	1.95	50
400 HR	±0.400	±10	0.15	0.25	0.35	0.60	0.79	32	1700	3000	−18	90	15	4.36	111	2.95	75
500 HR	±0.500	±12.5	0.15	0.25	0.35	0.75	0.70	28	460	375	−1	109	18	5.50	140	3.45	88
1000 HR	±1.000	±25	0.25	0.25	1.00	1.30*	0.39	16	460	320	−3	126	21	6.63	168	4.00	102
2000 HR	±2.000	±50	0.25	0.25	0.50*	1.00*	0.23	9.2	330	330	+5	168	27	10.00	254	5.30	135
3000 HR	±3.000	±75	0.15	0.25	0.50*	1.00*	0.25	10	115	375	+11	225	28	12.81	325	5.60	142
4000 HR	±4.000	±100	0.15	0.25	0.50*	1.00*	0.20	8.0	275	550	+1	295	36	15.64	397	7.00	178
5000 HR	±5.000	±125	0.15	0.25	1.00*	—	0.15	6.1	310	400	+3	340	36	17.88	454	7.00	178
10000 HR	±10.00	±250	0.15	0.25	1.00*	—	0.08	3.0	550	750	−5	580	43	30.84	783	8.50	216

*Requires reduced core length

ORDERING INFORMATION

(Fold out page 32 for instructions on how to use this chart.)

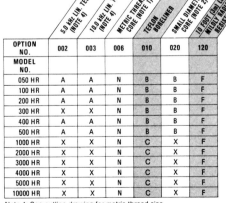

OPTION NO.	002	003	006	010	020	120
MODEL NO.						
050 HR	A	A	N	B	B	F
100 HR	A	A	N	B	B	F
200 HR	A	A	N	B	B	F
300 HR	X	X	N	B	B	F
400 HR	A	A	N	B	B	F
500 HR	A	A	N	B	B	F
1000 HR	X	X	N	C	X	F
2000 HR	X	X	N	C	X	F
3000 HR	X	X	N	C	X	F
4000 HR	X	X	N	C	X	F
5000 HR	X	X	N	C	X	F
10000 HR	X	X	N	C	X	F

Note 1: See outline drawing for metric thread size
Note 2: Consult factory for mass, dimensions, and thread size
Note 3: Withstands 10¹² NVT total integrated flux
Note 4: See frequency option note on page 32

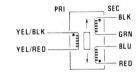

CONNECT GRN TO BLU FOR DIFFERENTIAL OUTPUT

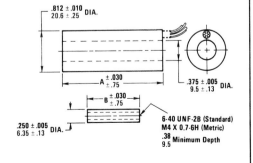

Figure 10-9 Specification sheet for an LVDT. (Data courtesy of Schaevitz Lucas, Pennsauken, NJ.)

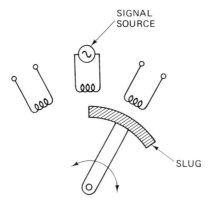

Figure 10-10   Rotary motion LVDT.

Rotating core design

Figure 10-10 shows the general design scheme of this particular system. The movable core rotates past the three coils, creating a variable-reluctance path between the two receiving coils. This system uses the same type of signal-conditioning circuitry as that used by the LVDT. A variation of this system is seen in Figure 10-11. Here we see the core in a form such that it inductively couples the primary coil to the two receiving coils through its length. Again, this type of configuration uses the same signal-conditioning circuitry as that used by the other variable-reluctance transformer types.

Rotating coil design

An example of the rotating coil design is shown in Figure 10-12. The primary or signal coil is rotated near the windings of the two secondary receiving coils, causing variable coupling to take place. The reluctance of this system in this case does not vary in

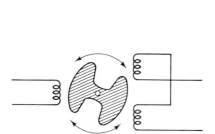

Figure 10-11   Another variation of measuring rotary motion using the LVDT.

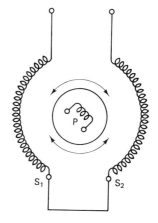

Figure 10-12   Rotating coil variable-reluctance sensor.

the sense that a core moves relative to the sensing coils. It could be argued that a certain amount of variable-reluctance action does take place due to the varying distances between the rotating primary and stationary secondary coils. However, that is not the primary reason for the behavior of this device. It is being included with this discussion only because it is customary to mention this type when the others are discussed.

Of the two rotating designs that we have discussed here, the rotating coil design is the more expensive of the two. The rotating primary coil requires a specially shaped armature to accommodate the windings, along with specially constructed flexible leads to facilitate partial rotation.

There are other rotating coil designs that fall under this particular category; we will discuss these in the sections that follow. For a detailed explanation of how the device pictured in Figure 10-12 works, see Section 10-4.4.

Signal conditioning

The only difference existing between the linear and rotary differential transformer systems is in the mechanical construction. Electrically, the two systems are virtually identical. Consequently, the signal-conditioning methods used by the rotary differential transformer systems are the same as those used by linear systems.

Practical applications

The RVDT is often used as an indicating device for angular rotation. Figure 10-13 shows a typical application where an RVDT is being used to generate a signal that operates a remote indicator. The indicator shows the amount of rotation taking place at the RVDT's rotor shaft. The shaft is coupled to a rotatable communications antenna.

As in the case of the LVDT, the RVDT is also used extensively in automatic control systems. In Figure 10-14 we see an RVDT being used as a flow rate device in a circuit similar to one used to describe the LVDT's flow control application,

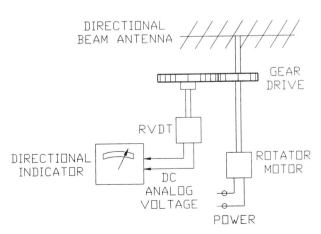

Figure 10-13 RVDT being used in the application of indicating antenna rotation direction.

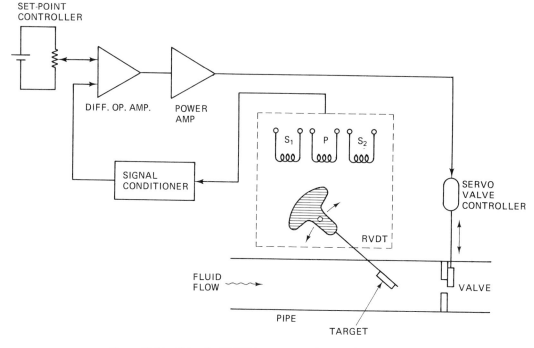

Figure 10-14   Using the RVDT in an automatic control system for controlling flow.

discussed earlier (see Figure 10-6). In place of the manometer flow transducer, we now have a flow-sensing device that has a pivoting target whose angular pitch varies with the pipe's flow rate. This pitch is detected by the attached RVDT.

10-4.3 Synchromechanism

The *synchromechanism* is an electromechanical device that contains a rotor supply voltage coil, a set of phasing stator coils, and a rotatable output shaft (Figure 10-15). Synchromechanisms usually operate in pairs, with one unit called the transmitter and the other unit referred to as the receiver. Synchromechanisms, or *synchros* as they are often called, can be divided into two general groups depending on their applications:

1. *Control synchros*, for indicating readings of position or location from a remote location.
2. *Torque synchros*, for performing work using remotely transmitted signals. Torque synchros usually involve the use of servomotors for their operation.

The only major difference between these two groups has to do with their internal construction. The torque synchro uses heavier wire for its coil windings. This is to allow for higher current flows associated with the higher torque capabilities. Also, the components used in its overall construction are usually much more ruggedly designed. The control synchro, on the other hand, usually is smaller in size

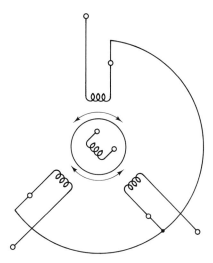

Figure 10-15 Coil configuration in a sychromechanism.

due to a lighter-weight design. Torque synchros are used for moving relatively massive system components, such as platforms, levers, wheels, and so on, whereas control synchros move only very lightweight items such as dial-type indicators, pointers, and graduated scales.

To understand how these devices work, we must study Figure 10-16. A stator

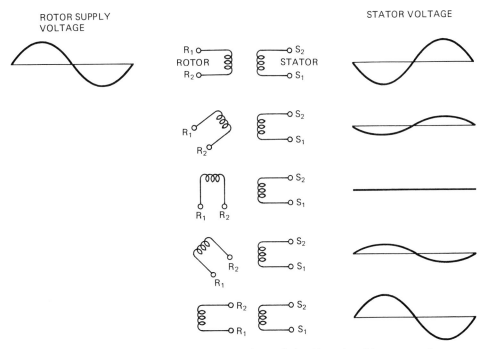

Figure 10-16 How the input and output phase relationship varies with a stator and rotor coil construction and their relative rotation.

coil and rotor coil are shown opposite each other, with the rotor coil supplied with an ac sinusoidal signal as shown. Initially, the coils' positions are shown. The stator coil's output is shown to the right of the coil. The two signal patterns generated for the given rotor position indicate a 180° phase shift. As the rotor is now rotated through a partial rotation, say 45°, the phase shift remains the same. However, the stator's amplitude is reduced as a result. As the rotor continues to a 90° position, the stator's output amplitude is reduced to zero. As the rotor continues rotating past 90°, to, say, 135°, the amplitude now begins to increase but with an in-phase phase shift with the rotor's supply voltage. As 180° is approached, the stator's output once again attains full amplitude output but with the in-phase signal relationship now existing. The rotor coil's winding sense is now reversed relative to the stator's sense to what it was earlier. We will see in the next discussion how this rotor-stator behavior is applied to the operation of a typical synchromechanism.

How it works

A schematic of a synchro pair is shown in Figure 10-17. Notice that the one pictured on the left is called a *CX*, for *control transmitter* (because the controlling signal originates with this unit), and the unit on the right is called a *CR*, or *control receiver*. Even though we use the control synchro systems for our explanation, we could just as well have used the torque synchros. In place of the CX unit we could have used a *TX* (*torque transmitter*) unit, and in place of the CR, a *TR* unit would have been used.

The theory of operation of the *CX–CR* system (or *TX–TR* system) is based largely on the action of a transformer. (If you are not too sure of your transformer

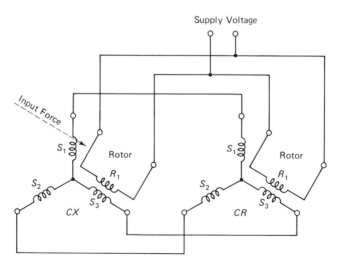

Figure 10-17 Schematic of a synchro pair system. (From James R. Carstens, *Automatic Control Systems and Components,* copyright 1990, p. 124. Reprinted by permission of Prentice Hall, Englewood Cliffs, NJ.)

theory, be sure to review the discussion above concerning the interaction between the rotor and stator windings of a synchro.) As you study Figure 10-17, it is important to realize that the stator voltage's magnitude and phase are determined by the angle of rotation of the rotor relative to the stator. This relationship permits the CX (or TX) to transmit its rotor position accurately to the matching CR (or TR). As a result, the receiver's rotor will respond to the voltages transmitted to its three windings and will rotate automatically due to motor action. It will continue to rotate until it finds a spot in its rotation where the three transmitted voltages and the supply voltage produce a torque that is essentially zero. This will, in turn, cause the CR synchro's shaft to come to rest at precisely the same location as the CX's shaft.

Another popular synchromechanism is the *control transformer*, or *CT*. Electrically, it is quite similar to the CX and CR. However, mechanically it operates quite differently. Like the CX, it has an input shaft and winding construction much like the CX, for indicating position. However, there is another input shaft that has a winding whose output is an ac voltage that is proportional to the difference between the rotational displacements of the two shafts (Figure 10-18). It may help to think of the CT synchro as being similar to a CX–CR synchro all housed in one unit, but with the common rotor connections removed. The CX rotor windings, instead, becomes the input connections for the CT where an ac voltage is applied, while the CR windings are the output connections. The output ac voltage is then proportional to the relative displacements of the two input shafts, the CX rotor shaft and the CR rotor shaft. This output voltage is called an *error voltage* and is frequently used in automatic control systems for controlling position. As a matter of fact, the control transformer is frequently referred to as an *error detector*.

Quite often when no control transformers are available for a particular design application, it is possible to simulate a CT circuit from a control transmitter and receiver. All that needs to be done is to make the appropriate wiring changes between the two rotor windings. Figure 10-19 is a photograph of a typical TX-TR synchro pair.

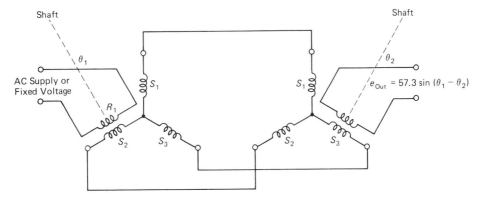

Figure 10-18 Control transformer (CT). (From James R. Carstens, *Automatic Control Systems and Components,* copyright 1990, p. 126. Reprinted by permission of Prentice Hall, Englewood Cliffs, NJ.)

Figure 10-19 TX–TR synchro pair.

Figure 10-20 shows a typical application of a CX–CR remote-reading synchro system. A CX is attached to the dial mechanism of a Bourdon tube pressure gage. The CR is located at a remote data-gathering site and is seen attached to a dial mechanism that would normally have been installed in the pressure gage itself. As the Bourdon tube on the gage responds to pressure, the CX transmits the amount of rotation of the Bourdon mechanism to the CR, where the actual pressure reading can then be read.

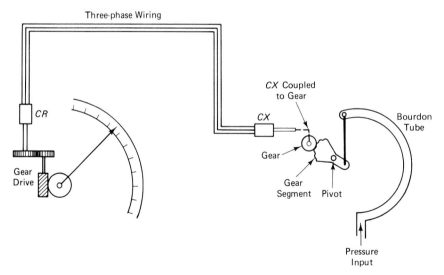

Figure 10-20 Application of CX–CR synchros in transmitting pressure data from a bourdon tube pressure gauge. (From James R. Carstens, *Automatic Control Systems and Components,* copyright 1990, p. 125. Reprinted by permission of Prentice Hall, Englewood Cliffs, NJ.)

The TX–TR synchro pair depicted in Figure 10-21 is being used to rotate a small microwave antenna system for a communications setup. For this application the torque-handling capabilities of a TX–TR are required over that of a CX–CR system. In this particular application two pairs of TX–TR synchros are used. One pair is used for adjusting the altitude of the antenna, while the other pair is for rotating the antenna for obtaining the proper horizontal or azimuth bearing.

10-4.4 Microsyn

The *microsyn* is a rotary reluctance device used in situations where the angular displacements being measured or controlled are very small, perhaps a few degrees or so. Figure 10-22 illustrates how a microsyn is generally constructed and wired. The rotor is nothing more than a solid bar of low-reluctance material such as iron mounted on a rotatable shaft. The stator is comprised of four equally spaced poles having windings that are wired as shown. One pair of poles contains one winding, while the other pair contains a similar winding. One pair is called the primary winding (the vertical pair) because it is supplied with an excitation voltage; the other (horizontal) pair, where the output occurs, is called the secondary. Flux lines generated by the excitation voltage are shown circulating throughout the stator's frame to which the poles and their windings are attached.

Looking at the rotor's vertical position as seen in Figure 10-22, there are no

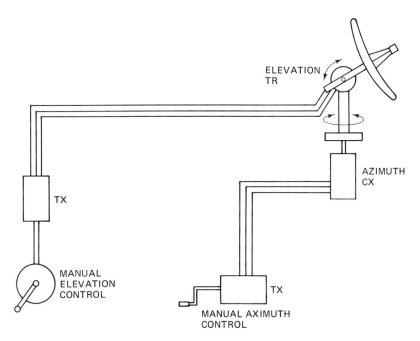

Figure 10-21 Using TX–TR sychros in controlling positioning of an antenna.

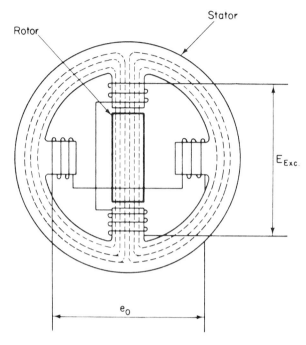

Figure 10-22   How a microsyn is constructed. (From Richard W. Miller, *Servomechanisms: Devices and Fundamentals,* copyright 1977, p. 39. Reprinted by permission of Reston Publishing Co., Inc., Reston VA.)

flux lines cutting the secondary windings as a result of this position. If the rotor were now to be rotated, say in a counterclockwise direction, some of the flux lines originating from the primary windings will cut through the secondary windings, causing a voltage to be induced as a result. Maximum induction takes place when the rotor is aligned horizontally with the secondary windings. If the rotor were rotated clockwise, similar voltage magnitudes but of opposite phase are induced into the secondary instead. Despite the typically small angles of rotation associated with a microsyn, the microsyn is noted for its good linearity and high accuracy.

Practical applications

An application of the microsyn is illustrated in Figure 10-23. A camera onboard a satellite often requires the use of a first-surface mirror to reflect into its lens system the planetary scene that is to be recorded. On a command signal transmitted from earth, a servomotor is ordered to move the mirror. The precise mirror angle is registered by an attached microsyn, and the information is sent to the automatic control circuitry onboard the spacecraft. The mirror's control system maintains the targeted image within sight of the camera regardless of small changes that might be made in the satellite's orientation.

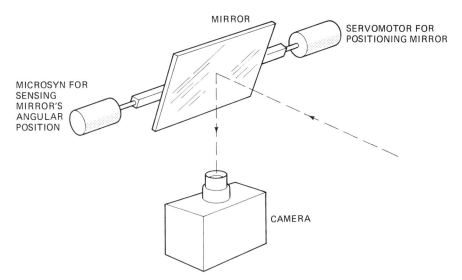

Figure 10-23   Application of a microsyn used for positioning a mirror for alignment of images in a satellite camera.

10-4.5 Resolver

Resolvers are rather unusual devices. In addition to transducing position, they have the ability to perform trigonometric calculations. In other words, as they perform a position-sensing function, they are also functioning as a small analog electrome-chanical computer. For example, they can add and subtract angles of rotation of two resolver shafts. They can also transform positional coordinate systems by converting the polar coordinates of an object's vector (in other words, its direction and displace-ment) to rectangular coordinates; this process is often referred to as *resolving*. In addition, resolvers can perform the reverse conversion. They can convert rectangular coordinates into polar coordinates; this is sometimes referred to as *composition*. Yet another function performed by the resolver is the *transformation* of coordinate axes. This process takes an existing set of axes and rotates them to a new angular orientation. To understand how all four functions are performed, we must look at a typical resolver's construction.

A resolver's construction is similar to that of a variable transformer—the same construction as that found in the synchros just described. However, unlike CX–CR systems, which require a three-phase coil system for a stator, the resolver can get by with only two coils. The same is true for its rotor. The coil pairs are positioned at right angles to each other, as shown in Figure 10-24. Ac voltages E_{S1} and E_{S2} are supplied to the two stator windings. The resultant rotor voltages are a function of the two stator voltages and the rotor's angular rotation, θ:

$$E_{R1} = E_{S1} \cos \theta + E_{S2} \sin \theta \qquad (10\text{-}2)$$

$$E_{R2} = E_{S2} \cos \theta - E_{S1} \sin \theta \qquad (10\text{-}3)$$

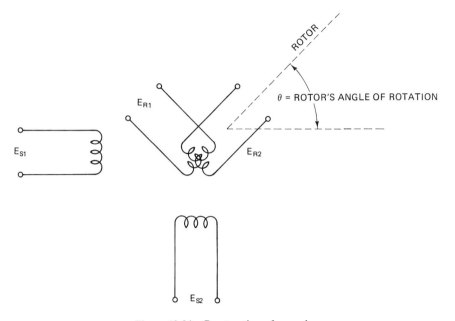

Figure 10-24   Construction of a resolver.

It should be pointed out here that eqs. (10-2) and (10-3) depend on the sense of the coils' windings in order to have the signs expressed in these two equations to be correct as shown. To understand why this is, we must first look at another diagram of the same system (Figure 10-25) but with the rotor oriented as shown. This rotor position is called the *electrical zero position*. At this position the rotor's angle of rotation is zero degrees. The winding sense of the two rotor coils is such that when stator currents i_{s1} and i_{s2} are in the directions shown, the induced rotor currents, i_{r1} and i_{r2}, are also as drawn in the figure. The polarities across each coil are also indicated. The induced rotor currents are the net results produced by the two stator currents. For instance, looking at rotor coil 1, the current induced in it is due solely to the current from stator coil 1 since stator coil 2 is at right angles and therefore cannot induce any current into it.

What happens when the rotor is now rotated through an angle of less than 90°? This is the situation shown in Figure 10-24. Both rotor coils now receive an induced current from both stator coil 1 and stator coil 2. The net amount received in each rotor coil depends on the rotational angle of the rotor and on the phase relationship of the current components received from both stator coils. This phase relationship, in turn, is dependent on which quadrant the shaft's rotation angle happens to be in as it rotates. This is because the relative sense of the windings between each stator and rotor pair also changes as the shaft goes through a full 360° rotation, as shown in Figure 10-16.

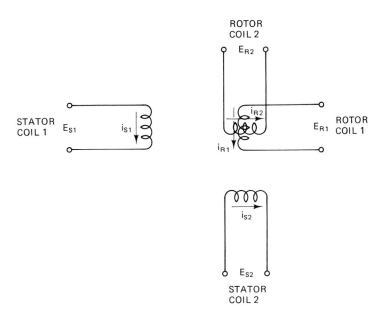

Figure 10-25   Resolver system of Figure 10-24 with rotor coils oriented 90° to stator coils.

How addition and subtraction are performed

Figure 10-26 shows how two resolvers may be wired to perform a mathematical addition of the two rotors' angles of rotation, α and β. A fixed ac supply voltage is applied across S_1–S_2 of rotor 1. We designate this voltage E_{S1-2}. Rotor 1 is then rotated through an angle α. The voltage induced into winding R_3–R_4 of this rotor then becomes $E_{S1-2} \cos \alpha - E_{S3-4} \sin \alpha$, where $E_{S3-4} = 0$ since that winding $(S_3$–$S_4)$ is shorted. The voltage induced into winding R_1–R_2 becomes $E_{S3-4} \cos \alpha + E_{S1-2} \sin \alpha$, where, again, because of the short across R_3–R_4, $E_{S3-4} = 0$. The voltage now appearing across S_3–S_4 of rotor 2 is the same voltage as that across winding R_3–R_4 of rotor 1, namely $E_{S1-2} \cos \alpha$; the voltage appearing across windings S_2–S_1 of rotor 2 is the same voltage as that across R_1–R_2 of rotor 1, or $E_{S1-2} \sin \alpha$.

Rotor 2 is now rotated through an angle β. What follows, then, is the same trigonometric process of combining voltages and angles at the output windings of rotor 2 as was followed at the output windings of rotor 1. Voltage $E_{S1-2} \cos \alpha$ appears at rotor 2's windings, S_3–S_4, and voltage $E_{S1-2} \sin \beta$ appears at rotor 2's winding, S_2–S_1. The voltage appearing at R_1–R_2 of rotor 2 now becomes $(E_{S2} \cos \alpha)(\cos \beta) - (E_{S2} \sin \alpha)(\sin \beta)$ or, after applying the difference-of-the-product-of-two-angles identity, $E_{S2} \cos(\alpha + \beta)$. The voltage appearing at R_3–R_4 becomes $(E_{S2} \sin \alpha)(\cos \beta) + (E_{S2} \cos \alpha)(\sin \beta)$ or, after applying the sum-of-the-products-of-two-angles identity, $E_{S2} \sin(\alpha + \beta)$.

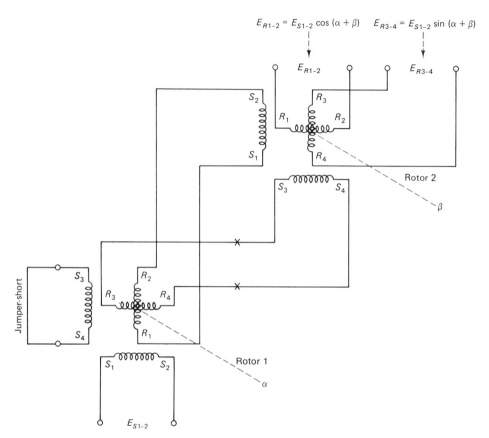

Figure 10-26 Resolver system used for the addition of two angles. (From James R. Carstens, *Automatic Control Systems and Components,* copyright 1990, p. 128. Reprinted by permission of Prentice Hall, Englewood Cliffs, NJ.)

If subtraction of these angles is desired instead of addition, merely cross the wires at the points marked with an × in Figure 10-26. The resultant voltage, E_{R1-2}, would then become $E_{S1-2} \cos(\alpha - \beta)$; the resultant voltage, E_{R3-4}, would become $E_{S1-2} \sin(\alpha - \beta)$.

Example 10-1

The resolver circuit in Figure 10-26 is shown wired for angle addition. If the supply voltage is 1 V ac (E_{S1-2}), find the two resolver output voltages, E_{R1-2} and E_{R3-4} if $\alpha = 30°$ and $\beta = 45°$. Discuss the calculated voltages' relationship to the input angles.

Solution: To find E_{R1-2} we use the equation shown at the output of E_{R1-2} in Figure 10-26:

$$E_{R1-2} = E_{S1-2} \cos(\alpha + \beta) = (1 \text{ V}) \cos(30° + 45°)$$

$$= 1 \cos 75°$$

$$= (1)(0.259)$$

$$= 0.259 \text{ V ac}$$

To find E_{R3-4} we use the equation shown at the output of E_{R3-4}:

$$E_{R3-4} = E_{S1-2}\sin(\alpha + \beta) = (1\ \text{V})\sin(30° + 45°)$$

$$= 1\sin 75°$$

$$= (1)(0.966)$$

$$= 0.966\ \text{V ac}$$

The numeric value of the voltage 0.259 V ac represents the cosine of 75°; therefore, taking the arccosine of 0.259 identifies the angle sum resulting from the addition of the two angles, 30° and 45°. Also, the value of the voltage 0.966 V ac represents the sine of 75°; therefore, the arcsine of 0.966 also represents the angle resulting from the addition of 30° and 45°.

Example 10-2

Using the same two angles of rotation and voltage input used in Example 10-1 but using a subtraction resolver circuit now instead, determine the output voltages and their relationship to the input angles.

Solution: According to the discussion in the paragraph immediately preceding Example 10-1, the output voltages for subtraction occurring at resolver 2 are $E_{R1-2} = E_{S1-2}\cos(\alpha - \beta)$, and $E_{R3-4} = E_{S1-2}\sin(\alpha - \beta)$. Therefore,

$$E_{R1-2} = E_{S1-2}\cos(\alpha - \beta) = (1\ \text{V})\cos(30° - 45°)$$

$$= (1)\cos(-15°)$$

$$= (1)(0.966)$$

$$= 0.966\ \text{V ac}$$

$$E_{R3-4} = E_{S1-2}\sin(\alpha - \beta) = (1\ \text{V})\sin(30° - 45°)$$

$$= (1)\sin(-15°)$$

$$= -0.259\ \text{V ac}$$

The arccosine of the numeric value of 0.966 V ac is 15° (or −15°) and the arcsine of the numeric value of 0.259 V ac is also −15°.

Converting polar coordinates to rectangular coordinates (resolving)

Figure 10-27 shows how a resolver is wired for resolving. Notice that one of the two stator windings is "shorted." The polar value R is applied to the other stator in the form of an ac voltage, whereas θ is applied to the resolver's shaft. The resultant values of x and y appear at the two rotor coils' output terminals in the form of ac output voltages. Except for the short across terminal E_{S1}, the system is similar to the one depicted in Figure 10-25. Referring once again to eq. (10-2), since $E_{S1} = 0$ because of the shorted coil, this equation may now be rewritten as

$$E_{R1} = y = E_{S2}\sin\theta \tag{10-4}$$

Similarly, eq. (10-3) can be rewritten as

$$E_{R2} = x = E_{S2}\cos\theta \tag{10-5}$$

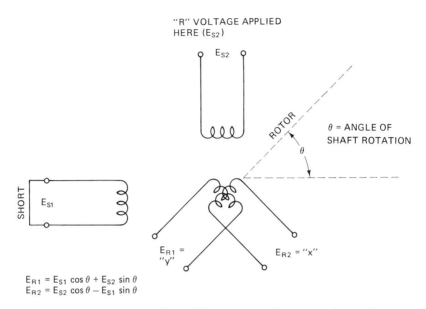

Figure 10-27   Resolver used for resolving polar angles into rectangular coordinates.

Example 10-3

A radar observer spots a ship at sea whose position makes an angle of 56° as measured from true north. The observer's radar indicates a distance of 3.5 miles from the observation site. If a resolver is used to resolve the x and y coordinates of this ship, what would the values of these coordinates be?

Solution: Referring to Figure 10-27, we apply a supply voltage to E_{S2} whose numerical value is equal to the ship's distance as determined by the radar; this is our R value. In this particular case $R = 3.5$. Therefore, we apply a voltage of 3.5 V ac to terminal E_{S2}. We then rotate the resolver's shaft through an angle of 56° to simulate the measured angle or bearing of the sighted ship. When this is done, then, according to eq. (10-2), we should be able to measure the following voltage at E_{R2}, which represents the x-coordinate value:

$$x = E_{S2} \cos\theta = 3.5 \cos 56°$$

$$= (3.5)(0.559)$$

$$= 1.96 \text{ V ac}$$

$$= 1.96 \text{ mi}$$

According to eq. (10-3), we should be able to measure the following voltage at E_{R1}, which represents the y-coordinate value:

$$y = E_{S2} \sin\theta = 3.5 \sin 56°$$

$$= (3.5)(0.829)$$

$$= 2.90 \text{ V ac}$$

$$= 2.90 \text{ mi}$$

Converting rectangular coordinates to polar coordinates (composition)

To perform the reverse conversion of resolving, some additional hardware is needed. A servo circuit must be assembled as shown in Figure 10-28. The short is removed from E_{S1} from the preceding example and the output of E_{R1} is fed to an amplifier, A. The amplifier's output is sent to a servomotor whose output shaft is mechanically coupled to the shaft of the resolver. Ac voltages representing x and y are applied to E_{S2} and E_{S1}, respectively. The induced voltages into rotor coil E_{R1} cause the amplifier to drive the motor, which, in turn, rotates the resolver's shaft until a "null" is found. This null will occur at an angle, θ, which is the vector angle formed by the combining of the x and y voltages. As a result, the shaft will rotate until the axis of the R_1 rotor coil is at right angles to this vector and E_{R1} becomes zero. The angle of rotation is the converted value, θ. The value of R is produced at the output of the second rotor coil, R_2, which is E_{R2}. Its value is the vector sum of E_{S1} and E_{S2}.

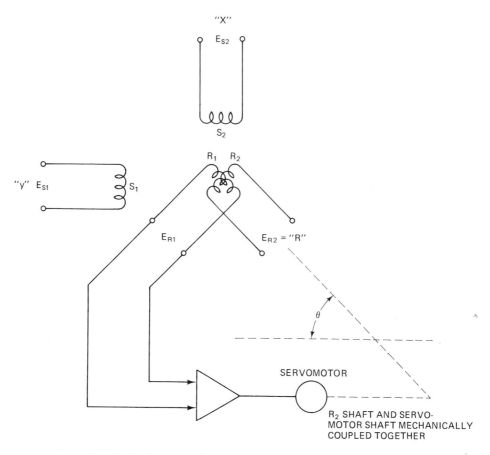

Figure 10-28   Resolver used for converting rectangular coordinates into polar values.

Rotation of coordinate axes (transformation)

Quite often it becomes necessary to convert one set of coordinates used by a moving vehicle into another set that is better understood. Examples are found in space vehicles, airplanes, and ships at sea. There are numerous ways to express a vehicle's location, depending on which set of coordinates you choose to use relative to a fixed reference mark. Since often more than one reference mark is used, conversions become necessary.

To illustrate how coordinate conversions are done, let's use the example of an airplane, A, that is flying in a direction relative to another airplane, B, as seen in Figure 10-29. According to the pilot in airplane A, he or she is flying at a zero-degree bearing according to relative coordinates X and Y. However, according to the true bear coordinates, airplane A is flying ϕ degrees east of north. Also, airplane A spots airplane B at an angle τ relative to A's X–Y coordinate system. The question is: What are the distances a and b?

By using the resolver circuit in Figure 10-27, we can, through "composition," determine a and b. But first, we must obtain the distance R through means of radar or some other ranging means. Having done that, we then feed this value and the relative bearing, τ, into our composition resolver so that we may determine the values of x and y at the stator coils. Next, these newly determined x and y values

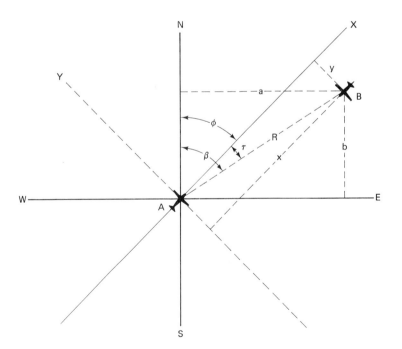

Figure 10-29 Practical application of resolving using the system described in Figure 10-28.

are then fed into the two stator coils of another resolver and its shaft rotated through an angle, ϕ. The output of the two rotor coils then represent the values of a and b predicted by eqs. (10-2) and (10-3).

Typical signal-conditioning methods

Resolvers usually do not require special signal-conditioning circuitry for their operation. However, it is not uncommon to convert the ac output signals from the resolver's coils to a dc signal to drive an analog meter. Also, more recently, the dc signal may then be converted to a digital signal for a digital readout display.

Practical applications

Resolvers are used extensively in guidance control systems where precise direction is required. A good example of an application is in the guidance system of a space satellite. It is necessary to know the satellite's exact space coordinates at all times in order to track the satellite from earth. It is also necessary to know the satellite's attitude relative to a "fixed" reference spot (usually, the sun or a bright star) for the precise aiming and control of imaging and other detection systems.

The military has used resolvers for years for the purpose of aiming platform-mounted firing devices and for obtaining ranging information on enemy targets. Industry is a large user of resolvers. Resolvers are used for the positioning of cutting tools used in CNC/DNC machining centers.

Typical design specifications

Figure 10-30 shows a typical design specification sheet for a resolver.

10-4.6 Inductance Bridge

Linear inductance bridge

The inductance bridge is a type of inductance coil circuit arranged in a Wheatstone bridge configuration. Figure 10-31 shows a linear inductance bridge in which the positioning core varies the inductance of L_1 and L_2, depending on the core's location relative to these two coils. The core's movement is followed by the produced imbalance of the bridge's center-leg current flowing through the current indicator. The bridge is balanced only when the core is at a center location relative to L_1 and L_2.

Rotary inductance bridge

The very same principle used in the linear inductance bridge may also be used to accommodate rotary motion. In place of the moving linear-motion core shown in Figure 10-31, a camlike rotary core may be installed in its place. The two induction coils, L_1 and L_2, are arranged in a manner to facilitate the core's rotation. Otherwise, the circuit's operation is basically the same as the linear configuration. As the

DIMENSIONS

	inches	mm		inches	mm
A	0·750	19·050	J	min	min
	0·747	18·974		0·260	6·604
B	0·692	17·577	K	min	min
	0·689	17·501		0·117	2·972
C	0·5000	12·700	L	0·231	5·867
	0·4995	12·687		max	max
D	0·015	0·381	M	0·1247	3·167
	0·005	0·127		0·1245	3·162
E	0·047	1·194	N	0·491	12·471
	0·043	1·092		0·471	11·963
F	0·065	1·651	O	0·1248	3·170
	0·059	1·499		0·1245	3·162
G	0·065	1·651	P	1·240	31·496
	0·059	1·499		max	max
H	0·491	12·471	Q	0·990	25·146
	0·471	11·963		0·980	24·892
I	0·294	7·468	R	min	min
	max	max		0·245	6·223

SPECIAL NOTES

Rotor Inertia: 0·005 oz in²   0·83 g cm²

Friction Torque (max): 0·04 oz in   3·0 g cm

Weight: 1·6 oz   44 g

Ambient Temperature Range: −55°C to +125°C

Function Error (max): ±0·1 % of max voltage

Inter-Axis Error (max): ± 5 mins of arc

Lead Wires: 7/·005, 18 in long, P.T.F.E. insulation ·039 in dia max

Notes:

(a) $E_{S1-3} = kE_{R1-3} \cos\Theta - kE_{R2-3} \sin\Theta$

(b) $E_{S2-4} = kE_{R2-3} \cos\Theta + kE_{R1-3} \sin\Theta$

(c) $E_{R1-3} = kE_{S1-3} \cos\Theta + kE_{S2-4} \sin\Theta$

(d) $E_{R2-3} = kE_{S2-4} \cos\Theta - kE_{S1-3} \sin\Theta$

*R = Rotor

S = Stator

SECONDARY	NOMINAL IMPEDANCE ohms					PERFORMANCE				
PHASE LEAD ON INPUT Degrees						EQUALITY OF TRANSFORMATION RATIO		NULL VOLTAGE mV rms		OUTPUT EQUATION
	Zro	Zrs	Zso Zco	Zss	Stator tuned	Rotor % Spread max	Comp mV	Fund	Total	MIL-R-23417
18	240 + j620		250 + j790	420 + j140	1·85k	1·0		30	46	SEE NOTE (a) SEE NOTE (b)
18	240 + j620		250 + j790	420 + j140	1·85k	1·0		30	46	SEE NOTE (a) SEE NOTE (b)
23	210 + j480	340 + j125	220 + j420 230 + j430		1k	1·0	50	25	30	SEE NOTE (c) SEE NOTE (d)
23	210 + j480	340 + j125	220 + j420 230 + j430		1k	1·0	50	25	30	SEE NOTE (c) SEE NOTE (d)
23	210 + j480	340 + j125	220 + j420 230 + j430		1k	1·0	50	25	30	SEE NOTE (c) SEE NOTE (d)
23	210 + j480	340 + j125	220 + j420 230 + j430		1k	1·0	50	25	30	SEE NOTE (c) SEE NOTE (d)

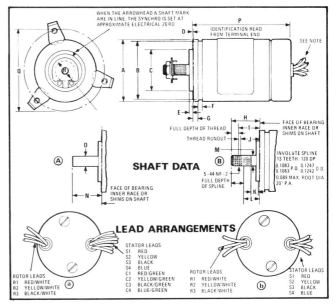

	DESCRIPTION					TRANSFORMATION RATIO					D.C. RESISTANCE at 20 C ohms (NOM)			PRIMARY		
Drawing Detail	TYPE DESIGNATION Muirhead	NATO NUMBER 5930 99	WINDING* (NUMBER OF PHASES)			(OUTPUT:INPUT)								RATED VOLTS	CURRENT (MAX)	POWER (MAX)
			Primary	Secondary	Compensator	Stator/ Rotor (k)	Rotor/ Stator (k)	Compensator/ Stator	Rotor/ Compensator (derived)	Rotor	Stator	Compensator		Amps	Watts
B–b	08M6H1		R2	S2		1·000:1 ±0·023				215	210		5 to 26	0·044	0·46
A–b	08M6J1		R2	S2		1·000:1 ±0·023				215	210		5 to 26	0·044	0·46
B–a	08M7B1		S2	R2	S2	0·913:1 ±0·027	0·913:1 ±0·027	1·000:1 ±0·010		175	195	200	5 to 26	0·038	0·32
A–a	08M7C1	−199−7033	S2	R2	S2	0·913:1 ±0·027	0·913:1 ±0·027	1·000:1 ±0·010		175	195	200	5 to 26	0·038	0·32
A–a	08M7E1		S2	R2	S2	0·913:1 ±0·027	0·922:1 ±0·027	0·990:1 ±0·010		175	195	200	5 to 26	0·038	0·32
B–a	08M7F1		S2	R2	S2	0·913:1 ±0·027	0·922:1 ±0·027	0·990:1 ±0·010		175	195	200	5 to 26	0·038	0·32

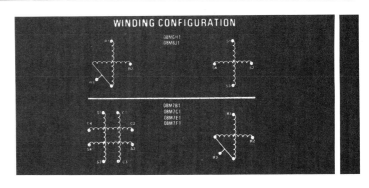

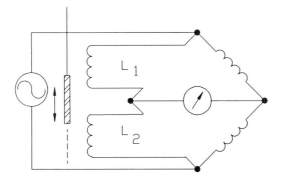

Figure 10-31 Linear inductance bridge circuit.

camlike core rotates between the two coils, the balancing current in the Wheatstone's center leg is varied accordingly.

Typical signal-conditioning methods

The signal-conditioning methods used for both the linear and rotary induction bridges are the same ones used for resolvers. The Wheatstone bridge circuit may be considered a signal conditioner by itself. It is an extremely precise method for obtaining a nulling current or voltage. This is a built-in advantage when comparing this type of system to the other resolver circuits.

Practical applications

Applications involving the inductance bridge are very similar to those of the other resolver types. In addition, the inductance bridge is used in those applications where a center null is advantageous, such as in wanting to maintain a zero position or location. Guidance systems in rockets or space vehicles are again a good example of this type of application.

REVIEW QUESTIONS

10-1. Explain the difference between the terms *reluctance* and *inductance*. How do *reluctance* and *permeability* differ?

10-2. Show how an LVDT can be used to measure the quantity acceleration.

10-3. Explain what a differential rectifier does in an LVDT signal-conditioning circuit.

10-4. What is the difference between a rotating induction potentiometer and the rotating core design in an RVDT?

10-5. Define the term *synchromechanism*.

10-6. What is the basic difference between a control synchro and a torque synchro? Describe an application of each.

10-7. Explain the purpose of the control transformer.

10-8. How would you use a CX–CR pair for the purpose of transmitting temperature information obtained from a remotely located bimetallic thermometer?

10-9. Explain how a microsyn works. Give a practical application for one.

10-10. Explain what a resolver is and describe the differences between its two functions, composition and transformation.

PROBLEMS

10-11. Design a sensing system using a RVDT for monitoring the changing depth of a liquid-filled tank. How would you modify your design to allow the RVDT automatically to maintain a particular depth in this tank? Show circuit details.

10-12. Refer to Figure 10-24. If $E_{S1} = 10$ V ac and $E_{S2} = 5$ V ac, determine the values of E_{R1} and E_{R2}. The angle of rotation for the resolver's shaft is 37°.

10-13. Referring to Figure 10-26, assume that a supply voltage of 12.6 V ac is furnished to winding S_1–S_2. Calculate the output voltages for E_{R1-2} and E_{R3-4} for a shaft rotation of 40° for rotor 1 and a shaft rotation of 63° for rotor 2. Check to see if these two angular rotations are properly added.

10-14. Again, referring to Figure 10-26, assume a supply voltage of 25 V ac applied to winding S_1–S_2. Calculate the output voltages for E_{R1-2} and E_{R3-4} for a shaft rotation of 83° for rotor 1 and 29° for rotor 2. However, configure this circuit now for angle subtraction. Check your results to see if the voltages obtained reflect this subtraction process.

10-15. Design a circuit that can be used in conjunction with a radar-ranging system so that launched spacecraft from a launch site can have their horizontal and vertical distances determined automatically from an observation site merely by noting the line-of-sight distance and the angle-of-sight (i.e., the angle subtended by the line of sight and the horizon) values.

REFERENCES

BOYLESTAD, ROBERT L., *Introductory Circuit Analysis*. Columbus, OH: Charles E. Merrill, 1983.

CARSTENS, JAMES R., *Automatic Control Systems and Components*. Englewood Cliffs, NJ: Prentice Hall, 1990.

DEROY, BENJAMIN E., *Automatic Control Theory*. New York: Wiley, 1966.

FINK, DONALD G., *Electronics Engineers' Handbook*. New York: McGraw-Hill, 1975.

MILLER, RICHARD W., *Servomechanisms: Devices and Fundamentals*. Reston, VA: Reston, 1977.

11

ELECTROMAGNETIC TRANSDUCERS

CHAPTER OBJECTIVES

1. To review electromagnetic theory.
2. To understand radio-wave detection.
3. To study methods used in detecting x-rays.

11-1 INTRODUCTION

It is interesting that in most textbooks dealing with sensors and transducers, the methods used for the detection of electromagnetic radiation is often overlooked or may be purposely omitted. Perhaps the reason has to do with the fact that the principles of detection are often simply relegated to the art and science of radio communications detection and are considered to be covered adequately in any communications textbook. However, there are instances in which communications does *not* play a role in this detection, but rather, the detection has a bona fide role in the conversion of radio-frequency electromagnetic radiation to a control signal for the necessary operation of a process. As a consequence, the detection methods certainly do deserve a spot in the discussion of sensors and transducers.

One of the confusing aspects of energy radiation is the following: We can describe this radiation as being in the form of *electromagnetic waves*, or we can describe it as being in the form of *particles* called *photons*. As a matter of fact,

282

radiation's dual role may lead you to conclude that radiation has a dual personality, or schizophrenic nature. (See again the discussion in Section 1-5.4 for additional comments concerning this interesting paradox.) This chapter is devoted to methods used for sensing radiation in its electromagnetic form, other than heat and light; these two radiation forms are covered in later chapters.

The detection of electromagnetic radiation covers a wide range of detection devices. In this chapter we discuss several methods that depend on electronic circuitry in one form or another for proper operation of these devices.

To understand fully how the detection of electromagnetic radiation takes place, you must first have a good foundation of understanding as to how the electromagnetic spectrum is put together. You should refer back to Figure 1-9 to review how the electromagnetic spectrum is laid out according to frequency and wavelength. You should also familiarize yourself with its various regions or "bands." What is interesting about the spectrum is that virtually every labeled region has its own unique electromagnetic characteristics, with many requiring their own unique methods of detection. We discuss these methods in detail and how they are implemented. However, as usual, we begin our discussion of detection methods with a review of the necessary theory.

11-2 REVIEW OF ELECTROMAGNETIC THEORY

Electromagnetic radiation is the energy that is emitted or radiated through space from either an energy source such as a thermonuclear or chemical reaction or from a source containing an oscillating electrical current. The intensity of this energy varies inversely with the square of the distance from the energy's source and is expressed by the following equation:

$$\frac{I_2}{I_1} = \frac{D_1^2}{D_2^2} \tag{11-1}$$

where I_2 = intensity of radiation measured at distance D_2 from the radiation source (any unit of radiation intensity)

I_1 = intensity of radiation measured at distance D_1 from the radiation source (any unit of radiation intensity that is the same as I_2)

D_1 = distance from radiation source at which I_1 is being measured (any unit of length)

D_2 = distance from radiation source at which I_2 is being measured (any unit of length that is the same as D_1)

Note that all variables in eq. (11-1) must have nonzero values.

The velocity of the emitted radiation is the speed of light or near the speed of light if the propagation takes place in any medium other than a vacuum. This speed (designated by the lower case letter c) is 299,792 km/s, or approximately 3×10^8 m/s.

The relationship that exists between the radiation's frequency and wavelength was described in Sections 1-5.4 and 1-5.5 and should be reviewed here before going on.

11-3 COMMONLY SENSED MEASURANDS

The most often sensed measurands using electromagnetic radiation are the following: (1) radio-wave radiation, (2) thermal radiation, (3) light radiation, (4) x-ray radiation, (5) atomic radiation, (6) position of a body, (7) velocity of a body, (8) acceleration of a body, and (9) fluid flow. As we go through the descriptions of how the various sensors work (again excluding heat and light radiation), enough examples will be given to explain how each of the foregoing measurands is detected.

11-4 TYPES OF TRANSDUCERS PRESENTLY AVAILABLE

There are basically three types of transducer systems that are presently used for sensing electromagnetic radiation (excluding heat and light). These fall under the broad categories of (1) radio detectors, (2) x-ray detectors, and (3) nuclear radiation detectors. With these three types of detectors we can sense all the measurands listed in Section 11-3. However, because of the somewhat specialized nature of the third category, nuclear radiation detectors, we will not cover these detectors in this chapter. Instead, we devote an entire chapter, Chapter 17, to these devices. Meanwhile, we begin our discussion and inspection of the first two categories in detail.

11-5 RADIO RECEIVERS

We now explore the different methods used for the detection of radio-frequency (RF) energy. There are a variety of methods used today but we look only at the more popular ones in the discussions that follow.

11-5.1 Radio (RF) Detector

There are a number of different radio detection circuits used to detect the presence of radio energy. The type of detector used is determined by the nature of the radiated radio waves being detected. In Chapter 2 we discussed the various forms of data transmission methods used in transmitting data and we discussed how they are received or detected. Here in Chapter 11 we are interested only in detecting the presence of radio energy and are not interested in demodulating a transmission to obtain data such as speech or encoded computer data pulses. This greatly simplifies the needed detection circuitry. We begin our discussion then with the simplest detection schemes used for this application, and then we finish with the more sophisticated ones.

Detection of AM and CW signals

The simplest method of RF detection is using a diode circuit, such as the one illustrated in Figure 11-1. When a diode is used in this particular application, it and its associated circuitry is referred to as a *demodulator*. The radio wave is essentially rectified much like an ordinary 115-V ac waveform is rectified in a power supply to produce dc voltages. In the case of RF voltages, the RF waveform is transformed into a pulsating dc signal by the receiving diode (D_1), the pulsating dc waveform being typical of the half-wave rectification process. This signal is then further filtered or smoothed by a filtering capacitor (C_2) and a blocking RF choke coil, inductor L_2. Both of these components prevent the RF frequency portion of the diode's output waveform from proceeding farther down the circuit, where it could interact with other stray signals to create distortion. The net result of C_2 and L_2 is an ac signal that has been stripped of all RF at the receiver's output. C_3 is a blocking capacitor that has the purpose of acting as a "curve-filling" device that serves to round out the half-wave ac signal at its input side and more or less "fills in" the missing bottom half of the rectified signal. The capacitor also blocks out any dc voltages that may have a tendency to surface within the circuit that could bias the receiver's output. The purpose of R_L is to act as a load resistor. Its purpose is to indicate simply that the circuit is usually connected to another circuit at this point. This circuit may be an indicating device to show the presence of an output signal. However, R_3 could also serve as a path for unwanted ac interference signals to travel to ground because of the blocking action of C_3.

The amplitude of the resultant output signal in this circuit is directly proportional to the amplitude of the RF waveform being varied at an audio-frequency rate. Notice that a tuned circuit comprised of L_1 and the variable capacitor C_1 is used to enable the receiving circuit to discriminate between different radio signal frequencies. To a great extent the antenna can also act as a tuned circuit to receive only a limited band or range of frequencies. Coupled with a tuned circuit, a certain amount of receiver frequency selectivity can be achieved with this system. (A receiver's selectivity refers to its ability to separate adjacent radio signals without one of the signals creating interference with the other.)

The receiving circuit above is called an *amplitude-modulation* (AM) receiver. At the receiver's output, a simple voltage indicator such as a voltmeter can be used across R_1 to determine if a desired radio signal is present or not.

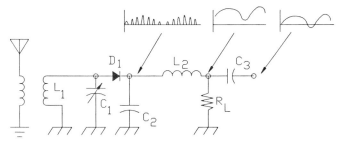

Figure 11-1 Typical CW and AM detector circuit.

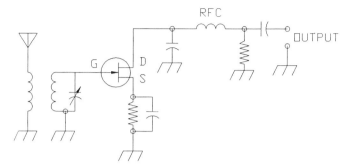

Figure 11-2   Improved version of circuit shown in Figure 11-1.

Better sensitivity for the receiver above can be achieved by using a method of detection that also furnishes some degree of amplification to the radio signal. Figure 11-2 shows such a circuit. The diode of Figure 11-1 has now been replaced with an FET (field-effect transistor). By using what is referred to as *plate detection* (an expression carried over from the days of the vacuum tube), it is possible to add gain to the receiver's detection circuit. The principle of operation is otherwise similar to that of the diode detector. The one other distinguishing difference between these two circuits is that the plate detector requires a dc power source, whereas the diode detector does not.

The problem with the two circuits above is that if the radio-frequency signal is not modulated, it is difficult to detect the signal's presence other than looking for an output voltage that is produced at the receiver's output. This output indication is strictly dependent on the radio signal's carrier strength. The term *carrier* in this sense refers simply to that portion of a transmitted radio wave having no modulation present at all. Figure 11-3 shows a carrier wave that is *modulated*. Therefore, an *unmodulated* wave would show *no* fluctuations; that is, there would be no *modulation envelope* present. The carrier would be of equal amplitude throughout its length.

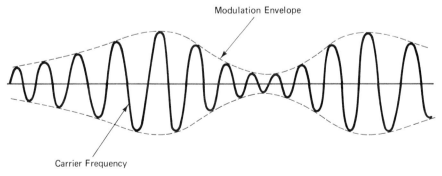

Figure 11-3   Modulated and unmodulated carrier wave. (From James R. Carstens, *Automatic Control Systems and Components,* copyright 1990, p. 351. Reprinted by permission of Prentice Hall, Englewood Cliffs, NJ.)

This type of unmodulated radio signal is often referred to as a *carrier-wave* (CW) signal.

Often, it is desirable to be able to hear the transmitted signal for troubleshooting purposes. Also, if more than one carrier is being transmitted at the same time and their carrier frequencies happen to be very close to each other, it is often much easier to separate the two signals by using audio filtering. It then becomes possible to filter out unwanted circuit noise and adjacent signal interference.

The circuit in Figure 11-4 allows for "listening" to unmodulated carrier signals and the filtering out of unwanted noise at audio frequencies. A process called *frequency beating* is used where a signal of slightly different frequency from that of the transmitted CW radio signal is injected into the receiving circuit's detection system to mix with the received signal (block 2 in Figure 11-4). The circuit that produces this second signal is called a *beat frequency oscillator* (BFO) (block 3). This mixing produces a third signal whose frequency is equal to the arithmetic difference between the original two, thereby creating an audio-frequency signal that is easily detected. This signal then causes an output indication to be produced at the receiver's output (block 4).

As an example, let's assume that we are trying to detect the presence of an unmodulated 4.000-MHz radio control signal. In our radio receiver we can adjust

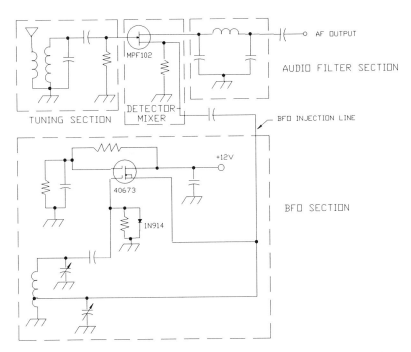

Figure 11-4 Circuit showing a beat frequency oscillator (BFO) for hearing unmodulated carrier signals.

the BFO circuit for a frequency of, say, 4.001 MHz. When this is combined with the 4-MHz signal, a tone of 4.001 to 4.000 MHz, or 0.001 MHz, or 1000 Hz is produced. This new frequency is easily detected and "heard" by our receiver. As a matter of fact, if a set of earphones were wired across the receiver's output, you could actually hear the 1000-Hz tone being generated. (It is interesting to note that the transmission and reception of "CW," as it is commonly referred to by military radio personnel and ham radio operators, is an extremely popular form of communications. The transmitted carrier is interrupted into a series of dots and dashes, or Morse code characters, which are then intercepted by a receiver containing a BFO and deciphered into a message form by the listener.)

Typical CW signal-conditioning method and how it works

The output of a CW receiver is usually fed into an audio filter network that has been specially designed to amplify a very narrow range of audio tones. The reason for this is to ensure that only one radio frequency, producing only one particular audio tone at the receiver's output for a given BFO frequency, is detected at a time. For instance, in the case of the example above, where only a 4-MHz signal is desired to be detected out of a field of several other frequencies that may happen to be present, only the 4.000-MHz frequency will produce the 1000-Hz tone. The other frequencies will produce tones other than 1000 Hz. In reality, a 3.999-MHz carrier frequency will also produce a 1000-Hz tone, but hopefully, the receiver will have enough selectivity to subdue that frequency when tuned to 4.000 MHz.

Because selectivity is an important factor in being able to receive individual radio signals for radio control and telemetry purposes, elaborate circuits have been devised to increase a receiver's selectivity. One such scheme is the heterodyne receiver, sometimes referred to as the superheterodyne receiver. In this receiver all incoming radio frequencies are first converted to a common frequency, often 455 kHz. This common frequency is often referred to as the receiver's *intermediate frequency* (IF). Then this frequency is combined with a BFO output to produce an audible CW signal. Figure 11-5 shows such a receiver. It is not unusual for a receiver to have two, perhaps even three stages of conversion or IFs before the signal is detected and mixed with a BFO. The advantage of having so many conversions is that following each conversion, since the frequencies are now all the same, each amplifier and filter can be "peaked" and "tweaked" for that one and only frequency rather than having to be able to amplify a relatively wide range of different frequencies. Filtering and amplifying efficiencies are so much greater for the multiconverted radio signals. Also, the BFO can be designed with greater stability now that it has to operate and beat against a much lower range of frequencies. In other words, in our 4-MHz receiver the BFO would have to operate in the 4-MHz range, whereas in our 455-kHz IF superheterodyne receiver, the BFO would have to operate only in the 455-kHz range.

The introduction of IF circuitry into our receiver design allows us to develop yet another method for RF carrier detection. The current drawn by the IF circuitry in this type of receiver varies proportionally with the signal intensity of the detected

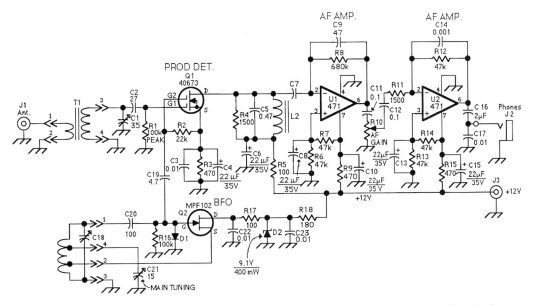

Figure 11-5 Single-conversion radio receiver with BFO. (From *Radio Amateur's Handbook*, 60th ed., p. 8-36. Reprinted by permission of the American Radio Relay League, Newington, CT.)

carrier. By monitoring this current it is possible to determine whether or not a carrier is present. Such a system is shown in Figure 11-6.

Practical applications

It was mentioned earlier that the radio receiver is used for detecting signals used for radio control and for telemetry. AM radio signals are rarely used for these applications because of AM's susceptibility to human-made and natural-made interference.

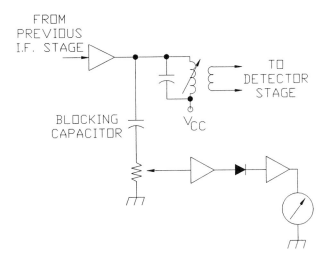

Figure 11-6 Signal strength indicating circuit.

On the other hand, CW is used extensively for simple radio control applications such as in the cases of controlling on–off switches. An example is in the case of a typical garage door opener (Figure 11-7). A radio signal is transmitted by the car owner's controller by the push of a button. The signal is received at the garage site by the receiver's antenna. The receiver detects the signal and converts the RF signal to an audio frequency, which is further converted to a signal of sufficient strength to operate a power relay. The relay, in turn, operates a motor device that opens or closes the garage door.

A more sophisticated control scheme, used where it is desired to open and close a control valve remotely on a pipeline, involves using the tone control system depicted in Figure 11-8. Because critical radio-control systems cannot depend merely on the presence of carrier waves for control purposes, since spurious signals could trigger the system inadvertently, transmitted tones can reduce this problem considerably. A transmitted audio tone, say 1450 Hz, could open the pipeline valve, while another tone, 1500 Hz, could close the valve. It is unlikely that these two tones would occur in spurious radio signals from other sources; consequently, fairly good control security is possible.

Detecting FM radio signals

A *frequency-modulated* (FM) radio wave or signal is a radio signal whose carrier frequency is varied (modulated) at an audio rate (Figure 11-9). This type of modulation has the distinct advantage of being relatively interference-free as compared

Figure 11-7 Radio controlled receiver for controlling a garage door opening and closing mechanism.

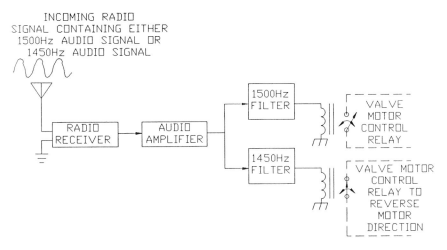

Figure 11-8   Radio remote control scheme involving two channels of audio control.

to an AM signal. AM signals are easily garbled by electrical interference created by static discharges such as lightning or by human-made radio noises created by relay contacts and motors. These interfering AM signals add to the amplitude or, in many cases, literally cover up the desired AM signals so that they cannot be detected. FM, on the other hand, is quite immune to these problems and consequently, has become very popular for telemetry applications.

Figure 11-10 shows the schematic of a typical FM receiver that can be used for telemetry. The output from an IF stage (see the discussion above concerning IFs) on a receiver is fed into the FM detector shown. This type of detector is called a *frequency discriminator* because it can discriminate between different modulation frequencies used to vary the carrier frequency. These frequencies are then converted into an amplitude-varied ac signal much like the output of the AM detectors discussed earlier.

As can be seen from its schematic, the FM detector is somewhat more complicated in design than the AM detector. The discriminator detector works like this: The transformer, T_1, in Figure 11-10 converts the incoming FM signal to an AM signal. This is done by introducing the FM signal through a center tap on the T_1's secondary winding and creating a phase relationship to develop between the voltages

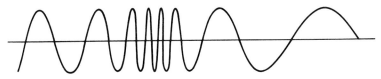

Figure 11-9   Frequency-modulated radio signal. (From James R. Carstens, *Automatic Control Systems and Components,* copyright 1990, p. 354. Reprinted by permission of Prentice Hall, Englewood Cliffs, NJ.)

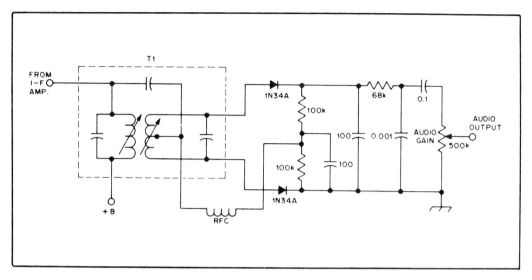

Figure 11-10 Discriminator stage for an FM receiver. (From *Radio Amateur's Handbook,* 60th ed., p. 13-8. Reprinted by permission of the American Radio Relay League, Newington, CT.)

on either side of the center tap, depending on whether or not there is frequency modulation present in the carrier wave received. No modulation causes the secondary voltages to be 180° out of phase with each other so that when rectified by D_1 and D_2, the two voltages cancel each other because of equal and opposite polarities. Any modulation present in the carrier will cause one or the other voltage component to dominate over the other, thereby producing an output voltage.

Another more recent form of FM detection is shown in Figure 11-11. This particular FM detector is called a *ratio detector*. This term arises from the fact that the ratio of the two rectified voltages appearing across R_1 and R_2 following the discriminator detection circuit is the same as the voltage amplitude ratio formed by the voltages on either side of the tapped secondary winding of the transformer. With this type of detection circuit the input FM signal may vary widely in signal strength, but it will produce little change in the output signal level. Only FM signals are detected with this system, but no AM.

There are other forms of FM detection that work as well or better than the two methods just discussed. However, these two systems have probably seen the heaviest use for FM detection.

Typical FM signal-conditioning methods

The methods used for conditioning FM signals, once they have been detected, are very similar, if not identical, to those used for CW detection since the output of each detector is a variable dc signal. One possible modification to conditioning an FM signal is to make provisions for receiving variable-frequency tones rather than just

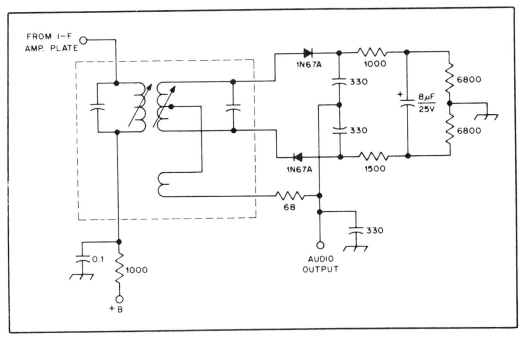

Figure 11-11   Ratio detector circuit for an FM receiver. (From *Radio Amateur's Handbook,* 60th ed., p. 13-9. Reprinted by permission of the American Radio Relay League, Newington, CT.)

receiving the one tone associated with CW reception. In this manner several channels of data may be transmitted simultaneously. The FM receiver then selectively responds to each individual channel since the receiver contains several tuned filters, one for each channel. Each filter is able to respond to its designed center frequency plus a narrow range of frequencies on either side of its center frequency to accommodate the transmitted data's range of values (Figure 11-12). Bandpass filters with relatively low Q values constructed from operational amplifiers work very well for this application. (The Q or *quality* of any frequency-sensitive circuit is an indication of its frequency-selective ability; the higher its Q, the greater its selectively, and vice versa.)

Practical applications

FM radio transmissions are used extensively in industry for remote control and data transmission applications. A typical data transmission system is shown in Figure 11-13, where a remote radio telephone employing touch-tone dialing is used. When a number is dialed, two tones are generated for each number. (Theoretically, single-tone generation would also work but lacks protection against accidental or unauthorized phone system access.) These tones are divided into two groups: one group (the low-frequency group) comprised of 697, 770, 852, and 941 Hz, the other

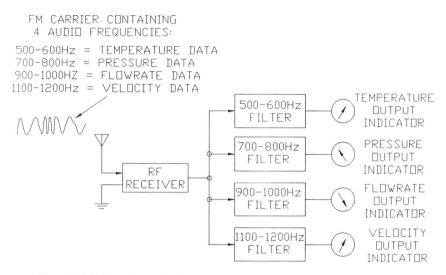

Figure 11-12   Four-channel audio-tone reception scheme involving radio detection.

group (the high-frequency group) comprised of 1209, 1336, 1447, and 1633 Hz. Each number (including the # and * symbols) is generated with a tone from each group. When a number is dialed at the remote site the tonal pairs are transmitted by FM radio to a distant receiver containing the appropriate audio filters for reception of these pairs. Each of the tonal pair's filters are designed to receive only those tone pairs and no others. The tones received then trigger the appropriate relay for dialing the phone. When a connection is made, the return phone audio is transmitted by a separate transmitter (not shown in Figure 11-13) back to the originator's receiver (also not shown).

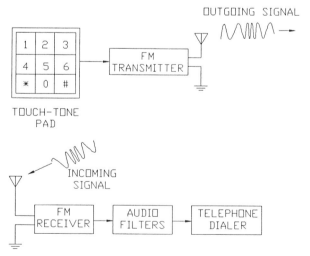

Figure 11-13   Transmission and reception of dial touch-tone signals for an FM radio telephone system.

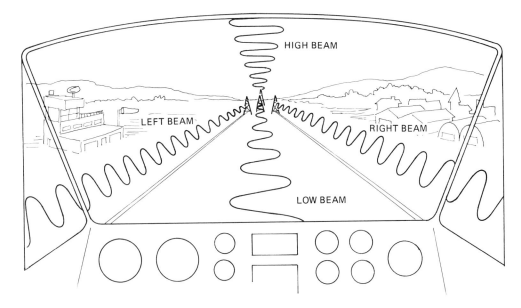

Figure 11-14 FM radio signals used in an aircraft landing system.

Another application of FM reception involves using a detection system for maintaining a course setting for landing aircraft (Figure 11-14). FM radio signals having very narrow beams of transmission are beamed down a runway toward landing aircraft. These beams are transmitted side by side, with each being modulated by a particular audio tone. An aircraft trying to land in weather with restricted visibility can receive these radio signals and determine whether the craft is to one side of the beam or to the other, or whether it is too high or too low. Figure 11-15 shows a very simplified block diagram for such a system. The audio tones transmitted

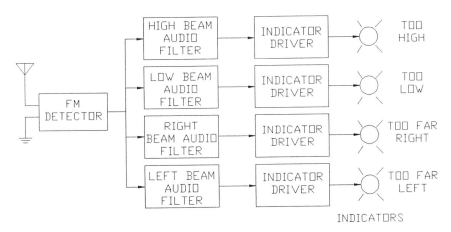

Figure 11-15 Simplified block diagram of an aircraft landing system.

within the received carrier will be detected and identified by the receiver depending on the aircraft's relative location to the transmitted beams. Referring to the block diagram in Figure 11-15, the appropriate audio filter will pass only its particular frequency. Any other frequency will go undetected. Each tone detected will trigger an appropriate onboard warning light or buzzer indicating the aircraft's relative location and notifying the pilot of the necessary corrective maneuvers needed to get back on the desired beam.

An application of FM data transmission is illustrated in Figure 11-16. A thermistor senses temperature, causing a voltage-controlled oscillator to vary its audio output according to the thermistor's varying output resistance, which, in turn, varies the input voltage to the VCO. The VCO, in turns, modulates a FM transmitter that transmits these data to a remotely located FM receiver. The receiver demodulates the FM signal and after passing the signal through an audio filter for conditioning purposes, a frequency counter identifies the audio frequency and interprets the thermistor's temperature.

Admittedly, the thermistor system just described may place the FM receiving system outside the realm of transducers. This is where the distinction becomes blurred between transducers and instrument systems. You are encouraged to pursue the subject of instrumentation and telemetry further by referring to books dealing with these subjects.

Other types of radio detectors

There are several other configurations of radio signals used for data transmission and communications. However, since we are interested primarily in detecting the presence of radio waves and not in deciphering the data they are carrying, we only mention here some of the other forms of radio transmissions.

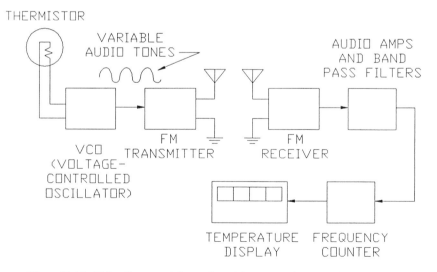

Figure 11-16   FM radio transmitting and receiving system for transmitting temperature data.

Single- and *double-sideband* (SSB and DSB) transmissions are rarely used for remote control transmission purposes other than for voice and some forms of complex data communications. A special detector called a *product detector* is used for receiving this type of transmission, although any AM receiver containing a BFO can be used with only some minor difficulty in tuning these signals.

The various pulsed modulation modes that are available can all be detected using the CW methods outlined earlier. Only the signal-conditioning methods may differ from one pulse form to another after signal detection takes place within the receiver. Chapter 2 covers all the various forms of pulsed transmissions; these should be reviewed having now discussed the various detection methods here in Chapter 11.

11-5.2 X-ray Detection

To understand x-ray detection, one must first understand the nature of x-rays themselves. X-rays are produced when very high-speed electrons that have been accelerated through potential differences in excess of 10,000 V or so strike a metallic surface. Secondary particles and x-rays are generated during this collision. X-rays behave much like light energy. In other words, when traveling through space, x-rays behave as electromagnetic waves do. However, when interacting with matter, x-rays have the characteristics of photons or particles of energy, called *quanta*. This wave–particle duality exhibited by x-rays is also common for the other forms of electromagnetic energy besides light.

Certain chemical coatings, when bombarded with x-ray radiation, emit electrons quite readily, as seen in Figure 11-17. These coatings are called *x-ray-sensitive phosphors*. The emitted electrons are accelerated to high velocities by a series of focusing or accelerator plates charged with high negative voltages. These voltages direct the dislodged electrons toward another phosphor-coated screen located at the opposite end of the converter tube. Here, the now concentrated electron beams collide with the phosphor's surface, causing the impact areas to emit light from this high-velocity collision and to produce a visible image of the otherwise invisible x-rays. This type of device is called an *x*-ray image converter tube. As we will discover

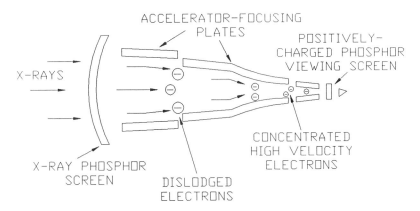

Figure 11-17   X-ray detection viewing system.

later, image conversion tubes are also used for converting other forms of electromagnetic energy into visible images. The phosphor used at the tube's front entrance determines the type of radiation that can be detected.

Instead of using a viewing phosphor at the far end of the converter tube, an anode plate can be installed to collect and allow measurement of the amount of electron flow emanating from the x-ray phosphor. This electron flow results in an immediate current readout so that the x-rays' intensity can be correlated precisely with the current flow and recorded as shown in Figure 11-18. The following equation describes the sensitivity:

$$\sigma = \frac{I_{out}}{\mathscr{R}_{int}} \tag{11-2}$$

where σ = converter tube sensitivity ($\mu A/\mathscr{R}$)
I_{out} = current output on anode (μA)

$\mathscr{R}_{int}$ = input intensity to x-ray phosphor (roentgen)

Note: The roentgen is defined as the quantity of x-ray radiation needed to produce 1 electrostatic unit of electrical charge in 1 cubic centimeter of dry air at standard conditions (i.e., 0°C, 760 mm barometric pressure), characteristics of the converter device above.

Typical signal-conditioning methods

Signal conditioning for the x-ray detector described in Figure 11-18 is comprised primarily of relatively high-gain dc amplifiers. The resultant signal is then of suffi-

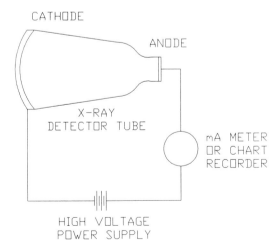

Figure 11-18   Recording circuit for x-ray detection system.

cient strength to operate warning lights, bells, servos, and so on, so that some sort of useful function can be initiated or completed.

Practical applications

X-ray detection is vital for personnel working around medical x-ray equipment such as the equipment found in clinics and hospitals. Because x-rays are a serious health hazard, it is very important for x-ray personnel to monitor their work area for x-ray leakage from the x-ray-producing machines. This can be done easily with the x-ray conversion device described in Figure 11-18. A current threshold circuit can be incorporated into the plate output circuit in place of the ammeter that is shown. When a certain current flow amount is exceeded, an alarm is triggered.

X-ray detection is particularly important in the science of astronomy. Recently, astronomers have intensified their studies of x-ray emissions coming from stars within our galaxy and the x-rays being emitted from the centers of distant galaxies. All of this is part of an overall effort by the astronomer to better understand the physical makeup of these objects. However, to do this it is necessary to scan fairly large portions of sky in the case of galaxies and certain stars. It becomes necessary to use a very narrow reception window on the detection device for the x-rays to enter through (Figure 11-19). This increases the resolution or selection of the receiver itself. This resolution is obtained by using a very small aperture in a metallic cover located at the front of the x-ray converter tube. When a sky area is scanned back and forth following a methodic pattern, then allows for the buildup of a composite electronic image comprised of varying current or voltage intensities that can be recorded on magnetic media such as tape or a floppy disk. Each voltage "pixel" signal is then played back in proper sequence and converted to a printed dot on paper. A total x-ray picture is then reconstructed in this manner, which then becomes a complete x-ray of that viewing area in the sky.

Another method extensively used for the detection of x-rays is the photographic method. Although this method is hardly considered a transduction method, much less an electrical transduction method, it is mentioned here merely to round out the discussion on x-ray detection.

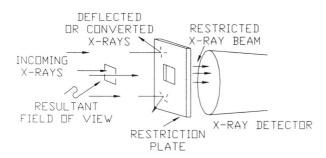

Figure 11-19 Restricting the field of view of an x-ray detector.

Ever since the discovery of x-rays late in the nineteenth century, it was well known that x-rays interacted with silver-sensitized films used in photography. This was, for many years, the classic method used for x-ray detection. A film's photosensitized surface becomes fogged or exposed when exposed to the rays, just as if light itself had fallen upon the surface. Again, by using a small x-ray aperture and scanning the suspected x-ray source several times, a mosaic is obtained on the photographic plate of the x-ray's intensity and location.

REVIEW QUESTIONS

11-1. Draw a circuit of a simple diode detector and explain how an AM radio signal is detected. Show at least three methods or steps that you would take to increase the selectivity of your receiver.

11-2. Show with a circuit sketch how you could transmit pressure data by means of variable audio tones on a carrier using an FM signal. Show both the encoding of the signal at the transmitter and the decoding at the receiver.

11-3. Explain how a superheterodyne receiver works. What is its purpose?

11-4. Explain the purpose of a discriminator circuit.

11-5. How are audio tones that are transmitted simultaneously on a carrier (each tone representing a channel of data) separated from each another at the receiver?

11-6. What is meant by a *quanta of energy*?

11-7. Draw a diagram of a typical x-ray detector-converter and show the relationship that exists between the x-ray phosphor cathode and the anode phosphor screen.

11-8. What is the purpose of the focusing plates used in an x-ray converter tube?

11-9. What is used in place of the anode viewing screen of an x-ray converter tube to create a measuring circuit for quantifying x-ray data?

11-10. How would the circuit in Review Question 11-9 be used to signal or warn personnel of exceeded levels of x-rays?

PROBLEMS

11-11. Assume that a radio telemetry system has been landed on the moon for the purpose of sending back certain recorded lunar data. The radio transmitter is known to have the capability to transmit a signal having the measured strength of 0.08 μW at a distance of 1 km using a particular receiving antenna. Assuming no path losses along the way, what would be the transmitter's theoretical, received signal strength here on earth? Assume that the earth-based receiving antenna is the same type of antenna as the one used for making the 0.08-μW measurement. The moon's average distance from earth is approximately 382,000 km.

11-12. Figure 11-20 shows the circuit and the design specifications for constructing an active audio filter. Using this information, design a filter for receiving a 650-Hz audio signal and having a Q of 2.5 and a gain of 2. (Note: $W_o = 2\pi f$; C is in farads, R is in ohms.)

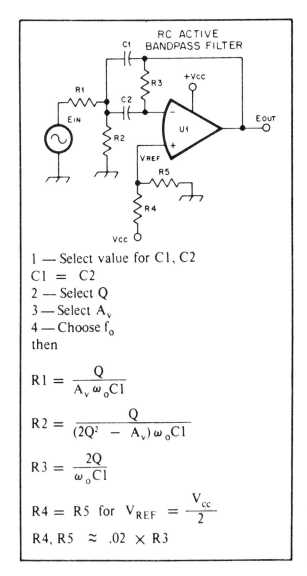

1 — Select value for C1, C2

C1 = C2

2 — Select Q

3 — Select A_v

4 — Choose f_o

then

$$R1 = \frac{Q}{A_v \omega_o C1}$$

$$R2 = \frac{Q}{(2Q^2 - A_v) \omega_o C1}$$

$$R3 = \frac{2Q}{\omega_o C1}$$

$$R4 = R5 \text{ for } V_{REF} = \frac{V_{cc}}{2}$$

$$R4, R5 \approx .02 \times R3$$

Figure 11-20 Circuit for Problem 11-12. (From *Radio Amateur's Handbook,* 56th ed., p. 4-40. Reprinted by permission of the American Radio Relay League, Newington, CT.)

REFERENCES

BUCHSBAUM, WALTER H., *Encyclopedia of Integrated Circuits*. Englewood Cliffs, NJ: Prentice Hall, 1981.

DEMAW, DOUGLAS, *ARRL Electronics Handbook*. Newington, CT: American Radio Relay League, 1976.

DUNCAN, FRANK R., *Electronic Communications Systems*. Boston: Breton, 1987.

MILLER, GARY M., *Modern Electronic Communication*. Englewood Cliffs, NJ: Prentice Hall, 1988.

SEARS, FRANCIS W., MARK W. ZEMANSKY, and HUGH D. YOUNG, *College Physics*. Reading, MA: Addison-Wesley, 1977.

SESSIONS, KENDALL W., *Master Handbook of 1001 Practical Electronic Circuits*. Blue Ridge Summit, PA: TAB Books, 1975.

VERGERS, CHARLES A., *Handbook of Electrical Noise: Measurement and Technology*. Blue Ridge Summit, PA: TAB Books, 1979.

12

THERMOELECTRIC SENSORS

CHAPTER OBJECTIVES

1. To review thermoelectric theory.
2. To study the behavior of thermocouple junctions.
3. To study the application of the thermopile in infrared pyrometers.

12-1 INTRODUCTION

Thermoelectricity is electricity that is generated through the application of heat. The key word in this statement is *generated*. There are many sensors that respond to heat by controlling a voltage or current source that has been physically attached to the sensor, but there are very few types of sensing materials that actually generate electricity by becoming the voltage or current source. In this chapter we discuss the characteristics of these materials and see how they are used in making sensors and transducers.

12-2 THERMOELECTRIC THEORY

Thermoelectric phenomena were observed as far back as the early eighteenth century, when T. J. Seebeck (1770–1831) discovered that an electric current could be produced in two dissimilar metal wires when one end of each of the wires were joined to form a junction and that junction heated. About this same time, another physicist, Jean Peltier (1785–1845), made a similar discovery to further establish this thermoelectric effect. Finally, a British scientist, William T. Thomson, Lord Kelvin (1824–1907), experimented further with the interaction of heat on charge flow in wire conductors, contributing further to the understanding of thermoelectricity. To better understand how this phenomenon works, we investigate the work of each of these three gentlemen and then formulate a conclusion.

In 1826, Seebeck discovered that by taking two pieces of wire, each wire being made from different metals and joining their ends together, he could heat one end and produce a very small current flow in the circuit. To illustrate this, Figure 12-1 shows two wires, wires A and B, with their two ends forming junctions J_1 and J_2. Junction J_1 is at a higher temperature (T_1) than junction J_2 (which is at temperature T_2). A net EMF, which can be measured with a millivolter, is generated between these two junctions; this is called the *Seebeck effect*.

The Seebeck effect results from the fact that the densities of free electrons (these are electrons that are free to "drift" from one metallic atom to another to constitute a current flow) are different for two different metal wires. Within each metal, the density of these free electrons can be further varied with temperature. Looking again at Figure 12-1, when heat is applied to J_1, the amount of electron flow into and out of this junction is different for each of the two wires. This immediately produces an EMF across that junction. The same effect also occurs at the other junction, J_2. If J_2 is at a different temperature, a net EMF will be produced between the two junctions; if the junction temperatures are the same, the net EMF will be zero.

Peltier discovered that as a result of an EMF existing across a junction, heat is either liberated at the junction or heat is absorbed, depending on the polarity of the EMF, that is, depending on the direction of charge flow across the junction. In other words, a single junction can either convert heat into electrical energy or convert electrical energy into heat. This is called the Peltier effect or Peltier EMF.

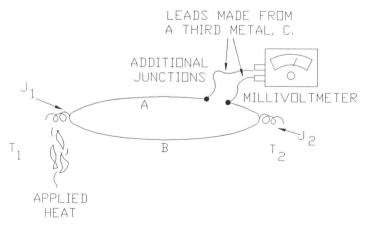

Figure 12-1   Seebeck effect.

As an interesting side note here, this effect implies that it is possible to cause a junction to act as a refrigerator and become cold if the current flow is in a particular direction. This is in fact exactly what can happen. (A number of manufacturers market small battery-operated refrigerators that operate on this very principle.)

The Peltier EMF is defined as the amount of heat energy that is absorbed or given off per unit quantity of charge flowing across a junction. That is,

$$E_P = \frac{W}{Q} \tag{12-1}$$

where E_P = Peltier EMF (V)

W = amount of Peltier heat expended (or absorbed) as result of charge moving across junction (J)

Q = amount of charge moved (C)

There are two Peltier EMFs for wires A and B because there are two junctions, E_{P1} and E_{P2} (Figure 12-2). The net EMF would then be the difference between these two voltages.

Thomson discovered that when a single wire is unevenly heated and that a current flow is taking place at the time, heat is either radiated or absorbed, depending on the direction of current flow and depending on the flow quantity. This Thomson effect is similar to what Peltier observed except that Thomson made his observations using only a single wire that had no junctions.

Like the Peltier effect, the Thomson effect can be described by a similar equation:

$$E_T = \frac{W}{Q} \tag{12-2}$$

where E_T = Thomson EMF (V)

W = amount of heat expended (or absorbed) as result of charge moving through wire (J)

Q = amount of charge moved (C)

Because we are dealing with two wire materials, A and B, there are two

Figure 12-2 Generating and measuring a voltage with a thermocouple and a sensitive voltmeter.

Thomson EMFs, E_{TA} and E_{TB}. As it turns out, the Seebeck effect is the sum of the two other effects. If we let E_S be the Seebeck effect, or Seebeck EMF, then

$$E_S = (E_{P1} - E_{P2}) + (E_{TA} - E_{tB}) \tag{12-3}$$

Equation (12-3) is a much oversimplified expression for accounting for the total voltage generated by a thermocouple when subjected to a temperature differential across its two junctions. Nevertheless, the overall concept of this equation is accurate. Keep in mind, however, that a thermocouple produces voltage only when there is a temperature differential that exists.

To measure the voltage output of a thermocouple created by a temperature differential, it is necessary to attach a voltmeter across one of the two junctions using a third metal, creating a third and a fourth junction in the process (refer again to Figure 12-1). However, we can nullify the effects of these additional nonintentional junctions by making very certain that they are both maintained at the same temperature as one of the other junctions. With that precaution the thermocouple's output can easily be measured. Typical voltages resulting from the thermoelectric effect are in the millivolt range.

When using thermocouples one of the two junctions formed by the two dissimilar metal wires is purposely maintained at a constant easily maintained temperature. In reality, virtually any temperature would do; however, the temperature most often used is 0°C since this is an easily obtained and held temperature. What is often done to maintain this particular temperature is to use a simple ice-bath arrangement, as shown in Figure 12-3. Notice how the voltmeter is connected to the thermocouple circuit. It is wired across the junctions J_1 and J_2. Unfortunately, in the process of connecting the voltmeter, additional junctions, J_3, J_4, J_5, and J_6, are also created. Fortunately, we can be sure to use the same wire material on both sides of the newly created junctions to prevent additional temperature-sensitive junctions from forming. Notice, for instance, that J_5 is comprised of metal wire, M_1, on both sides. This produces 0 V when exposed to any temperature amount. The same is true for the other new junctions also.

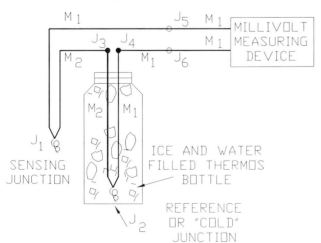

Figure 12-3 Using an ice bath for a reference temperature in a thermocouple setup.

As we will soon find out as we continue studying the thermocouple, there is a far more convenient method of obtaining a reference temperature for our single reference junction. But before we discuss this, let's investigate what materials are used in the fabrication of thermocouple wire.

12-2.1 Materials Used for Making Thermocouples

Thermocouple wires are made of many different kinds of metals. Considerable experimentation was done over the years to determine what pairs of dissimilar metals produce the best thermocouple readings. Various pairs of thermocouple wire were found very suitable and became so commonly used that eventually, the American National Standards Institute developed a set of letter symbols for referring to each of these frequently used pairs. Table 12-1 lists these symbols and the kinds of metals each symbol represents, along with the approximate temperature range that each thermocouple pair is capable of measuring. The first listing of each pair in Table 12-1 is the positive terminal when wired to a galvanometer or voltmeter; the second listing (at the right of the hyphen) is the negative terminal. Tungsten wire is also used as a thermocouple wire and when used with wire containing the element rhenium, this pair can read temperatures as high as 4200°F.

12-2.2 Reading Thermocouple Tables

The practice of reading thermocouple tables to obtain the temperature for a given thermocouple output voltage is gradually giving way to the newer solid-state digital readout systems. However, using these tables is an excellent way to better understand how the thermocouple operates.

Figure 12-4 shows an abbreviated form of a thermocouple table for copper–constantan (i.e., type T) thermocouple wire. More elaborate tables would enable

TABLE 12-1 POPULAR THERMOCOUPLE WIRE COMBINATIONS AND THEIR MAXIMUM TEMPERATURE RANGES

Letter designation	Metals used	Approximate temperature range[a] (°C)
T	Copper–constantan	−200 to 350
J	Iron–constantan	0 to 750
E	Chromel–constantan	−200 to 900
K	Chromel–alumel	−200 to 1250
R	Platinum–13% rhodium–platinum	0 to 1450
S	Platinum–10% rhodium–platinum	0 to 1450
B	Platinum–30% rhodium–platinum	800 to 1700
	Platinum–6% rhodium–platinum	800 to 1700

[a] Temperature ranges are based on a 0°C reference.

DEG C	0	1	2	3	4	5	6	7	8	9	10	DEG C

THERMOELECTRIC VOLTAGE IN ABSOLUTE MILLIVOLTS

DEG C	0	1	2	3	4	5	6	7	8	9	10	DEG C
-270	-6.258											-270
-260	-6.232	-6.236	-6.239	-6.242	-6.245	-6.248	-6.251	-6.253	-6.255	-6.256	-6.258	-260
-250	-6.181	-6.187	-6.193	-6.198	-6.204	-6.209	-6.214	-6.219	-6.224	-6.228	-6.232	-250
-240	-6.105	-6.114	-6.122	-6.130	-6.138	-6.146	-6.153	-6.160	-6.167	-6.174	-6.181	-240
-230	-6.007	-6.018	-6.028	-6.039	-6.049	-6.059	-6.068	-6.078	-6.087	-6.096	-6.105	-230
-220	-5.889	-5.901	-5.914	-5.926	-5.938	-5.950	-5.962	-5.973	-5.985	-5.996	-6.007	-220
-210	-5.753	-5.767	-5.782	-5.795	-5.809	-5.823	-5.836	-5.850	-5.863	-5.876	-5.889	-210
-200	-5.603	-5.619	-5.634	-5.650	-5.665	-5.680	-5.695	-5.710	-5.724	-5.739	-5.753	-200
-190	-5.439	-5.456	-5.473	-5.489	-5.506	-5.522	-5.539	-5.555	-5.571	-5.587	-5.603	-190
-180	-5.261	-5.279	-5.297	-5.315	-5.333	-5.351	-5.369	-5.387	-5.404	-5.421	-5.439	-180
-170	-5.069	-5.089	-5.109	-5.128	-5.147	-5.167	-5.186	-5.205	-5.223	-5.242	-5.261	-170
-160	-4.865	-4.886	-4.907	-4.928	-4.948	-4.969	-4.989	-5.010	-5.030	-5.050	-5.069	-160
-150	-4.648	-4.670	-4.693	-4.715	-4.737	-4.758	-4.780	-4.801	-4.823	-4.844	-4.865	-150
-140	-4.419	-4.442	-4.466	-4.489	-4.512	-4.535	-4.558	-4.581	-4.603	-4.626	-4.648	-140
-130	-4.177	-4.202	-4.226	-4.251	-4.275	-4.299	-4.323	-4.347	-4.371	-4.395	-4.419	-130
-120	-3.923	-3.949	-3.974	-4.000	-4.026	-4.051	-4.077	-4.102	-4.127	-4.152	-4.177	-120
-110	-3.656	-3.684	-3.711	-3.737	-3.764	-3.791	-3.818	-3.844	-3.870	-3.897	-3.923	-110
-100	-3.378	-3.407	-3.435	-3.463	-3.491	-3.519	-3.547	-3.574	-3.602	-3.629	-3.656	-100
-90	-3.089	-3.118	-3.147	-3.177	-3.206	-3.235	-3.264	-3.293	-3.321	-3.350	-3.378	-90
-80	-2.788	-2.818	-2.849	-2.879	-2.909	-2.939	-2.970	-2.999	-3.029	-3.059	-3.089	-80
-70	-2.475	-2.507	-2.539	-2.570	-2.602	-2.633	-2.664	-2.695	-2.726	-2.757	-2.788	-70
-60	-2.152	-2.185	-2.218	-2.250	-2.283	-2.315	-2.348	-2.380	-2.412	-2.444	-2.475	-60
-50	-1.819	-1.853	-1.886	-1.920	-1.953	-1.987	-2.020	-2.053	-2.087	-2.120	-2.152	-50
-40	-1.475	-1.510	-1.544	-1.579	-1.614	-1.648	-1.682	-1.717	-1.751	-1.785	-1.819	-40
-30	-1.121	-1.157	-1.192	-1.228	-1.263	-1.299	-1.334	-1.370	-1.405	-1.440	-1.475	-30
-20	-0.757	-0.794	-0.830	-0.867	-0.903	-0.940	-0.976	-1.013	-1.049	-1.085	-1.121	-20
-10	-0.383	-0.421	-0.458	-0.496	-0.534	-0.571	-0.608	-0.646	-0.683	-0.720	-0.757	-10
0	0.000	-0.039	-0.077	-0.116	-0.154	-0.193	-0.231	-0.269	-0.307	-0.345	-0.383	0

DEG C	0	1	2	3	4	5	6	7	8	9	10	DEG C
0	0.000	0.039	0.078	0.117	0.156	0.195	0.234	0.273	0.312	0.351	0.391	0
10	0.391	0.430	0.470	0.510	0.549	0.589	0.629	0.669	0.709	0.749	0.789	10
20	0.789	0.830	0.870	0.911	0.951	0.992	1.032	1.073	1.114	1.155	1.196	20
30	1.196	1.237	1.279	1.320	1.361	1.403	1.444	1.486	1.528	1.569	1.611	30
40	1.611	1.653	1.695	1.738	1.780	1.822	1.865	1.907	1.950	1.992	2.035	40
50	2.035	2.078	2.121	2.164	2.207	2.250	2.294	2.337	2.380	2.424	2.467	50
60	2.467	2.511	2.555	2.599	2.643	2.687	2.731	2.775	2.819	2.864	2.908	60
70	2.908	2.953	2.997	3.042	3.087	3.131	3.176	3.221	3.266	3.312	3.357	70
80	3.357	3.402	3.447	3.493	3.538	3.584	3.630	3.676	3.721	3.767	3.813	80
90	3.813	3.859	3.906	3.952	3.998	4.044	4.091	4.137	4.184	4.231	4.277	90
100	4.277	4.324	4.371	4.418	4.465	4.512	4.559	4.607	4.654	4.701	4.749	100
110	4.749	4.796	4.844	4.891	4.939	4.987	5.035	5.083	5.131	5.179	5.227	110
120	5.227	5.275	5.324	5.372	5.420	5.469	5.517	5.566	5.615	5.663	5.712	120
130	5.712	5.761	5.810	5.859	5.908	5.957	6.007	6.056	6.105	6.155	6.204	130
140	6.204	6.254	6.303	6.353	6.403	6.452	6.502	6.552	6.602	6.652	6.702	140
150	6.702	6.753	6.803	6.853	6.903	6.954	7.004	7.055	7.106	7.156	7.207	150
160	7.207	7.258	7.309	7.360	7.411	7.462	7.513	7.564	7.615	7.666	7.718	160
170	7.718	7.769	7.821	7.872	7.924	7.975	8.027	8.079	8.131	8.183	8.235	170
180	8.235	8.287	8.339	8.391	8.443	8.495	8.548	8.600	8.652	8.705	8.757	180
190	8.757	8.810	8.863	8.915	8.968	9.021	9.074	9.127	9.180	9.233	9.286	190
200	9.286	9.339	9.392	9.446	9.499	9.553	9.606	9.659	9.713	9.767	9.820	200
210	9.820	9.874	9.928	9.982	10.036	10.090	10.144	10.198	10.252	10.306	10.360	210
220	10.360	10.414	10.469	10.523	10.578	10.632	10.687	10.741	10.796	10.851	10.905	220
230	10.905	10.960	11.015	11.070	11.125	11.180	11.235	11.290	11.345	11.401	11.456	230
240	11.456	11.511	11.566	11.622	11.677	11.733	11.788	11.844	11.900	11.956	12.011	240
250	12.011	12.067	12.123	12.179	12.235	12.291	12.347	12.403	12.459	12.515	12.572	250
260	12.572	12.628	12.684	12.741	12.797	12.854	12.910	12.967	13.024	13.080	13.137	260
270	13.137	13.194	13.251	13.307	13.364	13.421	13.478	13.535	13.592	13.650	13.707	270
280	13.707	13.764	13.821	13.879	13.936	13.993	14.051	14.108	14.166	14.223	14.281	280
290	14.281	14.339	14.396	14.454	14.512	14.570	14.628	14.686	14.744	14.802	14.860	290
300	14.860	14.918	14.976	15.034	15.092	15.151	15.209	15.267	15.326	15.384	15.443	300
310	15.443	15.501	15.560	15.619	15.677	15.735	15.795	15.853	15.912	15.971	16.030	310
320	16.030	16.089	16.148	16.207	16.266	16.325	16.384	16.444	16.503	16.562	16.621	320
330	16.621	16.681	16.740	16.800	16.859	16.919	16.978	17.038	17.097	17.157	17.217	330
340	17.217	17.277	17.336	17.396	17.456	17.516	17.576	17.636	17.696	17.756	17.816	340
350	17.816	17.877	17.937	17.997	18.057	18.118	18.178	18.238	18.299	18.359	18.420	350
360	18.420	18.480	18.541	18.602	18.662	18.723	18.784	18.845	18.905	18.966	19.027	360
370	19.027	19.088	19.149	19.210	19.271	19.332	19.393	19.455	19.516	19.577	19.638	370
380	19.638	19.699	19.761	19.822	19.883	19.945	20.006	20.068	20.129	20.191	20.252	380
390	20.252	20.314	20.376	20.437	20.499	20.560	20.622	20.684	20.746	20.807	20.869	390
400	20.869											400

DEG C	0	1	2	3	4	5	6	7	8	9	10	DEG C

Figure 12-4　Abbreviated thermocouple table for copper–constantan. (REPRODUCED WITH THE PERMISSION OF OMEGA ENGINEERING, INC.)

the user to read temperatures to the nearest tenth of a degree. As do the majority of tables, ice water at 0°C is used for the reference temperature. To utilize the table, the millivolt reading obtained with the potentiometer which is generated for the particular temperature you are reading is read, and then the corresponding temperature is looked up in the table. The only thing to watch out for in using these tables is that you make certain that you are using the correct table for the particular thermocouple wire. An iron–constantan table will give significantly different and certain erroneous results. As a matter of fact, Figure 12-5 shows the relationship between temperature and the millivolt readings of the thermocouple wire pairs listed in Table 12-1.

12-2.3 Law of Intermediate Temperatures

The reference temperature for a thermocouple can be at any temperature other than 0°C and still produce temperature readings that have been obtained from 0°C thermocouple tables. This is due to what is called the *law of intermediate temperatures*. To understand how this law works, first consider the thermocouple circuit pictured in Figure 12-3. If we have an iron–constantan thermocouple with its sensing junction at, say, 100°C and its reference thermocouple at 0°C, according to a set of iron–constantan thermocouple tables, these temperatures produce a millivolt reading of 5.27 mV at their output measuring terminals (between J_5 and J_6 in Figure 12-3). Now, consider adding a third thermocouple junction, an intermediate junction that is at some intermediate temperature, say 18.3°C (Figure 12-6). The resultant millivolt readings are as shown. Notice that the output reading remains at the same 5.27 mV

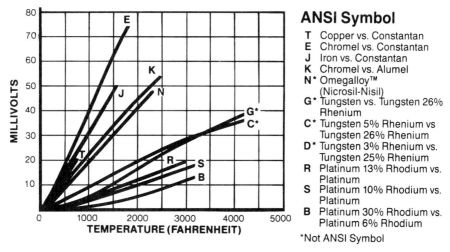

Figure 12-5 Millivolt output versus temperature graphs for several thermocouple types. (REPRODUCED WITH THE PERMISSION OF OMEGA ENGINEERING, INC.)

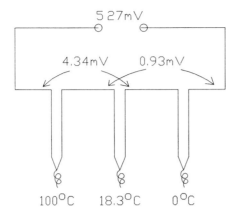

Figure 12-6 Showing the law of intermediate temperatures by adding a third junction.

reading as before. However, because the tables say that for a thermocouple at 18.3°C, its output is 0.93 mV (it is necessary to interpolate this value from the tables given in this book), the difference between 5.27 and 0.93 mV has to be 4.34 mV. This voltage must then occur across the 100°C and 18.3°C thermocouples. In other words, the output voltage for all three thermocouple junctions is the resultant voltage of the two individual voltages generated by all three thermocouples added together as shown in Figure 12-6. Another way to look at this is to remove the 0°C reference thermocouple leaving the circuit as shown in Figure 12-7. Doing this reduces the voltage to 4.34 mV, which is what appears across the output of the remaining two thermocouples. By adding the millivolt equivalent for 18.3°C (which is 0.93 mV as read in the iron–constantan tables) to the 4.34 mV reading, we get 5.27 mV, which is the reading we obtained earlier with all three thermocouples.

Let's look at an example now to see how we can take advantage of this intermediate temperature characteristic.

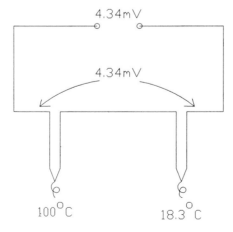

Figure 12-7 Result of using the law of intermediate temperatures.

Example 12-1

An iron–constantan thermocouple is used to read the temperature of a steam pipe at a thermonuclear power plant. The ambient temperature of 30°C is used as a reference junction temperature to obtain a total millivolt reading of 10.388 mV for the steam pipe. What is the pipe's temperature?

Solution: First, look up the millivolt reading in the iron–constantan tables for 30°C. We obtain a value of 1.536 mV. Second, we then add this value to the 10.388 mV reading obtained with the hot junction at the steam pipe to get a total value of 11.874 mV. Finally, we look up the temperature in our table for this total value to obtain a final temperature of approximately 219.8°C. This, then, is the actual temperature of the steam pipe.

12-3 COMMONLY SENSED MEASURANDS

Unlike many of the other sensors that we have discussed thus far, the thermocouple has only one measurand that it senses. This is its primary measurand of temperature; there are no secondary measurands that are sensed with this device.

12-3.1 Typical Signal Conditioning Methods

In reality, it is not too often that a thermocouple's millivolt output is utilized or read directly with a voltmeter. Instead, the output is often sent to a Wheatstone bridge–like circuit called a potentiometer, not to be confused with the variable-resistance potentiometer. The potentiometer is a device capable of reading very small voltages and being able to do it with extreme accuracy in the process.

Figure 12-8 shows a simplified schematic of one of these devices. A small standard voltage cell, which is a battery cell specifically designed to furnish a constant reference voltage over very long periods of time, is built into this device. The standard cell supplies its constant voltage across the slider resistor. This voltage is usually generated by a wet cell that requires little maintenance over periods of several years. The two slider contacts on the resistor are varied until the standard cell's voltage is nullified by the unknown voltage, as indicated by a current null reading at the galvanometer. (A galvanometer is a very sensitive current-indicating

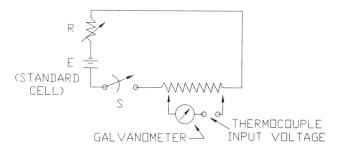

Figure 12-8   Potentiometer circuit for making millivolt readings from thermo-couples.

meter device found in many laboratory-grade precision instruments.) As in the case of a Wheatstone bridge, when a null condition is reached, there is no current flowing through the meter because of equal and opposite current flows taking place through the center leg of the circuit. In this case, the center leg is the slider resistor. Switch S is necessary to allow the user to check temporarily for null conditions to prevent sustained "pegging" of the galvanometer's needle during the nulling procedure.

The circuit for a commercial multirange potentiometer is shown in Figure 12-9. This system allows for measuring a very wide range of voltages and is capable of measuring thermocouple voltages out to several significant digits to the right of the decimal. As in the case with using any potentiometer, once the voltage is read, it becomes necessary to refer to a set of thermocouple tables published for the particular set of thermocouple wires being used. The temperature is then read from these tables. See Appendix A at the back of the book for typical thermocouple tables used with these instruments.

The potentiometer is rapidly being replaced by very sensitive and easily operated solid-state detection circuits that require no nulling. In addition, the thermocouple's reference ice-bath cell has been replaced by solid-state circuitry that simulates the constant-voltage output of the reference cell. Figure 12-10 shows the specifica-

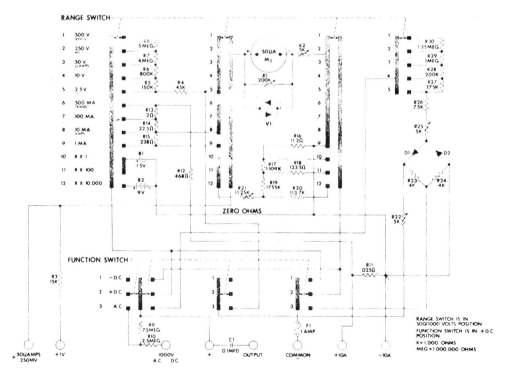

Figure 12-9   Circuit for a commercial multirange potentiometer.

SPECIFICATIONS (@ +25°C and V_S = 5V, Type J (AD594), Type K (AD595) Thermocouple, unless otherwise noted)

Model	AD594A Min	Typ	Max	AD594C Min	Typ	Max	AD595A Min	Typ	Max	AD595C Min	Typ	Max	Units
ABSOLUTE MAXIMUM RATINGS													
$+V_S$ to $-V_S$			36			36			36			36	Volts
Common-Mode Input Voltage	$-V_S - 0.15$		$+V_S$	$-V_S - 0.15$		$+V_S$	$-V_S - 0.15$		$+V_S$	$-V_S - 0.15$		$+V_S$	Volts
Differential Input Voltage	$-V_S$		$+V_S$	$-V_S$		$+V_S$	$-V_S$		$+V_S$	$-V_S$		$+V_S$	Volts
Alarm Voltages													
$+ALM$	$-V_S$		$-V_S + 36$	$-V_S$		$-V_S + 36$	$-V_S$		$-V_S + 36$	$-V_S$		$-V_S + 36$	Volts
$-ALM$	$-V_S$		$+V_S$	$-V_S$		$+V_S$	$-V_S$		$+V_S$	$-V_S$		$+V_S$	Volts
Operating Temperature Range	-55		$+125$	-55		$+125$	-55		$+125$	-55		$+125$	°C
Output Short Circuit to Common	Indefinite			Indefinite			Indefinite			Indefinite			
TEMPERATURE MEASUREMENT													
(Specified Temperature Range 0 to +50°C)													
Calibration Error at +25°C[1]		±3			±1			±3			±1		°C
Stability vs. Temperature[2]		±0.05			±0.025			±0.05			±0.025		°C/°C
Gain Error		±1.5			±0.75			±1.5			±0.75		%
Nominal Transfer Function		10			10			10			10		mV/°C
AMPLIFIER CHARACTERISTICS													
Closed Loop Gain[3]		193.4			193.4			247.3			247.3		
Input Offset Voltage		(Temperature in °C) × 51.70μV/°C			(Temperature in °C) × 51.70μV/°C			(Temperature in °C) × 40.44μV/°C			(Temperature in °C) × 40.44μV/°C		μV
Input Bias Current		0.1			0.1			0.1			0.1		μA
Differential Input Range	-10		$+50$	-10		$+50$	-10		$+50$	-10		$+50$	mV
Common-Mode Range	$-V_S - 0.15$		$+V_S - 4$	$-V_S - 0.15$		$+V_S - 4$	$-V_S - 0.15$		$+V_S - 4$	$-V_S - 0.15$		$+V_S - 4$	V
Common-Mode Sensitivity–RTO		10			10			10			10		mV/V
Power Supply Sensitivity–RTO		10			10			10			10		mV/V
Output Voltage Range													
Dual Supplies	$-V_S + 2.5$		$+V_S - 2$	$-V_S + 2.5$		$+V_S - 2$	$-V_S + 2.5$		$+V_S - 2$	$-V_S + 2.5$		$+V_S - 2$	Volts
Single Supply	0		$+V_S - 2$	0		$+V_S - 2$	0		$+V_S + 2$	0		$+V_S - 2$	Volts
Usable Output Current[4]		±5			±5			±5			±5		mA
3dB Bandwidth		15			15			15			15		kHz
ALARM CHARACTERISTICS													
$V_{CE(SAT)}$ at 2mA		0.3			0.3			0.3			0.3		Volts
Leakage Current		±1			±1			±1			±1		μA max
Operating Voltage at $-ALM$			$+V_S - 4$			$+V_S - 4$			$+V_S - 4$			$+V_S - 4$	Volts
Short Circuit Current		20			20			20			20		mA
POWER REQUIREMENTS													
Specified Performance	$+V_S = 5, -V_S = 0$			$+V_S = 5, -V_S = 0$			$+V_S = 5, -V_S = 0$			$+V_S = 5, -V_S = 0$			Volts
Operating[5]	$+V_S$ to $-V_S \le 30$			$+V_S$ to $-V_S \le 30$			$+V_S$ to $-V_S \le 30$			$+V_S$ to $-V_S \le 30$			Volts
Quiescent Current (No Load)													
$+V_S$		160	300		160	300		160	300		160	300	μA
$-V_S$		100			100			100			100		μA
PACKAGE OPTIONS													
TO-116 (D14A)	AD594AD			AD594CD			AD595AD			AD595CD			
CERDIP (Q14A)	AD594AQ			AD594CQ			AD595AQ			AD595CQ			

NOTES
[1]Calibrated for minimum error at +25°C using a thermocouple sensitivity of 51.7μV/°C. Since a J type thermocouple deviates from this straight line approximation, the AD594 will normally read 3.1mV when the measuring junction is at 0°C. The AD595 will similarly read 2.7mV at 0°C.
[2]Defined as the slope of the line connecting the AD594/AD595 errors measured at 0°C and 50°C ambient temperature.
[3]Pin 8 shorted to pin 9.

[4]Current Sink Capability in single supply configuration is limited to current drawn to ground through a 50kΩ resistor at output voltages below 2.5V.
[5]$-V_S$ must not exceed -16.5V.

Specifications subject to change without notice.

Specifications shown in boldface are tested on all production units at final electrical test. Results from those tests are used to calculate outgoing quality levels. All min and max specifications are guaranteed, although only those shown in boldface are tested on all production units.

OUTLINE DIMENSIONS
Dimensions shown in inches and (mm).

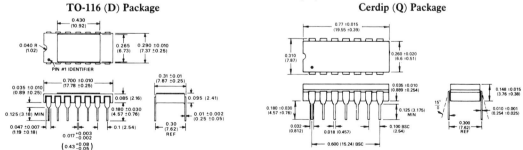

Figure 12-10 Specification sheet for a solid-state device used for thermocouple applications. (Courtesy of Analog Devices, Norwood, MA.)

tions for a solid-state device that is used for thermocouple temperature-sensing applications. A typical solid-state system has a digital temperature readout similar to the one shown in Figure 12-11. The user merely sets a dial or switch indicating the type of thermocouple that he or she intends to use and then attaches the thermocouple to a set of terminals. The temperature sensed by the sensing junction is then displayed automatically on the unit's front panel.

12-3.2 Radio-Frequency Interference

A fairly common problem with thermocouple wire has nothing to do with the thermocouple principle itself. Instead, it has to do with the fact that in many cases, long runs of thermocouple wire are necessary to transmit the very low amplitude voltages over large distances to the instrumentation site. Runs of many hundreds of feet are not at all unusual. Because of these lead lengths, the wires can behave like a radio antenna and as a result have induced into them significant amounts of RF energy from nearby radio broadcast stations. Because the Seebeck dc voltages within the thermocouple wire are so low, it is very possible to have these signals become masked with much larger amplitude RF radio signals. Consequently, it is always advisable to enclose long thermocouple runs in grounded conduit or other shielded containers to shield the dc signals from these interfering ac signals. It is also possible to purchase thermocouple wire containing a metallic sheath resembling a form of coax that is especially suited for these applications.

12-4 TYPICAL DESIGN SPECIFICATIONS FOR THERMOCOUPLE WIRE

Typical design specifications for thermocouple wire are shown in Figure 12-12.

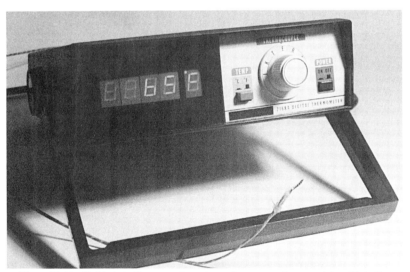

Figure 12-11   Solid-state thermometer using a thermocouple, complete with built-in reference junction voltage.

Insulation	AWG No.	Cat. No.	Price 1000'	Type Wire	Insulation Conductor	Insulation Overall	Max. Temp °F	Max. Temp °C	Nominal Size	Wt.† lb/1000'
Ceramic**	20	XC-K-20	$1275	Solid	Nextel Ceramic Refrasil	Nextel Ceramic Refrasil	2200	1250	.115 × .180	14
	20	XR-K-20	1260	Solid	Refrasil	Refrasil	1600	870	.115 × .180	14
High Temp. Glass**	20	HH-K-20	492	Solid	High Temp. Glass	High Temp. Glass	1300	704	.060 × .105	9
	24	HH-K-24	348	Solid	High Temp. Glass	High Temp. Glass	1300	704	.055 × .090	5
Glass	20	GG-K-20	410	Solid	Glass Braid	Glass Braid	900	482	.060 × .095	9
	20	GG-K-20S	600	7 × 28	Glass Braid	Glass Braid	900	482	.060 × .100	9
	24	GG-K-24	290	Solid	Glass Braid	Glass Braid	900	482	.050 × .080	5
	24	GG-K-24S	395	7 × 32	Glass Braid	Glass Braid	900	482	.050 × .085	5
	26	GG-K-26	270	Solid	Glass Braid	Glass Braid	900	482	.045 × .075	4
	28	GG-K-28	230	Solid	Glass Braid	Glass Braid	900	482	.045 × .070	3
	30	GG-K-30	240	Solid	Glass Wrap	Glass Braid	900	482	.045 × .070	3
	36	GG-K-36	320	Solid	Glass Wrap	Glass Braid	900	482	.045 × .070	2
Teflon® / Neoflon® Glass	30	TG-K-30	530	Solid	PFA	Glass Braid	500	260	.034 × .047	2
	36	TG-K-36	590	Solid	PFA	Glass Braid	500	260	.028 × .038	2
	40	TG-K-40	650	Solid	PFA	Glass Braid	500	260	.026 × .035	2
Teflon® Neoflon® PFA (High Performance)	20	TT-K-20	495	Solid	PFA	PFA	500	260	.068 × .116	11
	20	TT-K-20S	755	7 × 28	PFA	PFA	500	260	.073 × .126	11
	22	TT-K-22S	735	7 × 30	PFA	PFA	500	260	.065 × .133	9
	24	TT-K-24	370	Solid	PFA	PFA	500	260	.056 × .092	6
	24	TT-K-24S	515	7 × 32	PFA	PFA	500	260	.063 × .102	6
	30	TT-K-30*	335	Solid	PFA	PFA	500	260	.022 × .038	2
	36	TT-K-36*	375	Solid	PFA	PFA	500	260	.017 × .028	2
	40	TT-K-40*	500	Solid	PFA	PFA	500	260	.015 × .024	2
Teflon® Neoflon® FEP	20	FF-K-20	490	Solid	FEP	FEP	392	200	.068 × .116	11
	24	FF-K-24	350	Solid	FEP	FEP	392	200	.056 × .092	6
	30	FF-K-30	335	Solid	FEP	FEP	392	200	.022 × .038	2
	36	FF-K-36	375	Solid	FEP	FEP	392	200	.017 × .028	2
	40	FF-K-40	500	Solid	FEP	FEP	392	200	.015 × .024	2
Polymers	24	PR-K-24	230	Solid	Polyvinyl	(Rip Cord)	221	105	.046 × .086	5
	20	KK-K-20	850	Solid	Kapton	Kapton	600	315	.057 × .103	11
	24	KK-K-24	580	Solid	Kapton	Kapton	600	315	.045 × .079	6

Weight of spool and wire rounded up to the next highest lb.
Does not include packing material
* * Has color tracer

Some Wires available with special limits of error—Consult Sales Department for pricing and availability.
* Overall color: clear

Figure 12-12 Specification sheet for thermocouple wire. (REPRODUCED WITH THE PERMISSION OF OMEGA ENGINEERING, INC.)

315

12-5 PRACTICAL APPLICATIONS

Thermocouples can be used to obtain precise temperatures at specific locations. As an example, air temperature distribution across the discharge of a ventilation duct for a heating system can easily be obtained through application of a thermocouple grid similar to the one shown in Figure 12-13. This approach is used to determine the uniformity of distribution of the heated air and to determine if there are any dangerous hot spots. At each wire intersection a thermocouple is attached to monitor that particular location for a given test run. The thermocouple outputs are wired to a scanning and hard-copy recording device to record the results permanently. With these results it is then possible to reconstruct a temperature profile of the air discharge for that duct and to determine its temperature distribution characteristics.

During the test flights of the NASA space shuttle, the forward portion or bow of the shuttle's fuselage was heavily instrumented with thermocouples to monitor the heat of reentry (Figure 12-14). Specifically, the concern had to do with the heat-

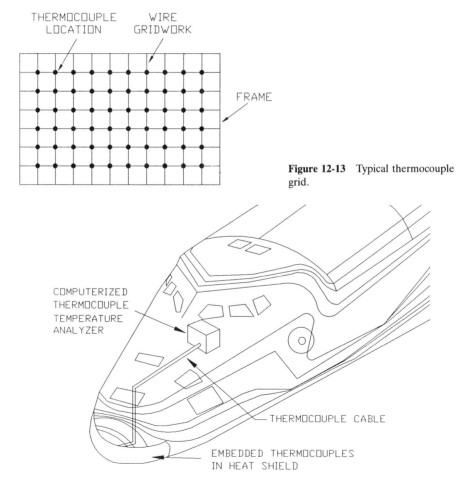

Figure 12-13   Typical thermocouple grid.

Figure 12-14   Thermocoupled front end of space shuttle.

dissipating qualities of the insulating tiles covering the nose of the fuselage. Again, each thermocouple that was installed was wired to a recording device onboard the spacecraft to create a recorded document for later data analysis.

Often, thermocouples are arranged in groups to create either higher voltages or greater current flows, depending on the application. In either case these thermocouple groupings are called *thermopiles*. When a thermopile consists of series-connected thermocouples the sensitivity of the thermocouple increases. In other words, for a given change in temperature a much larger voltage is generated than for a single thermocouple. This is the principle used in the case of a special heat-measuring device called an *infrared pyrometer* (Figure 12-15). An infrared pyrometer is a device that senses heat radiation in the infrared portion of the electromagnetic spectrum and is able, through proper calibration procedures, to indicate temperatures accurately within the range −50 to 3500°C. A schematic of a typical pyrometer containing a thermopile is illustrated in Figure 12-16.

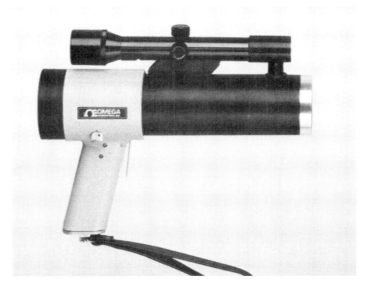

Figure 12-15   Infrared pyrometer. (REPRODUCED WITH THE PERMISSION OF OMEGA ENGINEERING, INC.)

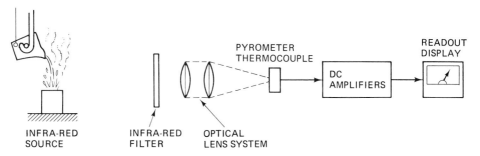

Figure 12-16   Block diagram of pyrometer using thermopile circuit.

12-6 TYPICAL DESIGN SPECIFICATIONS FOR THE INFRARED PYROMETER

Typical design specifications for an infrared pyrometer are shown in Figure 12-17.

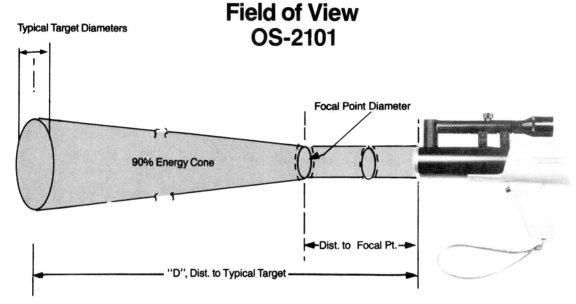

Field of View
OS-2101

Typical Target Diameters

90% Energy Cone

Focal Point Diameter

Dist. to Focal Pt.

"D", Dist. to Typical Target

Target Sizes

Model No.	6″	12″	18″	24″	30″	36″	42″	48″	60″	72″	84″	90″	8′	9′	10′	15′	20′	25′	50′	100′	>100″
OS-2101	1.4	1.2	0.9	0.7	0.5*	0.9	1.3	1.8	2.6	3.4	4.3	4.7	5.1	6.0	6.8	11.0	15.2	19.4	40.4	82.4	D/15
OS-2102S	1.4	1.3	1.2	1.1	1.1	1.0	0.9	0.8	0.6*	1.0	1.4	1.6	1.8	2.2	2.6	4.6	6.6	8.7	18.7	38.9	D/30
OS-2103S	1.5	1.5	1.5	1.4	1.4	1.3	1.3	1.3	1.2	1.1	1.0*	1.0*	1.0*	1.2	1.3	2.3	3.5	4.6	10.7	22.7	D/50
OS-2104S	1.1	0.7	0.5	0.25*	0.7	1.0	1.4	1.9	2.6	3.4	4.2	4.6	5.0	5.8	6.7	10.7	14.8	18.8	D/15		
OS-2105S	1.4	1.2	1.1	0.9	0.8	0.7	0.6	0.5*	0.8	1.2	1.6	1.8	2.0	2.4	2.8	4.9	7.0	9.0	D/30		
OS-2106S	1.6*	1.6*	1.7	1.8	1.8	1.9	2.0	2.0	2.2	2.3	2.4	2.5	2.6	2.7	2.8	3.5	4.2	4.8	8.2	14.8	D/90

*Minimum spot size

Specifications

Temperature Range: −20°F to 2500°F (−30°C to 1400°C)

Accuracy*: (NBS Traceable) Above 100°F (38°C) ± 1% of rdg. ±1° Below 100°F (38°C) ±1.5% of rdg. ±1°

Emissivity: 0.10 to 1.0 in 0.01 steps

Filter: 8 to 14 microns

Operating Temperature: 32°F to 120°F (0°C to 50°C)

Storage Temperature: −40°F to 150°F (−40°C to 65°C)

Battery: 9 V alkaline, supplied with each instrument

Battery Life: 8 hours, typically to 5000 uses. 3 months if simply maintaining memory and display

* Under standardized conditions: 1) instrument temperature well stabilized and within operating ambient range, 2) target emissivity is 1.0, 3) target fills 99% energy cone, 4) target distance is 24″, minimum.

C1-14

Dimensions

Length: 10.0″ (254 mm)
Height: 6.5″ (165 mm)
Width: 3.0″ (76 mm)
Weight: 2.0 lbs. (0.9 kg)
With Scope: 2.9 lbs. (1.3 kg)

Model OS-2101

Figure 12-17 Specification sheet for infrared pyrometer. (REPRODUCED WITH THE PERMISSION OF OMEGA ENGINEERING, INC.)

REVIEW QUESTIONS

12-1. Describe in detail the Peltier, Seebeck, and Thomson effects.

12-2. Explain how the law of intermediate temperature works. Why is it necessary, in general, to have a reference temperature for thermocouple work?

12-3. What is a thermopile? Where are they used?

12-4. List two advantages and two disadvantages in using thermocouples for temperature measurements.

12-5. What would you think the dc resistance would be for a thermocouple? Do you think the resistance would be in the hundreds of thousands of ohms, thousands of ohms, or just a few ohms?

12-6. Explain the function of a standard cell in a potentiometer.

12-7. What would you think the danger would be in using a welding process for obtaining a well-bonded junction for a thermocouple?

PROBLEMS

12-8. A copper–constantan thermocouple is used to read the temperature of a water bath. The ambient temperature in the room where the bath is located is 57°F. If a millivolt reading was taken of the thermocouple's output and was found to be 1.401 mV, find the water bath's actual temperature.

12-9. Obtain an extensive thermocouple table for, say, iron–constantan. Plot the temperature (*x*-axis) values versus the millivolt data readings (*y*-axis) on a large piece of Cartesian coordinate graph paper. What would you conclude from the resulting curve concerning its linearity or nonlinearity over an extensive temperature range? Over short temperature ranges?

REFERENCES

BLATT, FRANK J., *Principles of Physics*. Boston: Allyn and Bacon, 1986.

FRIBANCE, AUSTIN E., *Industrial Instrumentation Fundamentals*. New York: McGraw-Hill, 1962.

NORTON, HARRY N., *Sensor and Analyzer Handbook*. Englewood Cliffs, NJ: Prentice Hall, 1982.

O'HIGGINS, PATRICK J., *Basic Instrumentation*. New York: McGraw-Hill, 1966.

The Temperature Handbook, Vol. 27. Stamford, CT: Omega Engineering, 1990.

13

THERMORESISTIVE SENSORS

CHAPTER OBJECTIVES

1. To understand how the RTD works and to study the concept of self-heat.
2. To study the behavior of the thermistor.
3. To understand how the diode–junction sensor functions.

13-1 INTRODUCTION

A *thermoresistive sensor* is a device whose internal electrical dc resistance is a function of temperature. There are a number of different types of such devices and we attempt to discuss them all in this chapter. Unlike the thermocouple, however, the thermoresistive sensor does not generate its own voltage source; an external voltage source is necessary to "bring out" this change in resistance within the sensor. We begin our discussion, as usual, describing the measurands that are generally detected. We then describe the various resistivity detectors used today beginning with the earliest known thermoresistive device, the resistance temperature detector.

13-2 COMMONLY SENSED MEASURANDS

The primary measurand sensed by thermoresistive sensors is temperature. Only one secondary measurand has been detected with any practicality—fluid velocity.

13-3 RESISTANCE TEMPERATURE DETECTOR

The *resistance temperature detector* (RTD) operates on a very simple principle but requires the most sophisticated instrumentation for its proper operation. Although the term *resistance temperature detector* can actually refer to a relatively broad area of sensors whose dc resistances change with temperature, the term has recently become more closely associated with only a portion of all the resistive sensors. This portion comprises those sensors made from a pure wire-wound metal having a positive temperature coefficient. To understand what is meant by this term, however, we must next study a frequently observed relationship that exists between temperature and the electrical resistance for certain metals.

13-3.1 Temperature Coefficients of Metals

Virtually all the "earth-mined" metals are conductors of electricity, some being better than others. To state this in another way, virtually all metals display a certain amount of dc resistance to current flow. Some metals have more or less resistance than other metals, depending on their molecular structure. This resistance varies with the metal's temperature, and in almost all cases where we are dealing with metals in their most pure elementary form, this resistance varies directly with temperature. That is, as the temperature increases, the resistance also increases at a fairly linear, proportional rate, and vice versa. Hence the term *positive coefficient* has been affixed to these metals to describe their resistance behavior.

To understand why a metal's resistance increases with an increase in its temperature, we must first understand the nature of the metal's molecular structure and how its behavior relates to temperature. Temperature and energy are interrelated quantities. The thermal energy of a molecule, E_{th}, is related to its absolute temperature T through the equation

$$E_{th} = (1.5)(k)(T) \qquad (13\text{-}1)$$

where E = thermal energy (J)

$\quad k$ = Boltzmann's constant, 1.38×10^{-23} J/K

$\quad T$ = absolute temperature (K)

At absolute zero the molecules of a metal sample have no kinetic energy. The molecules are "stationary" in that they are not vibrating from outside heat energy sources contrary to those metals whose temperatures are above absolute zero and whose molecules do vibrate as a result. Because of this lack of vibrating electrons at the absolute zero condition, a current flow of electrons can pass relatively easily between these molecules without encountering any appreciable resistance (friction). However, with the addition of heat (energy) the metal's molecules vibrate about their locations. This, then, increases the chances of electron collisions from a current flow trying to move between the metallic molecules. Any collisions will produce friction, that is, resistance. The amount of resistance will vary proportionally with the amount of energy (temperature) input.

With some metals the linear rate of resistance change to temperature change is quite predictable. These few metals are generally very stable, chemically. That is, they react very little, if any, with other elements over time. Chemical activity is a problem with using metals for the purpose of temperature sensing. This activity causes the metals' dc resistance to change over a period of time for a given temperature. Figure 13-1 shows the linearity of various metals for a wide temperature range versus their dc resistance. The data acquired for producing this curve were obtained from metal samples having the same length and cross-sectional areas.

As stated earlier, for any given metal, its dc resistance varies directly with, and is proportional to, temperature. However, let's first investigate the relationship between resistance and a metal's cross-sectional area and length. We assume that the temperature is held at a constant room-temperature value. That relationship is

$$R = \frac{\rho l}{A_{CM}} \qquad (13\text{-}2)$$

where R = resistance of wire (Ω)

ρ = resistivity constant (Ω-mil/ft)

l = length (ft)

A_{CM} = cross-sectional area of wire (mil)

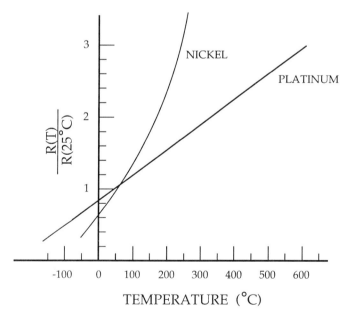

Figure 13-1 Graph of dc resistance for various metals versus temperature.

[Refer to Section 7-2 for a related discussion concerning the utilization of eq. (13-2). Compare this equation with eq. (7-3).] The value for ρ is good only for a specific temperature. The following two examples demonstrate this.

Example 13-1

Find the resistance or a coil of copper wire that is 750 ft in length and having a diameter of 25 mils. Assume a temperature of 20°C and a ρ value of 10.37 Ω-CM/ft for copper.

Solution: Using eq. (13-2) with the values given, we get

$$R = \frac{(10.37)(750)}{25^2}$$

$$= 12.44 \ \Omega$$

Keep in mind that this value is correct only for a 20°C temperature, however. To find the resistance of the coil for any other temperature, you must use the equation

$$R = \frac{\rho l (1 + \alpha T)}{A_{CM}} \tag{13-3}$$

where α = temperature coefficient between the operating temperature and 20°C (1/°C)

T = difference between the operating temperature and 20°C

Example 13-2

Continuing with Example 13-1, calculate the resistance of the same copper wire coil, but do this now for a temperature of 35°C. Assume an α value of 0.00392/°C.

Solution: Using eq. (13-3) with the values given above, we now get

$$R = \frac{(10.37)(750)[1 + (0.00392)(15)]}{25^2}$$

$$= (12.44)(1.0588)$$

$$= 13.17 \ \Omega$$

As can be seen from the two calculations above, the difference between the two calculated resistances is 0.73 Ω. This difference is the result of the 15° increase in temperature in Example 13-2.

13-3.2 Metals Used for Temperature Sensing

The metals most frequently used for temperature sensing, that is, metals that are used for their relatively linear resistance/temperature response characteristics, are nickel, platinum, and certain alloyed forms of copper. Of these three metals,

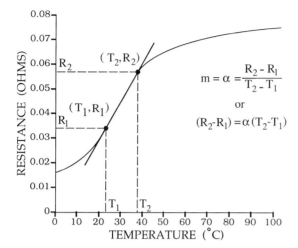

Figure 13-2 Temperature versus dc resistance for platinum.

platinum is probably used most frequently. Figure 13-2 shows platinum to have a very linear temperature versus resistance response over a narrow range of temperatures. Also, platinum is extremely stable in that it reacts with very few substances. This makes it a very reliable and predictable material to work with as a temperature sensor. However, it should be mentioned here that different platinum sensors may exhibit different resistance values for the same temperature. This is due to differing manufacturing and processing techniques used, and it also has to do with strain within the metal itself. Because strain also varies the resistance of metals, it is very important to be aware of this fact during the manufacturing and installation of the platinum sensor. This is true for any metal that is used for temperature detecting.

Because of the possible variations of resistance versus temperature for the same materials, attempts have been made to standardize the resistance/temperature ratios. Unfortunately, neither U.S. or foreign manufacturers have been able to agree on a ratio standard that satisfies everyone. For instance, one American standardizing organization stipulates that platinum sensors must have a resistance ratio, R_{100}/R_0, of 1.3924, where R_{100} is the resistance at 100°C and R_0 is the resistance at 0°C. On the other hand, a British standardizing organization requires a value for R_{100}/R_0 of 1.3850. These ratio figures are nothing more than an indirect means of referring to the slope of the sensor's resistance–temperature curve over a temperature range. While differing resistance ratios do not necessarily affect the performance of this kind of sensor, it does create somewhat of a problem. If you wanted to make a direct replacement, or if you wanted to swap a sensor between instruments, you must make certain that you are using sensors having the same R_{100}/R_0 ratios.

To summarize, to keep the temperature-indicating errors to a minimum, it is necessary to pick a sensor for a specific temperature range. Doing this will maximize the utilization of the sensor's resistance versus temperature linear response characteristics.

13-3.3 Typical Construction

RTDs come in two configurations. One is a wire-wound construction. The coiled construction allows for greater resistance variations for a given temperature change compared to single-strand detectors used many years ago. This type of construction increases the sensing device's sensitivity and resolution. The coil is formed around a nonthermally conductive material such as a ceramic to reduce temperature response time. A sheath is used to surround the coil to protect it from abuse and environmental reactions. The sheath is constructed from a highly conductive material so as not to create a thermal barrier between the sensing coil and the outside temperature. Usually, a stainless steel covering is used for this application. This sheath is hermetically sealed to prevent outside moisture and other elements from affecting the wire sensor inside. In addition, the sheath is filled with a dry thermal-conducting gas to increase thermal contact with the sheath; all of this is a further effort to reduce the sensor's response time.

The RTD's second configuration is in the form of a woven thin-film metallic layer that has been deposited on a nonthermally conductive substrate, such as a ceramic. This process allows for considerable miniaturization and also allows for doing away with an external sheathing material; the ceramic itself acts as the protective sheath. Figure 13-3 shows typical package arrangements used for both construction types.

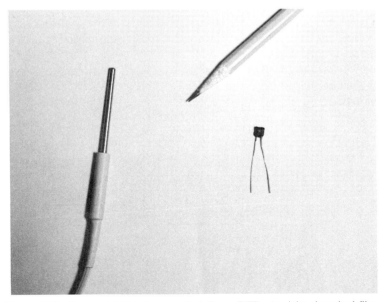

Figure 13-3   Shown at left, wire-wound platinum RTD. At right, deposited-film RTD.

13-3.4 RTD Signal-Conditioning Circuits

Because of the very small change in resistance for a given change in temperature, the Wheatstone bridge, or some variation of this circuit, is used for making these measurements.

Two-wire uncompensated RTD circuit

Figure 13-4 shows a basic two-wire RTD system for temperature measurement. The resistors, R_{L1} and R_{L2}, are the equivalent lead resistances for the two wire leads going to the RTD, R_3. When the bridge is balanced, the following condition is achieved:

$$\frac{R_1}{R_2} = \frac{R_{L1} + R_{L2} + R_3}{R_4} \tag{13-4}$$

where the R notations are as given in Figure 13-5. The circuit also possesses maximum sensitivity to a change in temperature when $R_1 = R_2 = (R_{L1} + R_{L2} + R_3) = R_4$. Notice, however, that the lead resistances play a significant role in determining the overall resistance of the RTD for a given temperature. This is true for lead lengths in excess of several feet or so, and it is certainly true for lengths of several hundreds of feet.

Two-wire compensated RTD circuit

One way to partially compensate for the additional lead resistances added by the wire leads going to the sensor is to add a similar resistance, R_C, to the leg containing R_4 of the Wheatstone bridge, as shown in Figure 13-5(a). Further temperature compensating can be obtained by placing this added compensating resistor with its own set of leads near the sensing leads of the RTD, as shown in Figure 13-5(b). In this way, any change in the RTD lead resistance due to temperature is partially nullified by a similar change in resistance in R_C. This type of circuit is called a *two-wire lead circuit*.

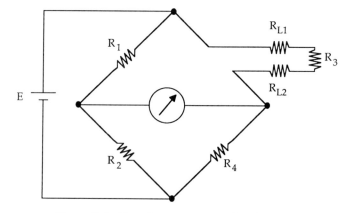

Figure 13-4 Two-wire uncompensated RTD circuit.

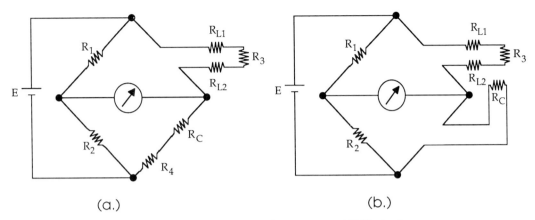

(a.)   (b.)

Figure 13-5   Two-wire compensated RTD circuits.

Three-wire RTD circuit

The *three-wire lead circuit* is one that should be used in those instances where the lead resistance is significant compared to the measured resistance of the RTD. Figure 13-6 demonstrates this particular hookup. Notice that lead resistance R_{L1} is in one leg of the bridge, while R_{L2} is in the other leg beneath the first. As a result of this configuration, the two resistances cancel each other out. Also, during the balanced condition of the bridge, there is no current flowing in R_{L3} since it is located in the center leg.

Four-wire RTD circuit

The *four-wire lead* RTD circuit is used for extremely sensitive temperature detection where precision laboratory-type measurements are necessary. The circuitry for this particular configuration is quite complicated and is not discussed further here only because of its extremely limited application.

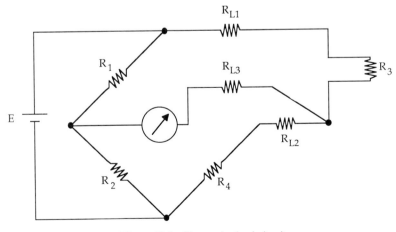

Figure 13-6   Three-wire lead circuit.

13-3.5 Self-Heat

One of the difficulties in using an RTD is to ascertain just how much current should be allowed to travel through it. After all, what is really being monitored when using an RTD to measure temperature is the voltage drop created by a current traveling through the RTD's internal resistance. The amount of current of course is a function of the voltage supply amount applied across the leads of the RTD. Obviously, there is a practical limit to the amount of voltage that can be supplied without damaging the RTD itself. Too-high voltage amounts would cause the RTD to generate excessive or possible damaging heat due to the power being dissipated by its internal resistance $(P = I^2R)$. As a matter of fact, for any voltage amount the resulting current flow will cause heat to be dissipated by the RTD. The RTD will not be able to differentiate between this *self-heat* and the heat energy to which it is supposed to respond. The question then becomes: What sort of temperature rise results from this self-heat characteristic?

Fortunately, manufacturers have anticipated this problem by publishing a self-heating error value in their catalog data literature for each RTD that they manufacture. A typical self-heat error figure may be 0.18°C/mW in still air, or 0.07°C/mW in moving air. In other words, for this particular RTD measuring a temperature of a heat source in still air, for every milliwatt of power dissipated by the RTD's internal resistance, there will be a 0.18°C increase in its temperature. This amount would have to be subtracted for each milliwatt of power consumed from the RTD's overall indicated temperature. Therefore,

$$T_{SH} = P_{SHE} \times P_{RTD} \qquad (13\text{-}5)$$

where T_{SH} = temperature rise due to self-heat (°C)

P_{SHE} = self-heat error (°C/mW)

P_{RTD} = power consumed by RTD (mW)

Let's investigate an example of self-heat with an RTD to become more familiar with how this quantity is handled.

Example 13-3

A platinum temperature sensor has a published self-heat error of 0.12°C/mW. The sensor's measured resistance for a particular temperature is found to be 129.78 Ω. The voltage supplied to the sensor is 1.88 V dc. According to a temperature versus resistance chart accompanying the sensor, the temperature for 206.66 Ω is supposed to be 280.0°C. What is the *actual* temperature existing at the 129.78 Ω condition after taking self-heating into account?

Solution: We must first calculate the amount of power dissipation in the sensor. This can be done by using the equation $P = E^2/R$. Therefore,

$$P = \frac{1.88^2}{129.78}$$

$$= 0.027 \text{ W}$$

$$= 27 \text{ mW}$$

Now, using eq. (13-5),

$$T_{SH} = (0.12)(27)$$

$$= 3.24°C \text{ rise due to self-heat}$$

We must next determine what the temperature *appears* to be based on the 129.78 Ω measurement. Because of the linear characteristics of an RTD, we can make a simple proportional ratio based on the supplied manufacturer's data to determine this apparent temperature. That is,

$$\frac{206.66 \ \Omega}{280.0°C} = \frac{129.78 \ \Omega}{x}$$

$$x = \frac{(129.78 \ \Omega)(280.0°C)}{206.66 \ \Omega}$$

$$= 175.84°C$$

Because this is a temperature that includes the self-heat of the RTD; in other words, this is a temperature obtained with data (the RTD's measured resistance) influenced by self-heat. We can now compensate for this effect by performing the following step. We subtract the temperature amount calculated with eq. (13-5) from our apparent temperature. That is,

$$\text{actual temperature} = \text{apparent temperature} - T_{SH} \qquad (13\text{-}6)$$

Therefore,

$$\text{actual temperature} = 175.84°C - 3.24°C$$

$$= 172.60°C$$

13-3.6 Typical Characteristics and Design Specifications

Typical temperature ranges for the RTD are the following. For copper RTDs, the range is -200 to $260°C$; for nickel, the range is -80 to $300°C$, and for platinum, the range is -260 to $630°C$. Most RTDs require some sort of protective shield to protect the sensor wire from corrosion and general abuse. As a result, the RTD tends to be somewhat fragile in its overall construction, more fragile than the thermocouple. Because of the required sheathing (see Figure 13-3) the thermal response time is significantly increased. A typical response curve for a platinum wire RTD is illustrated in Figure 13-7. Compared to other temperature sensors, this response is somewhat slow.

13-3.7 Practical Applications

Because of the compactness furnished by modern solid-state circuitry, it has recently been possible to construct hand-held RTD transducers complete with signal conditioning and digital temperature readout displays. These devices employ three- and four-wire balanced bridge circuitry for excellent temperature reading accuracy. Figure 13-8 shows a complete hand-held RTD thermometer using a platinum sensor and having a temperature range from -220 to $630°C$. This is a remarkable feat when

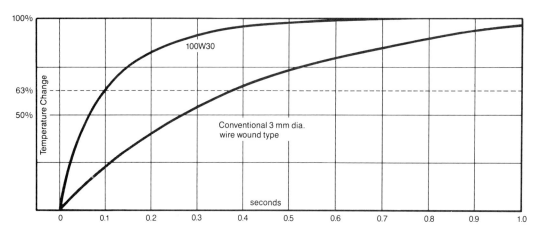

Figure 13-7 Time-response curve for a platinum wire RTD. (REPRODUCED WITH THE PERMISSION OF OMEGA ENGINEERING, INC.)

you compare this device to a typical RTD setup of, say, 15 years ago. At that time the instrumentation needed would have covered the area of roughly 4 ft^2 and its bulk would have weighed in excess of 30 lb or so.

RTDs have found their way into many laboratory applications where extreme temperature accuracy is needed along with their very linear, durable, and high-temperature-range characteristics. In industry a typical application would be in

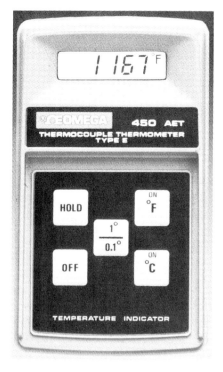

Figure 13-8 Hand-held RTD thermometer. (REPRODUCED WITH THE PERMISSION OF OMEGA ENGINEERING, INC.)

monitoring and controlling temperature in a hydraulic press used for molding plastics. Certain plastics require a particular curing temperature and curing rate. By implanting RTDs into the press' platens (Figure 13-9) it becomes possible to monitor and control these temperatures. Each RTD is located in a separately controlled heating zone so that each zone is controlled by that particular RTD.

13-4 THERMISTOR

We now look at another thermoresistive sensor, the *thermistor*. This device has entirely different characteristics compared to the RTD. For one thing, the thermistor is made from a human-made substance rather than being produced from naturally occurring materials, as in the case of the RTD. This human-made substance is called a semiconductor. Second, the thermistor's behavior with respect to temperature is entirely different from that of the RTD. These and other characteristics are discussed extensively in the sections that follow.

13-4.1 Theory of Operation

To understand how thermistors operate, we must first understand what a semiconductor is. Recalling our discussion concerning why metals conduct better at lower temperatures rather than at higher temperatures, there are certain types of materials that conduct very little, if any at all, at extremely low temperatures. The reason for this is because of the degree of bonding of the outer electrons in the shells of these materials. In metals these electrons experience very little attraction from the nucleus and, consequently, have little difficulty in being dislodged from the parent atom.

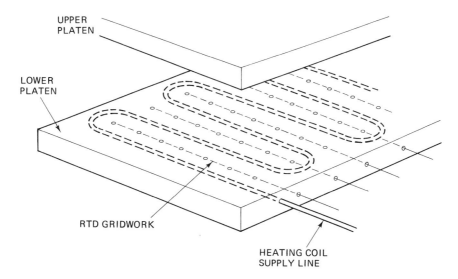

Figure 13-9 RTD gridwork in a hydraulic press platen.

However, in certain nonmetallic compounds, the bonding is much greater, and unless the outer electrons are agitated by heat, the bonding is so high that no free electrons are free to roam at the coldest temperatures. Only through the addition of heat can these electrons become agitated enough because of their vibrating that they can break their bonds and travel on to similarly ionized atoms. Compounds that show this type of behavior are called semiconductors.

13-4.2 Characteristics of the Thermistor

Because a semiconductor's resistance decreases with an increase in temperature, which is a behavior pattern opposite to that of metals, semiconductors are said to possess a *negative temperature coefficient*. Figure 13-10 depicts a typical resistance versus temperature curve for a thermistor semiconductor. You can conclude from observing this curve that a thermistor is not considered a linear device, especially when compared to the RTD curve also shown in Figure 13-10. Both curves shown are only typical curves. Thermistors can be manufactured so that different resistance values can be formulated to occur at a particular temperature, say at 0 or 25°C. These values can range from 100 Ω to as high as 1 MΩ for these temperatures. If the resistance versus temperature curves of these other thermistors were plotted alongside the one in Figure 13-10, you would observe a family of curves paralleling the one shown in Figure 13-10.

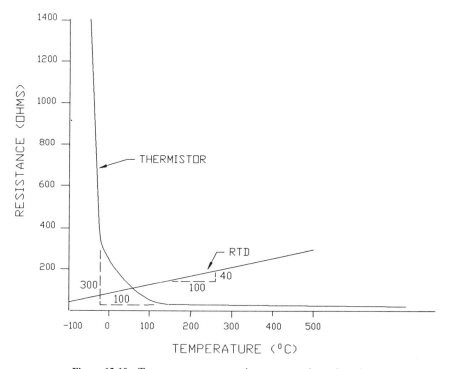

Figure 13-10   Temperature versus resistance curve for a thermistor.

Because the thermistor is a semiconducting material, the temperature range of this type of sensor is somewhat limited. Semiconductors can rarely withstand temperatures above about 300°C before melting. However, at this temperature the thermistor conducts current very well, approaching the conduction qualities of a poor conductor. In addition, the slope of the response curve approaches zero, meaning that the thermistor's sensitivity becomes almost nonexistent. The thermistor loses its effectiveness as a sensing device at this point.

Below −50°C the thermistor's resistance approaches that of a poor insulator. Resistances of several megohms are likely to be found. At −100°C the practical sensing limit of a thermistor is probably reached. However, again noting the shape of a thermistor's curve, you can see that the thermistor is most linear in the cold region of operation, say below 0°C. Also, its sensitivity is highest in this region.

The thermistor is generally a more sensitive sensor than the RTD. Since the sensitivity of any sensor is found by *dividing its output signal by its input signal*, this quantity is reflected in the amount of the slope in its plotted response curve. In other words, for the thermistor, we can observe the slope of its temperature versus resistance response curve (refer again to Figure 13-10), compare it to the slope of the RTD curve, and see immediately that the thermistor is far more sensitive than the RTD to a given change in temperature (assuming that we neglect the very high temperature end of the thermistor's response curve).

Example 13-4

Using the slope information shown in Figure 13-10, compare the sensitivities of the thermistor and RTD.

Solution: The sensitivity amount for the thermistor obtained from the slope information in Figure 13-10 is the following [using eq. (1-30)]:

$$\text{sensitivity} = \frac{\text{change in output indicator}}{\text{change in measurand that produced change in indicator}} = \frac{300\ \Omega}{100°F}$$

$$= 3\ \Omega/°F$$

The sensitivity amount for the RTD obtained from the slope information in the same figure is the following:

$$\text{sensitivity} = \frac{\text{change in output indicator}}{\text{change in measurand that produced change in indicator}} = \frac{40\ \Omega}{100°F}$$

$$= 0.4\ \Omega/°F$$

Comparing the two sensitivity results we see that the thermistor is 7.5 times more sensitive than the RTD in this particular example.

13-4.3 Typical Construction

Thermistors come in a variety of physical sizes and shapes. The various types of shapes or styles are the following: (1) beads, (2) disks, and (3) rods or probes. There are other shapes and styles, but the ones listed above are probably the most common. Figure 13-11 shows an illustration of each example.

Description	Configuration
GENERAL PURPOSE. Vinyl tipped, most rugged probe. Used for short-term water and sub-soil readings. Waterproof construction now standard.	
SMALL FLEXIBLE. Vinyl sheath and tip. Cuvette temperatures. General purpose measurement.	
GENERAL PURPOSE. Non-immersible, epoxy tipped probe. Can be potted in place. Probe is suitable for temperature measurements on surfaces.	
"BANJO" SURFACE TEMPERATURE. Skin, oral, axillary, water bath, air, surface temperatures. Stainless steel.	
ATTACHABLE SURFACE TEMPERATURE. Stainless steel cup, epoxy backed. Easy to tape on flat surfaces. Good for heat loss or compression efficiency study of piping.	
SMALL SURFACE TEMPERATURE. Cuvette, water bath, leaf and other surfaces. 24" Teflon-covered flexible wire. Stainless steel disc with epoxy back.	
AIR TEMPERATURE. Stainless steel probe suitable for test rooms, incubators, remote air readings, monitoring of gas streams, etc.	
TUBULAR. Stainless steel probe for rugged duty. Often used for liquid immersion. Probe is immersible only to cap.	
TUBULAR-GLASS. Chemically inert for liquid immersion use. Thermometric titration. Freezing point determination. Pyrex, 5" long.	
TUBULAR WITH FITTING. Rugged, stainless steel probe with pipe fitting. Suitable for taking readings in pipes or inside closed vessels.	

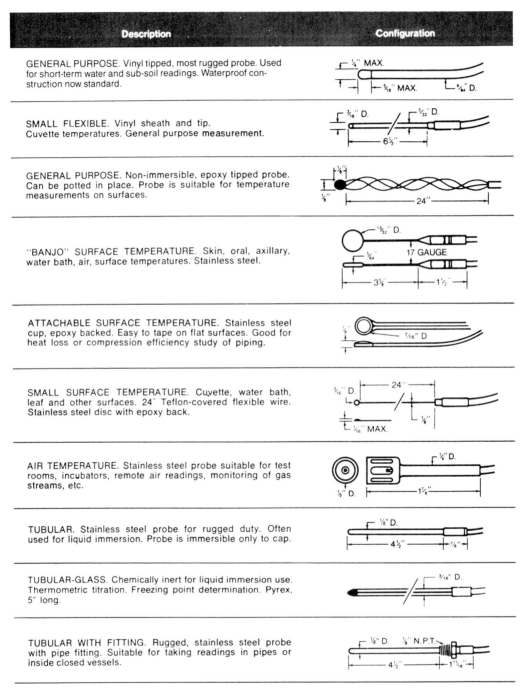

Figure 13-11 Various forms of thermistors. (REPRODUCED WITH THE PERMISSION OF OMEGA ENGINEERING, INC.)

In some cases the thermistor is coated with a protective substance, usually glass, to prevent oxidation of the solid-state material used in the thermistor's construction. The glass furnishes a hermetic seal that wards off oxidation and corrosion that may occur due the thermistor's environment. Another purpose of the coating is to bond the leads of the thermistor more firmly to the thermistor's body, thus acting as a strain relief for the leads.

The glass-coated beads on thermistors range in size from approximately 0.1 to 2 mm in diameter. On the other hand, the disk-type thermistors typically have diameters ranging from 1 to 3 mm and are often not coated. These are often cheaper in cost. During manufacturing, the thermistors are easily adjusted to the desired resistance by grinding away a portion of the thermistor's disk. Thermistor rods are typically manufactured in diameters of about 0.5 mm to as much as 50 mm.

The most common application for thermistors is for thermometry (Figure 13-12). Because of the wide variety of shapes available, it is possible to find a thermometer probe that can fit into the tightest of temperature-measuring physical restrictions. This is discussed further in the next section concerning applications.

Figure 13-12 Thermistor temperature probe being used in conjunction with its temperature readout unit. (REPRODUCED WITH THE PERMISSION OF OMEGA ENGINEERING, INC.)

13-4.4 Thermistor Self-Heat

Like the RTD the thermistor is susceptible to the problem of self-heat. Unlike the RTD, however, the thermistor is placed into a potential self-destructive mode when self-heat becomes excessive. That is, as the thermistor becomes warmer, its internal resistance becomes lowered, allowing an increase in current flow which creates additional heat, and so on. It is especially critical to limit the current flow through a thermistor to prevent this situation from happening.

13-4.5 Linearizing Thermistors

Because of the nonlinear characteristic of thermistors, various circuit schemes have been devised to make the thermistor behave in a linear fashion. One such scheme is shown in Figure 13-13. Two thermistors of identical response are so wired that one is parallel to another having a resistor in series with it. Through considerable experimentation it is found that this particular circuit achieves at least a 10-fold decrease in the amount of temperature deviating from linearity. For instance, for a temperature range of, say, 0 to 100°C, a maximum deviation of only 0.15°C from linearity is achieved, whereas normally, a single thermistor system yields an amount of 1.5°C. By adding the additional matching parallel thermistor with its series resistor, a dramatic linearization for the entire system can be obtained.

13-4.6 Practical Applications

Thermistors are usually associated with thermometry-type instrumentation, that is, instrumentation used solely for the reading of temperature. Figure 13-12 in Section 13-4.3 is a good example. Despite their nonlinear characteristics, thermistors still enjoy widespread use. The reason for this is due to the relative ease in linearizing an otherwise nonlinear sensor through the use of lookup tables that have been stored

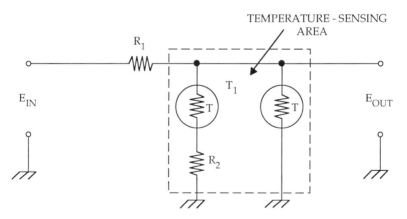

Figure 13-13   Linearizing circuit for a thermistor.

in firmware. Many thermistor-equipped electronic thermometers contain IC chips that can convert a thermistor's nonlinear output into a very linear response, resulting in extremely accurate output data. The temperature probe containing the thermistor is also made so that it is replaceable in case it is damaged.

There are at least two major forms of signal-conditioning circuitry that are used in converting a thermistor's resistance change to a signal that eventually drives a digital or analog readout. One form uses a resistance-measuring circuit such as a Wheatstone bridge (discussed in Section 7-4) to determine the temperature (refer to Figure 13-5). In this circuit R_1 is adjusted for a null according to the center-leg indicator. Instead of calibrating R_1 in ohms to read the resistance of R_3, in this case the thermistor, it is calibrated in units of temperature instead.

A less sophisticated resistance-measuring circuit employs a basic ohmmeter circuit (Figure 13-14). This is where the voltage drop occurring across resistance, R_{IND}, produces the necessary voltage signal for creating the desired temperature reading.

The second type of circuit used for thermometry purposes is illustrated in Figure 13-15. In this example we see a thermistor, R_{TH}, whose varying resistance controls the frequency of oscillation of an oscillator (the 555 timer IC, R, and C). The output of this system is in the form of a square wave with varying frequency. The frequency of this wave can then be measured and easily converted to a temperature display. In this circuit the frequency of oscillation varies inversely with the change in resistance of R_{TH}.

Example 13-5

In Figure 13-15 the frequency of oscillation of the circuit shown is determined by

$$f = \frac{0.722}{(R_{TH} + R)C} \tag{13-7}$$

Assuming that capacitor C has a value of 0.005 μF and resistor R a value of 1200 Ω, determine the values of f for the following three values of R_{TH}: 750 Ω, 1500 Ω, and 300 Ω.

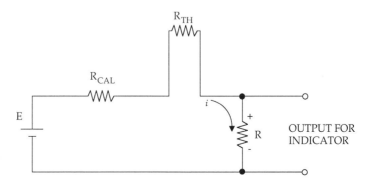

Figure 13-14   Using a simple ohmmeter circuit to read a thermistor's resistance.

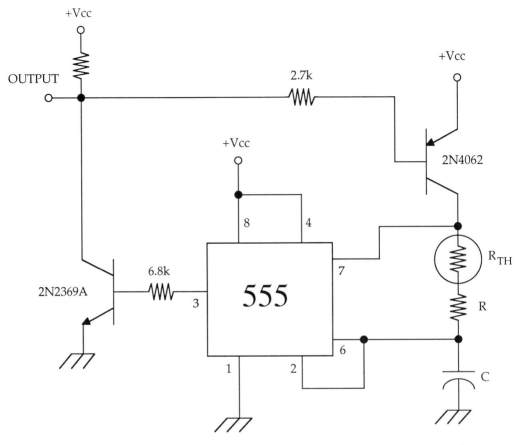

Figure 13-15 Using a thermistor to control the frequency output of an oscillator. (From Howard M. Berlin, *The 555 Timer Applications Sourcebook,* copyright 1976, p. 65. Reprinted by permission of the author.)

Solution: Placing the values of C, R, and the values of R_{TH} into eq. (13-7), we get

$$f_1 = \frac{0.722}{(750 \; \Omega + 15000 \; \Omega)(0.005 \times 10^{-6} \; F)} = 9168 \; Hz$$

$$f_2 = \frac{0.722}{(1500 \; \Omega + 15000 \; \Omega)(0.005 \times 10^{-6} \; F)} = 8752 \; Hz$$

$$f_3 = \frac{0.722}{(3000 \; \Omega + 15000 \; \Omega)(0.005 \times 10^{-6} \; F)} = 8022 \; Hz$$

In addition to being used in thermometry, thermistors are also often used for temperature control applications. A typical application is seen in Figure 13-16. In

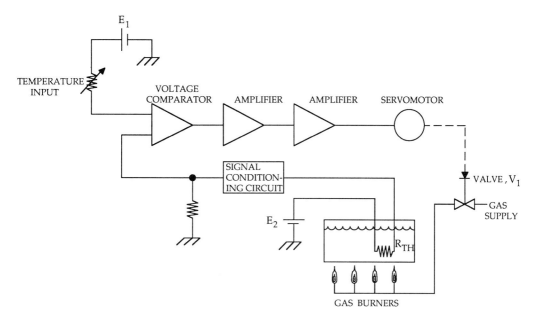

Figure 13-16 Thermistor being used in a process control system for controlling the temperature of a liquid in a heated container.

this circuit we see a controller being used to maintain a desired water temperature in a food-processing system. The operator adjusts the temperature controller to the desired temperature, while the thermistor detector senses the water's temperature. The sensed temperature is then compared to the desired temperature. Any difference in the two values (this difference is called the *error signal*) causes the gas valve, V_1, either to close or to open depending on the arithmetic sign of this difference.

It is the self-heat characteristic of thermistors that has created this next application. Thermistors are often used as level-indicating alarms, as depicted in Figure 13-17. A thermistor is adjusted for a particular height in a liquid-filled container that is being filled. The thermistor's height represents the maximum liquid height that is desirable. Since the thermistor is suspended in air, out of the liquid, its self-heat is sufficient enough to create a current flow in a current-sensing network to keep a holding coil energized on an alarm system. The alarm is off in this position. Should the liquid cover the thermistor, the self-heat will become dissipated within the liquid, causing the thermistor to cool, thereby increasing its resistance and lowering the current flow through it. This lower current is not enough to keep the holding coil to the alarm system energized; therefore, the relay opens, triggering the alarm circuit. The alarm circuit may also be wired to a control valve that opens a dump valve to the container so that the liquid will partially drain, causing the thermistor once again to become "self-heated," causing the alarm's relay coil to close once again.

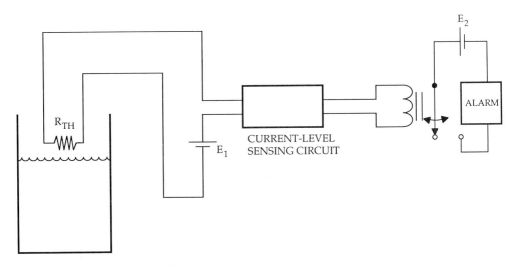

Figure 13-17   Thermistor being used as a level-indicating device.

13-5 DIODE JUNCTION SENSORS

A diode junction sensor is a solid-state device that contains a junction formed by two crystals that have been fused or otherwise grown together to form a semiconducting material. Such junctions are common in the fabrication of transistors and diodes. As in the case of the thermistor, the diode junction device has a negative temperature coefficient. However, it has the advantage of having a very linear voltage drop versus temperature response, approaching the linearity of an RTD. This type of sensor has a temperature-sensing range of approximately -200 to about $300°C$ depending on the semiconducting substances used for the junction fabrication.

Often, rather than manufacturing a special junction device just for temperature sensing, existing stock transistors may be used instead. Since bipolar transistors are comprised of two junctions of alternating p- or n-type silicon material (Figure 13-18), one of these two junctions may be used as a temperature sensor. Typically, what is done when this is the case is to cut off the unused third lead to the transistor. Physically small transistors are preferred over the larger ones, so that the thermal response time is minimized. The smaller units use a lesser mass of the epoxy encasement material, which is normally used to protect the silicon chip from the environment and to act as a heat sink.

13-5.1 How They Are Used

Diode junction sensors are used just like thermistors. As a matter of fact, their signal-conditioning circuitry is quite similar. Both the ohmmeter and Wheatstone bridge circuits are used for measuring the sensor's resistances. The only major difference between these two sensors is in the temperature ranges being sensed.

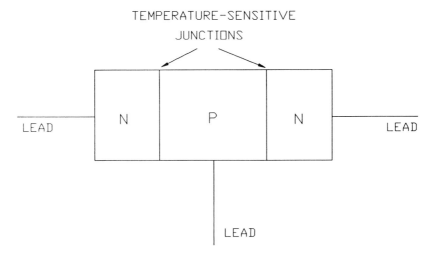

Figure 13-18   Temperature-sensitive junctions of a semiconductor.

The diode junction sensor does not have as high a temperature tolerance as the thermistor.

13-5.2 Typical Construction

The outward appearance of a junction diode is very similar to that of the thermistor. It's quite difficult to look at the packaging and determine if the installed sensor is a thermistor or a junction diode. A typical junction diode transducer system is displayed in Figure 13-19, where a flexible probe containing the junction diode is

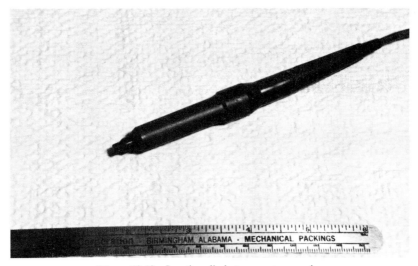

Figure 13-19   Junction diode temperature probe.

shown attached to a signal-conditioning unit. In this particular instance, however, rather than using a junction diode for the sensor, this particular probe uses the diode junction of a transistor. The base-to-emitter voltage variations created by varying temperature values are monitored and converted to a proportional dc voltage by the signal conditioner. This signal is then either displayed or sent to a recording device.

REVIEW QUESTIONS

13-1. Explain the difference between a substance having a positive temperature coefficient and a negative temperature coefficient.

13-2. What is the advantage of using a platinum RTD versus one made from nickel? What is the advantage of a nickel RTD?

13-3. Explain how strain inadvertently placed on an RTD's sensor metal could affect its temperature indication.

13-4. When comparing the thermistor and the RTD, which has the greater self-heat problem for the same current input, in your opinion?

13-5. Explain the designation R_{100}/R_0 applied to RTDs. Why are these designations so useful?

13-6. What is the advantage of a coiled RTD versus a single-strand RTD?

13-7. Explain how the protective sheath surrounding an RTD increases that RTD's time constant.

13-8. Explain the circuit for a three-wire RTD and explain its advantages over a standard two-wire uncompensated circuit.

13-9. Show a circuit that is typically used to linearize a thermistor's output.

13-10. What is an advantage in using a junction diode rather than a thermistor for a temperature-measuring application?

PROBLEMS

13-11. Find the resistance of an alloyed wire having a length of 15 in., whose temperature coefficient, α, is 0.00277/°C, whose resistivity constant is 9.89 Ω-CM/ft at 20°C and has a diameter of 25 mils. The ambient temperature under which the wire is to be used for a resistance measurement experiment is 47°C.

13-12. In a two-wire uncompensated RTD system an RTD is attached to leads that are 145 ft in length. The wire is No. 22 AWG copper wire. Design a Wheatstone bridge having maximum temperature sensitivity. Explain how you would compensate your system for the lead lengths.

13-13. An alloyed RTD sensor has a self-heat error of 0.09°C/mW. At a particular temperature the sensor's dc resistance was measured and found to be 192.67 Ω. The sensor operates from a 2.15-V supply. If the manufacturing data state that the dc resistance for this sensor is supposed to be 223.56 Ω at 270°C, what is the actual temperature after compensating for self-heat?

REFERENCES

ALLOCCA, JOHN A., and ALLEN STUART, *Transducers: Theory and Applications*. Reston, VA: Reston, 1984.

BERLIN, HOWARD M., *The 555 Timer Applications Sourcebook*. Indianapolis, IN: Howard W. Sams, 1979.

HORDESKI, MICHAEL F., *Microprocessor Sensor and Control Systems*. Reston, VA: Reston, 1985.

SESSIONS, KENDALL W., *Master Handbook of 1001 Practical Electronic Circuits*. Blue Ridge Summit, PA: TAB Books, 1975.

The Temperature Handbook, Vol. 27. Stamford, CT: Omega Engineering, 1990.

14

PHOTOCONDUCTIVE SENSORS

CHAPTER OBJECTIVES

1. To review the physics of light and photometry.
2. To study the characteristics of the photoresistor.
3. To study the characteristics of junction diode sensors.
4. To understand the characteristics of the phototransistor and the photoFET.
5. To study the characteristics of the LASCR.

14-1 INTRODUCTION

The photoconductive sensor is certainly a product of very recent development in solid-state electronics. We are speaking of those special materials that when exposed to light, alter their internal electrical resistance in proportion to the amount of light falling on them. This chapter, then, is devoted to the study and application of these devices. However, as usual, before we tackle this subject matter, we must first understand the physics that control the behavior of light. This will then help you develop a firmer understanding of what light is and help you better understand why the photoconductive sensor works the way it does. Following the review section we discuss the first of three types of photoconductors, the *photoresistor*. The other two, the photodiode and phototransistor, follow.

14-2 PHYSICS OF LIGHT AND REVIEW OF PHOTOMETRY

The behavior of light was covered in Section 1-5.4. Additional discussion concerning electromagnetic radiation detection was presented in Sections 11-1 and 11-2. Recalling these discussions, light can take on one of two behavior modes: (1) it can appear to behave as an electromagnetic waveform whose frequency of oscillation can be calculated by the equation $f = c\lambda$ [eq. (1-13), transposed], or (2) it can behave as a photon, which is a tiny energy-laden particle. The amount of energy contained by this particle can be calculated from the equation $E = hf$ [eq. (1-15)].

Photometry is the science of measuring the intensity of visible light. Unfortunately, the units of measurement used in photometric studies can be rather confusing and difficult for many people to understand. Consequently, the discussion here will focus first on just one unit of light measurement, the *lumen*. The lumen, or *luminous flux,* is a unit of illumination having to do with the "flow rate" of light. In other words, if you were to think of light as being comprised of many lines of flux, similar to the approach used in dealing with a magnet and its surrounding magnetic lines of force or lines of flux, the lumen is the quantity or concentration of light. The lumen is a unit of measure found in the SI system.

When luminous flux strikes a surface, that surface becomes *illuminated* by that flux. If the amount of luminous flux is F and the illuminated area is A, the amount of *illuminance* is defined as the density of luminous flux or flux per unit of illuminated area. That is,

$$E = \frac{F}{A} \tag{14-1}$$

where E = amount of illuminance (lm/m^2)

F = luminous flux (lm)

A = area of illuminated surface (m^2)

Example 14-1

Determine the illuminance of a light source capable of producing a luminous flux of 12.57 lm at a distance of 1 m.

Solution: At a distance of 1 m the area through which the 12.57 lm must penetrate and therefore illuminate is determined by calculating the inside area of a sphere having a radius of 1 m. Since the area of any sphere is found by the equation $A = 4\pi r^2$, then the area of a 1 m radius sphere is 4π, or 12.57 m^2. Then, using eq. (14-1),

$$E = \frac{F}{A} = \frac{12.57 \text{ lm}}{12.57 \text{ m}^2}$$

$$= 1 \text{ lm/m}^2$$

We must next define the unit of *luminous intensity*. If a point source of light were placed at the center of a sphere of radius R (Figure 14-1), and that source

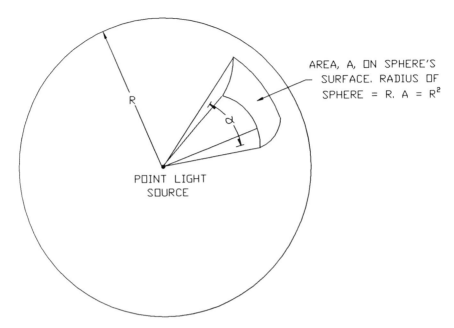

AREA, A, ON SPHERE'S
SURFACE. RADIUS OF
SPHERE = R. A = R²

R

α

1

POINT LIGHT
SOURCE

Figure 14-1   Defining luminous intensity using a point source of light at the center
of a sphere of radius R.

illuminated an area on the inside of that sphere equal to R^2, the solid angle (α)
subtended by the illuminated area and originating at the point source of light would
form a solid angle of *1 steradian*. Now the *luminous intensity, I,* is determined by

$$I = \frac{F}{\alpha} \qquad (14\text{-}2)$$

where I = luminous intensity [lumen/steradian (lm/sr) or candela (cd)]

F = luminous flux (lm)

α = number of steradians in sphere (12.57 sr)

The following example will help illustrate the explanation above.

Example 14-2

The light source in Example 14-1 illuminates a surface that is located at a distance of
1 m. Determine the luminous intensity, in candelas, on that surface.

Solution: In Example 14-1 the light source has a luminous flux of 12.57 lm. At 1 m
assume that this flux is spread out over the inside surface of the same imaginary spherical
surface depicted in Example 14-1. Because luminous intensity is based on the steradian,
and because we want to know this intensity at 1 m from the light source, we must now
determine the number of steradians in our 1m radius sphere. By definition, *1 steradian*
is the solid angle (which we call α) subtending an area on the inside surface of our sphere
equal to r^2 and extending back to the light source. Since the area of a sphere = $4\pi r^2$

and since 1 steradian in this sphere is r^2, the total number of steradians $= 4\pi r^2/r^2$, or 4π steradians. The luminous intensity, I, falling on 1 sr then becomes

$$I = \frac{F}{\alpha} = \frac{12.57 \text{ lm}}{12.57 \text{ sr}}$$

$$= 1 \text{ lm/sr}$$

$$= 1 \text{ cd}$$

Now consider this next example.

Example 14-3

What would be the illuminance, in lm/m^2, experienced at a 2-m distance from the same light source as in Example 14-1? Compare this result with the results obtained in Example 14-1.

Solution: The area of a sphere that has a 2-m radius is equal to $4\pi(2 \text{ m})^2$ or 16π. Consequently, the illuminance, or lm/m^2, is, using eq. (14-1),

$$E = \frac{F}{A} = \frac{12.57 \text{ lm}}{16\pi \text{m}^2}$$

$$= 0.25 \text{ lm/m}^2$$

Comparing this result with the results of Example 14-1, we see that as a result of doubling the distance from the light source, we have reduced the illuminance by one-fourth. This result could also have been determined by using eq. (11-1). This equation is a statement of the inverse-square law for electromagnetic radiation intensity in general and would have given us the same results.

14-3 THEORY OF THE PHOTORESISTOR

In its simplest form a photoresistor is made from a piece of semiconducting material that is placed between two conducting end plates, forming a sandwich. Each of the end plates has a wire lead for connection to other circuitry. The semiconducting material is usually made of cadmium sulfide (CdS), cadmium selenide (CdSe), germanium, or silicon. Certain metals having halides (i.e., compounds of hydrogen) are also sometimes used for this application.

There is a similarity between the semiconductor slab in Figure 14-3 and the semiconducting material of the thermistor. In the case of the thermistor, heat supplies the energy to free-up electrons to allow the conduction of a current (refer to Section 13-4.1). In a photoresistor that energy is supplied from the "$h \times f$" energy content of the photon. Although this is a much oversimplified explanation of what actually occurs inside the semiconducting material, it nevertheless is a fair representation of the actual process.

As a consequence of the event above, one can assume that when a photoresistive material is in the dark, there is little or no current conduction; its resistance to current flow is extremely high. On the other hand, when the photoresistor is exposed to light, its resistance decreases, thereby allowing the passage of larger amounts of current (see Section 14-3.3). This should be understandable when one compares the behavior of this device to that of the thermistor. The only real difference between these two sensors is that the thermistor is designed to be more sensitive to longer wavelengths of electromagnetic radiation (heat) as compared to the photoresistor's designed sensitivity.

14-3.1 Commonly Sensed Measurands

The primary measurand sensed by the photoresistor is, of course, light. There are numerous secondary measurands. As a matter of fact, most of the measurands mentioned in the list in Section 1-4 can eventually be detected by a photoconductor. This is why this device has such a universal appeal among applications engineers in industry in the design of products. The following discussion shows how the photo-resistor can be used to detect other measurands.

Position

Figure 14-2 shows how position is detected using a photoconductor. A sliding glass grating containing finely ruled lines is designed to cut a light beam that is intercepted by a resistor. The number of light-beam interruptions made by the opaque lines is counted by a frequency counter. By knowing the spacing distance used by the ruled lines on the glass slide, it is easy to calculate the slide's amount of movement and therefore the position of the slide itself.

Motion

Motion is simply a modification of the position-monitoring scheme just outlined. By timing the change in position, you can easily obtain the velocity of the attached rod to the slide in Figure 14-4. This can be taken a step further by measuring any change in velocity in order to obtain acceleration.

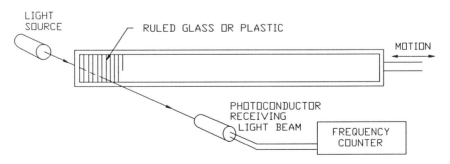

Figure 14-2 Detecting position using a photoconductor and light source.

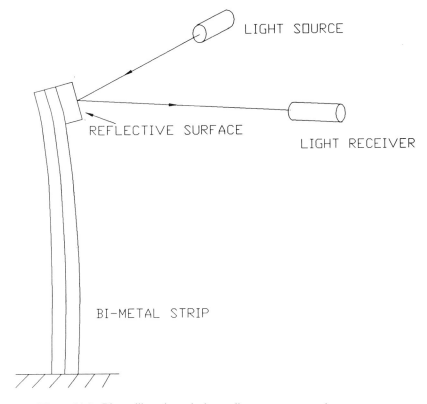

Figure 14-3 Bimetallic strip and photocell arrangement used as a temperature indicator.

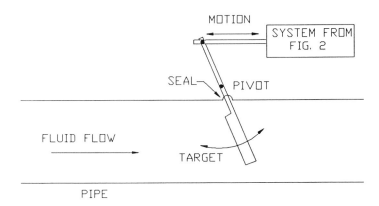

Figure 14-4 Flow rate measured with target device and photocell.

Sound

Sound may be detected by using a vibrating reflector diaphragm that replaces the diaphragm of a microphone. A light source is used to furnish a light beam that is reflected off the diaphragm onto a photoresistive cell. The varying reflections that are sensed by the photocell are an exact reproduction of the sound waves striking the diaphragm.

Temperature

The bimetallic temperature-sensitive strip shown in Figure 14-3 has a reflective surface attached to it, so that when the strip moves due to temperature, a varying portion of the light beam coming from a light source is reflected into the photoresistor's sensor. The beam's varying intensity is calibrated so that a temperature scale can be used in place of intensity.

Flow

Flow rate measurements can be made with the device illustrated in Figure 14-4. A pivoted target is placed in the middle of a pipe carrying a fluid. The fluid's flow creates a drag (force) on the target's cross-sectional area, causing a displacement of the pivoted arm on the opposite end to occur. The amount of displacement will be detected by the photocell and ruled grating system and will be in direct proportion to the amount of drag created by the flowing fluid.

Force

A simple modification of the system just described for measuring flow can also be used for measuring force. The target can be incorporated into a weighing mechanism by attaching the optical displacement system with its ruled displacement grating to the target and applying a force or weight to the target (Figure 14-5). Any displacement detected by the optical displacement system can be correlated directly with the applied force.

14-3.2 Typical Construction

As is true with many sensing devices, photoresistors are manufactured in a variety of sizes and shapes. This is to accommodate a wide variety of installation needs. Basically, photoresistive cells come in four different package styles, shown in Figure 14-6: (1) coated, (2) plastic-encased, (3) metal-encased, and (4) glass-encased units. The photo-sensitized surface in all four cases is usually comprised of a serpentine pattern woven between two electrodes, with the structure bonded to a layer of ceramic for a substrate. The least expensive unit and the weakest, mechanically, of the four described here is the coated unit. The coating is a plastic that is applied through dipping. However, this particular unit is one of the most popular because of its very small size. Sizes less than a millimeter in diameter are possible with this type. Figure 14-6 shows some of the various styles of cells in use today.

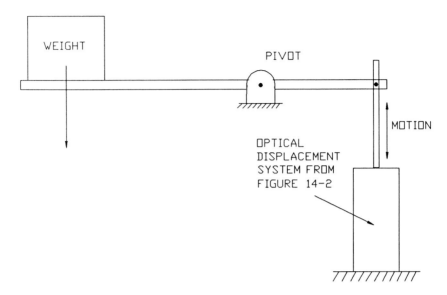

Figure 14-5   Measuring force with a photocell.

14-3.3 Photoresistor Characteristics

There are at least four attributes that are important in describing the characteristics of a photoresistor: (1) illumination sensitivity, (2) spectral response sensitivity, (3) response time, and (4) allowable power dissipation (self-heat).

Figure 14-6   Various photocell types available today.

Illumination sensitivity

A photoresistor's sensitivity is determined by comparing the change of its output resistance with the amount of change of luminous intensity falling onto the cell. In other words,

$$K_{\text{photo}} = \frac{\Delta I}{\Delta R} \tag{14-3}$$

where K_{photo} = sensitivity of photoresistor (cd/Ω)

ΔI = change in luminous intensity (cd)

ΔR = change in resistance (Ω)

A typical resistance versus illumination curve is shown in Figure 14-7. The following example illustrates how eq. (14-3) is used.

Example 14-4

Referring to Figure 14-7, determine the approximate sensitivity of the photoresistor that produced this curve.

Solution: Looking at the curve in Figure 14-7, we choose its middle to make our determination. At an illumination of 30 cd we see that the photocell's output resistance is 30 Ω. At 50 cd the cell's output is 10 Ω. According to eq. (14-3), then, the photocell's sensitivity is

$$K_{\text{photo}} = \frac{\Delta I}{\Delta R} = \frac{(50 - 30) \text{ cd}}{(30 - 10) \text{ } \Omega}$$

$$= \frac{20 \text{ cd}}{20 \text{ } \Omega}$$

$$= 1 \text{ cd/}\Omega$$

This calculated sensitivity value is valid only for the illumination range used in the calculation, namely for the range 30 to 50 cd.

To produce the data used for generating the curve in Figure 14-7, it is necessary to use a standard light source that has a high spectral purity in its illumination. This type of light source is referred to as a *monochrome* or *single-frequency light source*. For this type of light source a tungsten lamp is used whose filament temperature is closely monitored and maintained. Also, the photocell is operated in a constant ambient temperature, usually 25°C. As we will soon find out, the reason for this is that a photoresistor's resistance is somewhat temperature dependent. Again, this is because of the similarities between thermistor and photoresistors and their semiconductor chemistry.

It is obvious from the curve in Figure 14-7 that the response characteristics for a photoresistor is quite nonlinear. This is one of the drawbacks in using this type of device.

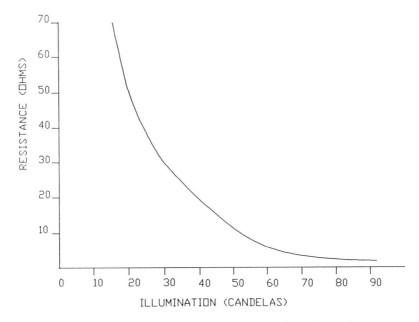

Figure 14-7   Illumination versus resistance curve for a photoresistor.

Spectral response sensitivity

Recalling that the sensitivity curve in Figure 14-7 was produced using a monochrome light source, the resulting curve is valid only for that particular color or wavelength of light. Other color portions of the light spectrum produce different curve responses. As a result of this color-sensitive behavior it is necessary to generate a spectral response curve that covers the entire visible spectrum. What can be expected from any given photocell device is a frequency response curve that peaks at a particular spectral frequency, the frequency value being determined by the type of photosensitive material used in the cell's construction and the type of light-admitting lens used. Figure 14-8 shows the curves resulting from the different types of cells that are presently available.

Response times

Many electrical switching devices that are made to have a fast turn-on and turn-off switch time possess a certain characteristic response-time curve. This curve is shown in Figure 14-9. Photoresistors are no different. Most photoresistors possess a time-of-response curve shape similar to the one shown in this figure. The *rise time* is the time it takes a photoresistor (or any other switching device) to rise to 63.2% of its full conductive value as a result of placing the cell instantly into full illumination from total darkness. The decay time is the time it takes for the cell to decrease from 100% full conduction to 36.8% of that value. This condition is obtained through the instant withdrawing of full illumination to total darkness.

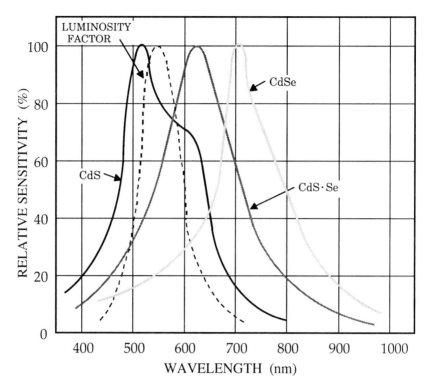

Figure 14-8 Spectral response curves for photoresistors. (From John A. Allocca and Allen Sturart, *Transducers: Theory and Applications,* copyright 1984, p. 259. Reprinted by permission of Reston Publishing Co., Inc., Reston, VA.)

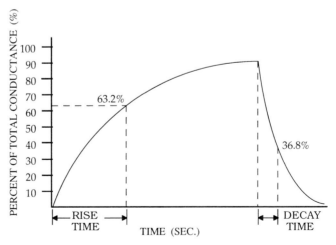

Figure 14-9 Response time for a photoresistor. (From John A. Allocca and Allen Sturart, *Transducers: Theory and Applications,* copyright 1984, p. 261. Reprinted by permission of Reston Publishing Co., Inc., Reston, VA.)

Example 14-5

Through precise measurements it was determined that for a particular photoresistive cell it took 52 ns for the cell to obtain its 100% conductive state coming from total darkness. What is its rise time?

Solution: Since rise time for a cell $= 63.2\% \times$ total response time, the rise time $= 0.632 \times 52$ ns $= 32.9$ ns.

A photoresistor's response time can be affected by several things. A cell's illumination amount can change the response time significantly. A brighter illumination can actually decrease the response time, as do higher temperatures. Also, higher loads placed on the photoresistor tend to decrease its response time.

14-3.4 Typical Design Specifications

Typical specifications for a photoresistor are seen in Figure 14-10.

14-3.5 Practical Applications

The photoresistor has so many practical applications that there are far too many to mention all of them here. Only a few applications and circuits will be mentioned.

Phototransistors are used in security systems where a light beam is used to activate a photoresistor that maintains the holding voltage on a relay circuit (Figure 14-11). If the light beam is interrupted by an unauthorized person or by an object, the holding voltage is interrupted, causing the relay circuit to open and to set off an alarm system. Figure 14-12 shows a typical amplifier circuit used in conjunction with a photoresistor to activate a relay circuit. This circuit could very easily be used in the alarm system described in Figure 14-11.

Another frequently used application for the photoresistor is in the construction of a photosensitive tachometer. A tachometer is used to measure the rpm of rotating mechanical members in machinery. A typical example is using a tachometer to check the rpm of a fan blade in a cooling system. A shiny surface or reflector is initially installed on the fan blade's rotary shaft. A light source is used to shine a beam of light onto the reflective surface, which is then reflected into a photocell. The number of reflected light pulses detected by the photocell is a direct indication of the rotating shaft's rpm. A flowchart circuit for this device is suggested in Figure 14-13. In the system shown, we see the output of Figure 14-12's circuit being sent a Schmitt trigger circuit. The purpose of this circuit is to "square" the output of the previous circuit and to eliminate any circuit noise. The Schmitt trigger's output is then sent to a one-shot multivibrator circuit which generates one output pulse for every input pulse that it receives. This further ensures that only one pulse is generated for every relay switch closure that is made in the photocell circuit. Finally, each pulse is then counted by means of a digital counter display that counts each of the pulses sent to it by the previous circuit.

CdS & CdSe PHOTOCELLS

Cell Type No.	Typical 1.0 Ftc Res. KΩ	Min. Dark Res. MΩ	Max. Peak Volt.	Max. Power MW.	Description
NSL-3120	75	100	70	50	CdSe Photoconductive Cell
NSL-3130	30	100	70	50	in Hermetic Glass and Metal
NSL-3140	14	100	70	50	TO-18. CdS versions also
NSL-3150	550	1000	250	50	available.
NSL-3160	260	1000	250	50	
NSL-3170	110	1000	250	50	
NSL-3180	50	500	120	50	
NSL-3812	43	10	80	100	CdSe Photoconductive cell
NSL-3822	18.6	10	80	100	in TO-5 plastic coated "open
NSL-3832	8	10	80	100	pill" package. Also available
NSL-3842	3.5	10	80	100	in CdS material.
NSL-3852	330	100	320	100	
NSL-3862	155	100	320	100	
NSL-3872	66	100	320	100	
NSL-4440	120	50	1000	1000	1" CdS Photoconductive
NSL-4450	26.0	15	1000	1000	Cell in Hermetic Glass and
NSL-4460	11.4	5	1000	1000	Metal case 1.25" diameter.
NSL-4470	5.5	0.5	1000	1000	
NSL-4442	120	10	1000	500	1" CdS Photocoductive
NSL-4452	26	1.5	1000	500	Cell in plastic coated
NSL-4462	11.4	0.5	1000	500	"open pill" package.
NSL-4472	5.5	0.2	1000	500	
NSL-4610	65	80	80	100	CdS Photoconductive Cell
NSL-4620	28	20	80	100	in Hermetic Glass and
NSL-4630	12	2.7	80	100	Metal ¼" dia. case.
NSL-4640	5.2	0.7	80	100	Hermetic CdSe versions
NSL-4650	550	160	250	100	also available.
NSL-4660	230	20	250	100	
NSL-4670	100	4	250	100	
NSL-4810	43.0	65	120	200	CdS Photoconductive Cell
NSL-4820	18.6	15	120	200	in Hermetic Glass and
NSL-4830	8	1.8	120	200	Metal TO-5. Plastic coated
NSL-4840	3.5	0.25	120	200	"open pill" versions also
NSL-4850	330	200	320	200	available.
NSL-4860	155	130	320	200	
NSL-4870	65	20	320	200	
NSL-2930	1.7	0.5	80	500	TO-8 Hermetic CdS/Se
NSL-2940	.75	0.1	80	500	Photoconductive Cell.
NSL-2970	7.5	5.0	250	500	
NSL-3921	4.0	2	80	200	TO-8 CdSe Photoconductive
NSL-3931	1.7	2	80	200	Cell in Plastic Case. Glass
NSL-3941	0.75	2	80	200	Lens and Plastic Coated
NSL-3951	40.0	5	320	200	"open pill" and CdS
NSL-3961	17.0	5	320	200	versions also available.
NSL-3971	7.5	5	320	200	
NSL-4920	4.0	2.0	80	500	CdS Photoconductive Cell
NSL-4930	1.7	0.4	80	500	in Hermetic Glass and
NSL-4940	0.75	0.05	80	500	Metal TO-8. Hermetic CdSe
NSL-4950	40	30	320	500	version also available.
NSL-4960	17	4.0	320	500	
NSL-4970	7.5	0.5	320	500	
NORP-11	20	2.5	320	250	CdS Photocell in 0.5"
NORP-12	9	1	320	250	lensed Plastic case.
NORP-13	4	0.5	320	250	
NSL-4512	43.0	5	120	100	CdS Photoconductive cell in
NSL-4522	18.6	1.5	120	100	Hermetic TO-5 package. Two
NSL-4532	8	0.50	120	100	plastic encapsulated
NSL-4542	3.5	0.10	120	100	versions also available.
NSL-4552	330	20	320	100	
NSL-4562	155	15	320	100	
NSL-4572	65	2	320	100	

Figure 14-10 Typical photoresistor specifications.

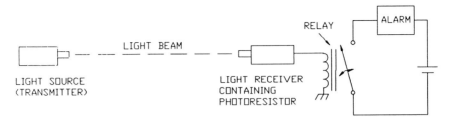

Figure 14-11   Photocell security system.

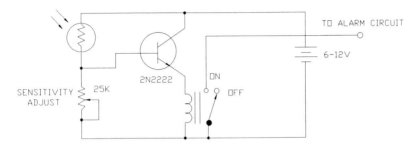

Figure 14-12   Amplifier circuit for relay-operated photocell system.

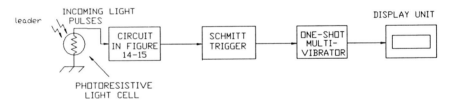

Figure 14-13   Block diagram for RPM-counting system.

Light-intensity data may be sent by telemetry with the circuit shown in Figure 14-14. The 555 timer chip functions as an astable multivibrator whose frequency of oscillation is determined by the equation

$$f = \frac{0.722}{(R_{pr} + R_1)\, C} \tag{14-4}$$

where F = frequency of oscillation (Hz)

  R_{pr} = photocell resistance (Ω)

  R_1 = limiting resistor (Ω)

  C = charging capacitor (F)

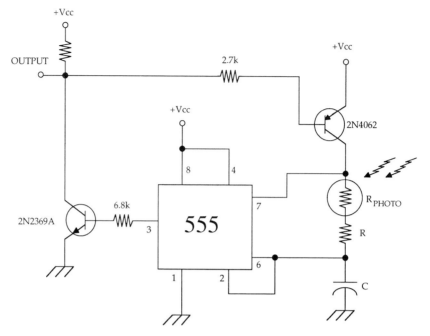

Figure 14-14 Light-intensity transmitter using astable multivibrator. (From Howard M. Berlin, *The 555 Timer Applications Sourcebook,* copyright 1976, p. 65. Reprinted by permission of the author.)

Example 14-6

Determine the frequency of oscillation in hertz for the circuit in Figure 14-17 if $C = 0.05$ μF, $R_1 = 10$ kΩ, and $R_{pr} = 1500$ Ω.

Solution: Placing the given values into eq. (14-4) to determine the frequency of oscillation, we get

$$f = \frac{0.722}{(R_{pr} + R_1)\, C} = \frac{0.722}{(1500\ \Omega + 10{,}000\ \Omega)(0.05 \times 10^{-6}\ \text{F})}$$

$$= 1256\ \text{Hz}$$

The output of his circuit is a variable-frequency square wave whose duty cycle is determined by the transistor Q_1. It is possible to obtain a square wave with a 50% duty cycle with this type of arrangement. Data may then be sent in the form of pulse-modulated square waves over great distances either by radio or by wire using this system.

14-4 DIODE JUNCTION SENSORS

Diode junction sensors are manufactured from semiconducting material which may be either of the p-type or n-type. That is, this material may be equipped for the supporting of current flow through the process of "hole conduction" (the p-type material), or this material may be suitable for supporting current flow through the process of free-electron "carriers" (the n-type material). Whichever type is used, the semiconducting material is sensitive to light. This is true more or less for all semiconducting materials. It is for this reason that most semiconducting materials are encased in opaque packages to prevent this light interaction from occurring. In the case of the photoresistor, however, this interaction is, of course, encouraged— hence the transparent window that is always found molded into the otherwise opaque container.

There are three basic kinds of diode junction devices: (1) silicon and germanium photodiodes, (2) the phototransistor, and (3) the photoFET and photoSCR. We discuss all three categories in the following sections.

14-4.1 Silicon and Germanium Photodiodes

The silicon photodiode is a silicon diode that is light sensitive. Otherwise, it functions much like an ordinary diode. That is, it allows current flow to take place in one direction only. Taking a look at the characteristic curve for a typical diode in Figure 14-15, we can see that when a voltage is applied across the diode in what is referred to as the *forward-biasing direction,* we obtain a relatively large current flow through the diode. That is, the current flow is large relative to the current obtained when the diode is *reversed biased.* The application of this forward-biasing voltage is illustrated in Figure 14-16(a), where we see the current traveling in the direction of the diode's symbolic arrow. If the voltage supply's polarity is reversed as shown in Figure 14-16(b), the current suddenly drops to a very small value, as can be seen by studying the curve in Figure 14-15. Again, this is a typical characteristic of the diode.

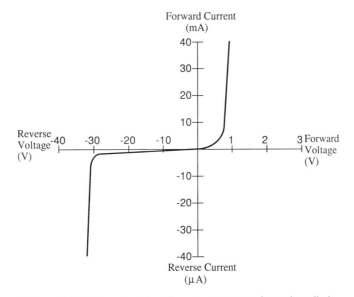

Figure 14-15   Characteristic voltage–current curve for a photodiode.

FORWARD CURRENT FLOW

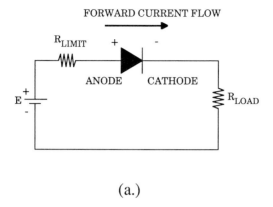

(a.)

REVERSE CURRENT FLOW

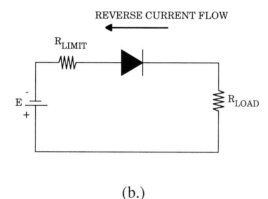

(b.)

Figure 14-16 Forward and reverse biasing a diode.

To explain how the diode is affected by the photons of light, we must first understand how the ordinary junction diode operates. Figure 14-17 illustrates a typical silicon (or germanium) diode with its one n-type wafer and one p-type wafer joined together at a common border called the p-n junction. The diode is shown being reverse biased by the attached power supply, E. The electrons in the n-material side are attracted away from the pn junction border by the power supply's positive terminal. The "holes" in the p-material side are attracted to the negative terminal of the same power supply. What occurs consequently is a depleted area on each side of the p-n junction that is almost void of either electrons or holes, depending on whether it is n- or p-type material. This mostly emptied zone is called the *depletion region* for that diode. In this depleted state very little, if any, current can flow across the junction. What little current flow there is is referred to as the diode's *leakage current.*

In a photodiode the light-admitting window is located either over the p-n junction area or at either the p end or n end. Let's assume that the window is located over the p-n junction area. When light strikes this area the photons are immediately converted to photoelectrons. These photoelectrons are forced by the negative side

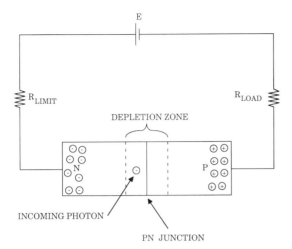

Figure 14-17   PN junction of a photo-diode and the photo-conversion process.

of the voltage source to flow across the junction into the material and eventually, into the negative terminal of the supply. This means that an increase in flow of leakage current is produced across the junction into the p-side of the diode. (Remember that conventional current flow is opposite in direction to electron flow.) As a result, then, we have a device that produces a leakage current that varies in proportion to the amount of light striking that device.

Now let us asume that the window is located over either end of the diode. In either case the released photoelectron is attracted to the positive side of the supply voltage (or repelled by its negative side), which again constitutes a current flow.

An interesting side characteristic of the photodiode is the following: If a rather high reverse bias voltage is used to operate the photodiode, a photo-multiplying affect can be achieved. The converted photodiode–electron is suddenly accelerated when it comes into the vicinity of the much higher electric fields created by the higher biasing voltage. Traveling at a higher energetic level, the electron collides with other electrons in the n-type material. This collision causes even further collisions to take place, resulting in several electrons being attracted by the bias supply's positive terminal instead of the one single electron. In effect, this type of photodiode operation produces a gain in the overall current. As we will discover later in the discussion of photomultiplier circuits, this *avalanche gain mechanism* in our photodiode is not the only device that produces gain by this type of electronic chain reaction.

14-4.2 Photodiode Characteristics

Photodiodes are noted for their very high sensitivity to light. Their sensitivity approaches that of certain light-sensing circuits that have to use amplifiers to attain the same amount of sensitivity. The photodiode's sensitivity is due mainly to the avalanching characteristics described earlier.

The silicon photodiode's spectral or frequency response is represented by the curve in Figure 14-18. The curve's peak response favors the red end of the visible spectrum and in many cases actually favors the infrared portion. As a matter of fact, the germanim photodiode has a typical response curve that favors the infrared end of the visible spectrum even more so. A typical range of operation for this device is found in the range 14,000 to almost 20,000 angstroms. This places this type of photodiode well beyond the *near*-infrared portion of the spectrum and into the *far*-infrared portion.

Certain silicon photodiodes have also been made to function at extremely short wavelengths when operated in combination with certain other materials that have been bonded to its pn structure. Crystals made from either sodium iodide or cesium iodide are joined to the silicone photodiode's body to give this device the ability to detect wavelengths that extend into the x-ray portion of the electromagnetic spectrum. As a result, this modified photodiode can function solely as an x-ray frequency detector rather than a detector of light frequencies.

Silicon photodiodes are also known for their fast response time to high-frequency pulsed light. Some devices have response times measured in nanoseconds. These units have special applications in high-speed data communications and are discussed further in Section 14-4.4.

Silicon photodiodes also possess very low *dark current* characteristics. The dark current of any photo-sensitive device is the current that can pass through that device under total darkness conditions. In the case of the silicon photodiode, typical dark current values are measured in microamperes and nanoamperes. As a matter of fact, it is not unusual to find dark currents of less than 1 nA in these units. Low dark current is advantageous because any current produces electrical noise within a circuit. Under dark conditions with very little dark current to generate this noise, the sensor is more likely to respond properly to very weak light conditions.

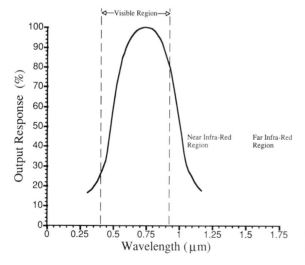

Figure 14-18   Silicon photodiode's frequency response curve.

14-4.3 Typical Construction

Typical silicon photodiodes are pictured in Figure 14-19. Physical dimensions are shown in Figure 14-20.

14-4.4 Design Specifications

Specifications for a photodiode are shown in Figure 14-21.

14-4.5 Practical Applications

Silicon photocells find many applications in the field of fiber optic communications. Because of their very high siwtching speeds they are suited for the reception of the high-speed data rates that are often used by computer circuitry. Silicon photodiodes are often found in fiber optic systems where the transmitted data which are normally in the form of electrical pulses and carried by wires are converted to light pulses and then transported by light "pipes" or fibers to photodiode receivers. These photodiodes then convert the rapid light pulses back into electrical pulses that can then be interpreted by the data processor circuits or computer (Figure 14-22).

A similar application is found in the hand-held laser voice communicators being developed by NASA. Built into these units are small very low power laser units with extremely narrow, highly collimated light beams capable of many tens of miles of transmission. The same units also contain photodiode receivers for the reception of these light beams. The laser beam is voice modulated by a person speaking into one of these units, whereas the diode receiver in the unit being held at the receiving

Figure 14-19   Typical silicon photodiodes. (Courtesy of Hamamatsu Corp., Bridgewater, NJ.)

Unit: mm

❶ S1226-18BQ etc.

❷ S2386-18L etc.

❸ S1190-01

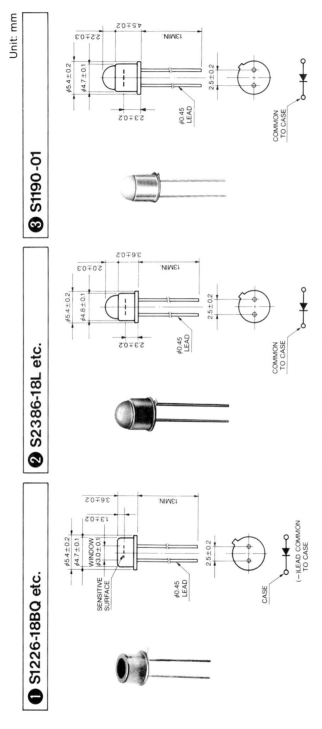

Figure 14-20 Physical dimensions of photodiodes. (Courtesy of Hamamatsu Corp., Bridge-water, NJ.)

S1226 Series (Metal Case)

Type No.	Outline	Photosensitive Surface Package (mm)	Size (mm)	Effective Area (mm²)	Spectral Response Range (nm)	Peak Wavelength (nm)	Peak Wavelength Typ.	200nm Min.	200nm Typ.	633nm He-Ne Laser Typ.	930nm GaAs LED Typ.	Short Circuit Current I_{sh} 100 lux Min. (µA)	Typ. (µA)
S1226-18BQ	①	TO-18	1.1×1.1	1.2	190~1000	720	0.35	0.08	0.1	0.33	0.17	0.5	0.65
-18BK					320~1000			–	–				
-5BQ	⑦	TO-5	2.4×2.4	5.7	190~1000			0.08	0.1			2.2	2.8
-5BK					320~1000			–	–				
-44BQ			3.6×3.6	13	190~1000			0.08	0.1			4.4	5.5
-44BK					320~1000			–	–				
-8BQ	⑩	TO-8	5.8×5.8	33	190~1000			0.08	0.1			12	15
-8BK					320~1000			–	–				

S1227 Series (Ceramic Case)

Type No.	Outline	Photosensitive Surface Package (mm)	Size (mm)	Effective Area (mm²)	Spectral Response Range (nm)	Peak Wavelength (nm)	Peak Wavelength Typ.	200nm Min.	200nm Typ.	633nm He-Ne Laser Typ.	930nm GaAs LED Typ.	Short Circuit Current I_{sh} 100 lux Min. (µA)	Typ. (µA)
S1227-16BQ	⑭	2.7×15	1.1×5.9	5.8	190~1000	720	0.35	0.08	0.1	0.33	0.17	2	2.5
-16BR	⑮				320~1000		0.42	–	–	0.39	0.20	2.2	2.8
-33BQ	⑯	6×7.6	2.4×2.4	5.7	190~1000		0.35	0.08	0.1	0.33	0.17	2	2.5
-33BR	⑰				320~1000		0.42	–	–	0.39	0.20	2.2	2.8
-66BQ	⑱	8.9×10.1	5.8×5.8	33	190~1000		0.35	0.08	0.1	0.33	0.17	11	14
-66BR	⑲				320~1000		0.42	–	–	0.39	0.20	13	16
-1010BQ	⑳	15×16.5	10×10	100	190~1000		0.35	0.08	0.1	0.33	0.17	32	40
-1010BR	㉑				320~1000		0.42	–	–	0.39	0.20	36	45

Figure 14-21 Specifications for a photodiode. (Courtesy of Hamamatsu Corp., Bridgewater, NJ.)

Figure 14-22   CNC machining operation using fiber optics for data handling.

end of the conversation demodulates the incoming signal. Both units are capable of transmitting and receiving. The photodiode converts the transmitted light signals into audio-rate electrical signals.

A similar application is found in the design of a compact disk player (Figure 14-23). A photodiode is frequently used to demodulate the stored pulse information on the CD disk, containing voice and music information. This information is stored in the form of reflective and nonreflective spots that have been embedded into the plastic disk. As the disk's surface is scanned by a laser light source, a photodiode is used to receive and demodulate the varying intensities of reflections created by the spots.

14-5 PHOTOTRANSISTOR

The *phototransistor* is another type of diode junction photo-sensitive device, in addition to silicon and germanium photodiodes, that utilizes the light-sensitive qualities of pn material. Unlike all of the other photo-conducting devices, however (with the single exception of the high-voltage-biased silicon photoconductive diode), the phototransistor possesses the ability to detect not only the presence of a light signal, but it can amplify its signal strength at the same time. In the next section we see how this is done.

Figure 14-23   Photodiodes being used in a CD player.

14-5.1 How the Phototransistor Works

To understand how a phototransistor works, we must first review how an ordinary transistor functions. In Figure 14-24 we see the schematic of a typical npn transistor (we could just as easily have used a pnp transistor for our discussion instead). We see the transistor in our circuit being used to amplify the very weak output voltages, of a microphone to drive a speaker. (This circuit, of course, has been highly

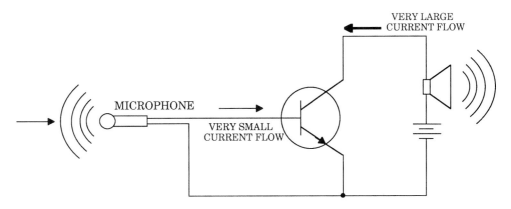

Figure 14-24   Highly simplified transistor amplifier.

simplified to emphasize the purpose of our discussion here.) Recalling that in a transistor, when an electron flow is drawn from the base of an npn transistor, this allows the much larger flow of electrons to take place from the power supply to pass through the transistor's emitter and collector circuits. To describe this in another way using conventional current flow, a very small insertion of current into the transistor's base causes a much larger flow of current to pass from the collector to the emitter. Assuming the transistor to be a proportional amplifier, we can then expect true reproduction of the sound coming from the microphone to be emitted from the speaker.

We now apply the same principle of operation to a phototransistor. In place of the microphone, we let light fall onto the base–emitter junction (Figure 14-25) instead. What with each photon being converted into an electron and the electron allowed to flow into the emitter region, this constitutes a withdrawal of electrons from the base just as in the case of the microphone circuit. As a result, a much larger current, a current that is proportional to the light's intensity falling onto the phototransistor, flows in the collector–emitter circuit. It is possible to obtain current gains that result in many hundreds of electrons being controlled by just one photoelectron. As a result of this electron control multiplication ability, the phototransistor is capable of very weak light detection.

14-5.2 Typical Construction

Phototransistors come in a variety of housing packages that often resemble ordinary discrete transistors. However, they can also be disguised in other packages. Figure 14-26 shows one such form that they can assume. The physical size and shape of the phototransistor's package is usually determined by the design of the phototransistor's optical input system. This is the system that focuses or concentrates incoming light onto the phototransistor's light-sensitive p-n junction area.

14-5.3 Phototransistor Characteristics

To fully appreciate the operating characteristics of the phototransistor we must first review the characteristics of a regular non-phototransistor. Remembering that a transistor is a current-sensitive device, that is, it amplifies current rather than voltage

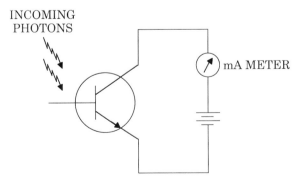

Figure 14-25 Phototransistor circuit.

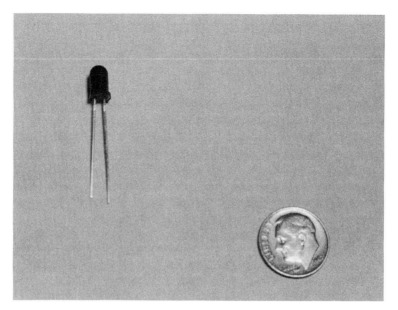

Figure 14-26   Phototransistor.

as was the case of its older predecessor, the vacuum tube, the gain, β, of a transistor is defined as the ratio of its collector current, I_c, to its base current, I_b. Looking at Figure 14-27, we see a typical family of base current curves for a low-power, low-gain transistor operating at various values of collector voltages. This information is plotted against the collector currents resulting for that transistor. In calculating a

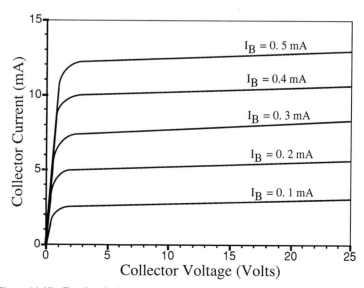

Figure 14-27   Family of collector voltage versus collector current curves for various base-current values for a transistor.

value for β using the data graphed in Figure 14-27, we obtain a value of approximately 25. This is done by dividing the transistor's collector current by its base current. For this particular case, since all the curves shown in this figure appear to have the same slope, we chose any one of these curves and divide the collector current indicated on the graph's *y*-axis (at any collector voltage value read on the linear portion of that curve) by the curve's labeled base current. For instance, choosing the $I_B = 0.3$ mA curve at a collector voltage of 5 V, we obtain a collector current of 7.5 mA. Therefore, β = 7.5 mA/0.3 mA = 25. This value verifies that we are dealing with a low-gain transistor. Silicon transistor gains typically range from about 20 to as high as 500.

Let us now look at the curve characteristics for a typical phototransistor (Figure 14-28). Immediately we see the similarity between its characteristic curves and those of the non-phototransistor. Perhaps the only difference in performance between the two transistors is in the amount of collector current controlled by each. In the case of the phototransistor the I_c operating values are usually much less than what is generally experienced in non-phototransistors. Because the base current values are not given for the various light exposure amounts, we cannot calculate the β-gain figure. However, β values take on the same range of values as the non-phototransistors.

As in the case of non-phototransistors, phototransistors are also subject to thermal runaway problems. Recalling for a moment what this problem is, *thermal runaway* in a transistor is a condition where the transistor becomes increasingly warmer by its own self-heat. This results in the transistor tending to conduct larger amounts of current. This is especially true for common emitter–type transistor

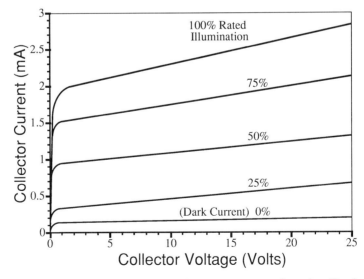

Figure 14-28 Collector voltage versus collector current curves for various illuminations for a photo transistor.

amplifier circuits. Unfortunately for the transistor, this produces a "snowballing effect"—the more conducted current, the warmer the transistor becomes. If allowed to go unchecked, this generally leads to the device's own self-destruction. This problem can be reduced if not entirely eliminated simply by inserting a resistor, R_{TH}, in series with the transistor's emitter circuit, as shown in Figure 14-29. What R_{TH} does is to produce a voltage drop that reduces the transistor's gain. As the current tries to increase through the emitter due to an increase in operating temperature, R_{TH}'s increasing voltage drop cuts back the transistor's desire to allow more current to flow by having its gain reduced by this increased voltage drop.

In the case of the phototransistor, an additional problem occurs. The ratio of light current to dark current flow in the phototransistor's collector is significantly affected by the value of R_{TH}. How the total base-to-emitter resistance affects the light-to-dark collector current ratio in a typical phototransistor is shown in the curves of Figure 14-30. As can be expected, the phototransistor's operating temperature also has a significant effect on this current ratio. Notice in Figure 14-29 that *relative illumination levels* are used to indicate the amounts of light exposed to the phototransistor. These levels are simply ratios of illumination used for relative comparison purposes only. For a relative illumination level value of say, 6, a change of 20°C in operating temperature could produce a change as large as 20-fold in the light-to-dark current ratio.

Spectral response characteristics

The spectral response for a typical phototransistor is shown in Figure 14-31. The peak response occurs in the infrared portion of the visible spectrum; the blue end of the response curve really does not exist. The cutoff frequency appears to be in

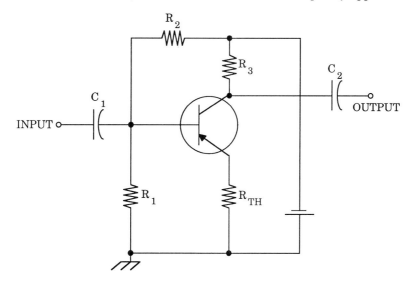

Figure 14-29 Correcting thermal runaway in a transistor circuit.

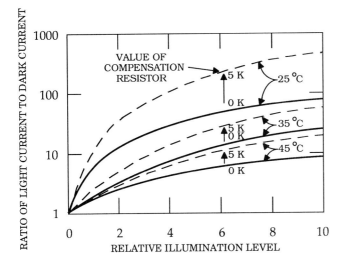

Figure 14-30 How the collector current is affected by the base–emitter resistance in a phototransistor. (From Walter H. Buchsbaum, *Practical Electronic Reference Data,* copyright 1978, p. 214. Reprinted by permission of Prentice Hall, Englewood Cliffs, NJ.)

the region of green light. However, just like the other photosensitive devices, they can be manufactured to be sensitive in virtually any portion of the visible spectrum, if desired. This can be done through special doping of the silicon or germanium material used in the transistor's construction and through the use of selective optical filtering in manufacturing the lens at the phototransistor's optical input.

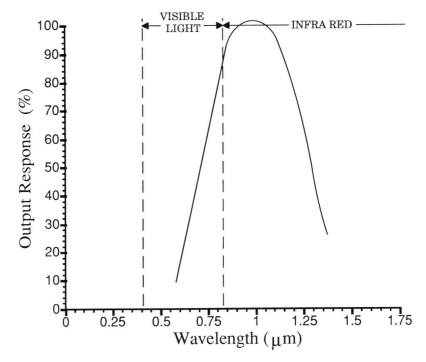

Figure 14-31 Spectral response for a typical phototransistor.

Responses time

The response time for a phototransistor is characteristically very short. Typical values are in the area of 1×10^{-8} s. This makes this device not quite as responsive as the photodiode.

14-5.4 Design Specifications

Typical catalog design specifications for a phototransistor are shown in Figure 14-32.

14-5.5 Practical Applications

The phototransistor's inherent ability to amplify its photoelectronic signal and to accomplish this inside a very small package makes the phototransistor a popular device in the area of industrial and consumer electronics. One very large application is found in the manufacturing of *opto-isolators* (Figure 14-33). These are devices that contain a light source and a photoconductive device, usually a phototransistor, used for the purpose of electrically isolating portions of electronic or electrical circuits from other portions. The opto-isolator may be considered a contactless high-speed switch having its input and output separated by an infinitely high impedance. The light source (the input) is most often a LED (light-emitting diode), while the output is the phototransistor's collector circuit. The voltage input supplies the necessary voltage (usually around 5 V or so) to fire the opto-isolator's LED. The phototransistor responds to the light by producing a replica of the voltage pulse that triggered the LED. This replica is of a higher magnitude than the original pulse, due to the phototransistor's β gain. Often, additional amplification is used to further increase the signal's amplitude, as in the case shown here.

In general, then, it can be said that the opto-isolator is used for isolating circuits for the purpose of reducing, if not eliminating, certain forms of circuit noise altogether. Also, opto-isolation is a very good solution to an interfacing problem where there is an impedance-matching problem. A LED is not sensitive to impedances and it only becomes necessary for the opto-isolator's output to match the impedance of the circuit it is "looking into." This can be done with a simple resistor network; often a single matching resistor will suffice.

Figure 14-34 shows another practical application for the phototransistor. In the schematic shown here we see the output of a phototransistor, Q_1, controlling the holding current for the coil of an electromechanical relay, K_1. When Q_1 is energized by a light source, K_1 is pulled in as a result of the collector current passing through Q_1. At the same time this current passes through D_1, an SCR, which is held in the conductive state as a result of its gate voltage being generated by the voltage drop across variable resistor, R_1. By pushing the reset switch, SW_1, the SCR may be switched off again. R_1 may be used as a sensitivity adjustment for the system. While this circuit represents the simplest circuit for this type of control, some modifications are necessary to accommodate the inductive problems created by the relay's holding coil. Usually, a diode of appropriate rating can be placed across the coil's terminal to suppress switching spikes and improve the circuit's performance.

 **MOTOROLA**

MFOD200

PHOTOTRANSISTOR FOR FIBER OPTICS SYSTEMS

. . . designed for infrared radiation detection in medium length, medium frequency Fiber Optic Systems. Typical applications include: medical electronics, industrial controls, security systems, M6800 Microprocessor systems, etc.

- Spectral Response Matched to MFOE100, 200
- Hermetic Metal Package for Stability and Reliability
- High Sensitivity for Medium Length Fiber Optic Control Systems
- Compatible with AMP Mounting Bushing #227015

MAXIMUM RATINGS (T_A = 25°C unless otherwise noted)

Rating (Note 1)	Symbol	Value	Unit
Collector-Emitter Voltage	V_{CEO}	40	Volts
Emitter-Base Voltage	V_{EBO}	10	Volts
Collector-Base Voltage	V_{CBO}	70	Volts
Light Current	I_L	250	mA
Total Device Dissipation @ T_A = 25°C Derate above 25°C	P_D	250 1.43	mW mW/°C
Operating and Storage Junction Temperature Range	T_J, T_{stg}	-55 to +175	°C

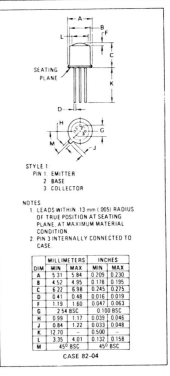

STYLE 1
PIN 1. EMITTER
    2. BASE
    3. COLLECTOR

NOTES
1. LEADS WITHIN .13 mm (.005) RADIUS OF TRUE POSITION AT SEATING PLANE, AT MAXIMUM MATERIAL CONDITION.
2. PIN 3 INTERNALLY CONNECTED TO CASE.

DIM	MILLIMETERS MIN	MAX	INCHES MIN	MAX
A	5.31	5.84	0.209	0.230
B	4.52	4.95	0.178	0.195
C	6.22	6.98	0.245	0.275
D	0.41	0.48	0.016	0.019
F	1.19	1.60	0.047	0.063
G	2.54 BSC		0.100 BSC	
H	0.99	1.17	0.039	0.046
J	0.84	1.22	0.033	0.048
K	12.70	–	0.500	–
L	3.35	4.01	0.132	0.158
M	45° BSC		45° BSC	

CASE 82-04

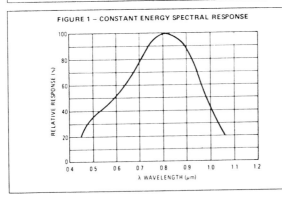

FIGURE 1 – CONSTANT ENERGY SPECTRAL RESPONSE

Figure 14-32 Typical specifications for a phototransistor. (Copyright of Motorola, Inc. Used by permission.)

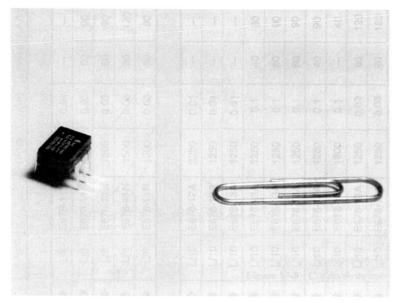

Figure 14-33   Opto-isolator.

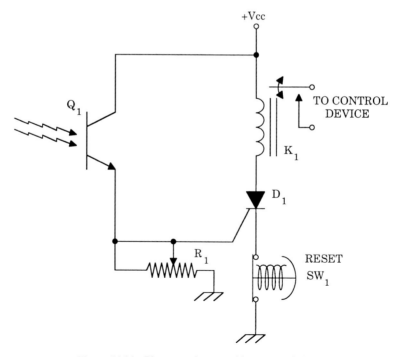

Figure 14-34   Phototransistor used in a control circuit.

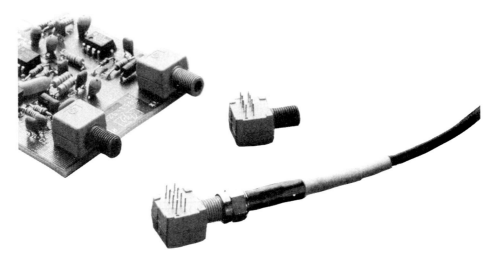

Figure 14-35 Fiber optic cable and receiver. (Courtesy of Hewlett Packard Co., San Jose, CA.)

Because the phototransistor is a linear device, it is well suited for handling modulated light frequencies of rather wide range. Consequently, phototransistors find extensive use in those applications where modulated light is transported over long distances requiring light signal relaying (i.e., repeating) or requiring final demodulation at the end of a lengthy transmitted light path. This type of application is particularly common where fiber optics are used for communications purposes (Figure 14-35).

14-6 PHOTOFET

The *photoFET,* or *photo field-effect transistor,* is similar to the phototransistor in many ways as far as operating characteristics are concerned. The only differences between the two devices are in the basic characteristics of the germanium or silicon bipolar transistor and the field-effect transistor. The FET is a voltage-actuated device, whereas the bipolar transistor is current-actuated. The FET is characterized by a high input impedance, whereas the bipolar transistor is characterized by a low input impedance. All in all, it can be said that the FET has many of the characteristics of the old pentode vacuum tube.

There are two common types of FETs: the junction field-effect transistor or *JFET,* and the metal-oxide-semiconductor field-effect transistor or *MOSFET.* Both types are used in making photoFETs. The only basic differences in their operating characteristics are that the MOSFET has a higher input impedance and the

MOSFET is more flexible in circuit applications since either a positive or a negative voltage can be applied to its gate. Their internal structures are quite different, but we are speaking here only of operating characteristics.

14-6.1 How the PhotoFET Works

To understand how the photoFET works, we must first review the operation of the ordinary FET. Figure 14-36 is a conceptual cross-sectional drawing of a FET. The flow of current within a FET is from the source through the channel to the drain side of the device. When voltage is applied to the gate, the channel may or may not form a restriction or resistance to current flow within this channel, depending on how the FET is designed. The current restriction is created by an electric field produced by the gate voltage, hence the term "field-effect transistor." FETs may be constructed as either p-type or n-type devices, just like the bipolar transistor. In Figure 14-36 the device shown is an *n-channel device*.

   Like the vacuum tube, depending on the FET's construction, one can obtain a wide variety of characteristic curves to fulfill a desired device behavior. Figure 14-37 shows the characteristic operating curves for two FET devices. These curves tells us that depending on the type of FET device, the gate voltage signal is the ultimate controlling signal for that device, hence making it a voltage-oriented device. For very small changes in this voltage (1 V or less), very large changes in drain–source current are produced. Also, the MOSFET [Figure 14-37(a)] has a gate that can take either a positive or negative voltage to produce a drain current. The JFET shown in Figure 14-37(b), on the other hand, requires a negative gate voltage to cut off the drain current. The JFET, by the way, is called a *depletion-mode* FET, where a negative voltage is required to deplete the FET's controlling channel causing the channel to become "pinched" off. Some MOSFETs are *enhancement-mode* devices in which a positive gate voltage enhances the conduction characteristics of the FET's

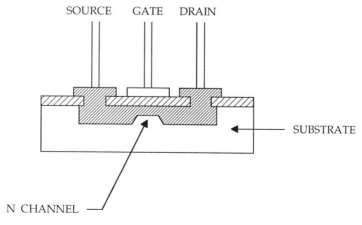

Figure 14-36   Internal construction of a FET.

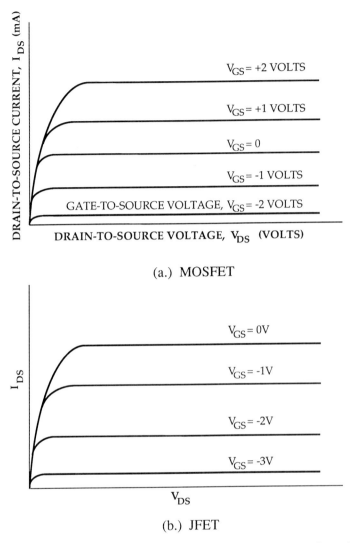

Figure 14-37 Drain-to-source voltage versus drain-to-source current for various gate voltages in a MOSFET and JFET.

controlling channel, thereby allowing a source–drain current to flow; a negative gate voltage restricts this channel, thereby reducing the source–drain current.

In making a JFET into a light-sensitive device, a transparent window is constructed over the one gate's pn junction. In the case of the MOSFET this window is placed over one of the gate's insulated terminal connections to the device itself. The incoming photons are converted to photoelectrons in very much the same manner that is experienced in the pn-type photo-sensitive semiconducting devices.

These electrons, in turn, cause the "squeezing" action or resistance within the p- or n-channel of the FET, which, in turn, controls the source–drain current flowing through this channel.

14-6.2 Typical Construction

It is difficult to distinguish photoFETS from phototransistors, just as it is difficult to tell non-photoFETs from non-phototransistors. The only distinguishing feature in both photo devices is the transparent lens construction for admitting light into the unit. This is built into each unit's top. Figure 14-38 shows a typical photoFET device.

14-6.3 Operating Characteristics

Figure 14-39 shows a typical set of characteristic curves for a photoFET. The amount of drain current variation between light-level changes, in other words the sensitivity, will vary markedly from one FET type to another. However, in principle, they all behave somewhat the same.

Spectral response and response-time characteristics

The spectral response and response-time characteristics for photoFETs of all types closely resemble those of the bipolar phototransistor. That is, the spectral response of photoFETs tend to favor the red end of the visible spectrum. Because of this,

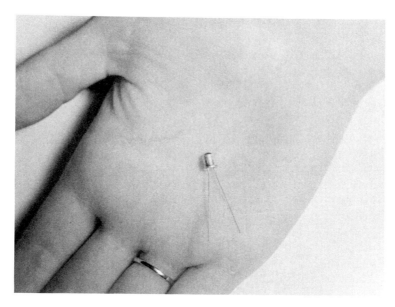

Figure 14-38 Typical photoFET.

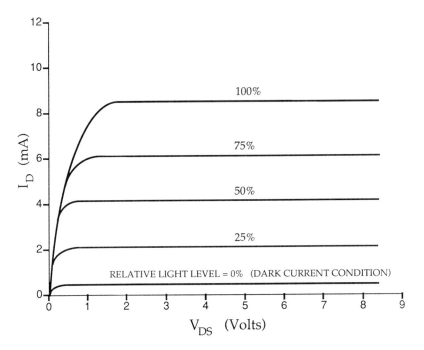

Figure 14-39   Typical characteristic curves for a photoFET.

photoFETs lend themselves to applications in the security detection industry, where invisible infrared light paths are not visible to the eyes of unauthorized personnel. The response times of the photoFET are approximately the same as those of the phototransistor: typically in the range 10^{-6} to 10^{-8}.

14-6.4 Practical Applications

As mentioned in Section 14-6.3, because of its spectral response characteristics, the photoFET lends itself to security detection or warning applications where an unauthorized person would not be able to see the infrared detection light beam paths used for protecting secured areas. A typical circuit using the photoFET is illustrated in Figure 14-40. In this circuit an infrared light source shines on the photo-sensitive surface of a FET device. As long as the beam is uninterrupted, the FET's circuit produces an output voltage at test point *a* in the circuit. This voltage is amplified by the power amplifier, whose output holds in relay *R*. As soon as the light beam is interrupted, the voltage at *a* drops to a very low value, causing the voltage at the relay coil to do the same. The relay then drops out, thus completing the alarm circuit. Other applications involve using the photoFET in the construction of opto-isolators similar to the one illustrated in Figure 14-33.

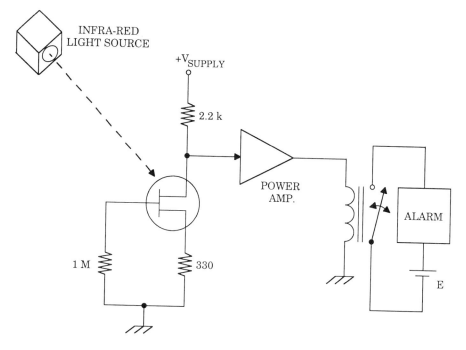

Figure 14-40   PhotoFET circuit.

14-7 LASCR

LASCR is an acronym for *light-activated silicon-controlled rectifier*. The LASCR is also called a *photoSCR*. The LASCR is an SCR device having its gating junction sensitive to light. This means that the SCR can be triggered into conduction by exposure to light, in addition to being triggered by an application of a small triggering voltage or current. We discuss the characteristics and behavior of this device in the following sections.

14-7.1 How the LASCR Works

In an ordinary SCR we have a diode that functions like any other diode as far as rectifying ac signals. However, the distinguishing feature of an SCR is that this diode can be turned on into its conduction state by an attached circuit called a *gate*. In other words, the gate control circuit tells the diode when to act as a rectifier. This gate control voltage is usually fairly small, usually 1 to 10 V or so, with a gate current measured in milliamperes.

   In the LASCR this gating circuit is replaced by a light-admitting window (Figure 14-41). This construction allows photons to strike a photo-sensitive silicon

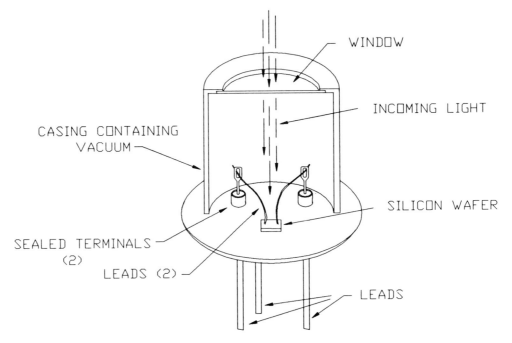

Figure 14-41 Light-activated SCR (LASCR).

circuit so that, when activated by light, the photoelectrons thus generated replace or augment the gating electrons that originally performed the diode switching.

14-7.2 Typical Construction

The photoSCR looks very much like an ordinary SCR except for the addition of its light-admitting window, pictured in Figure 14-41. The light-activating circuit has been constructed to assist the normally operated gating circuit, so that now, when the gate is exposed to light, very little additional gating current is needed to trigger the SCR. (See the additional discussion in the next section.)

14-7.3 Operating Characteristics

Because the photoSCR is yet another solid-state photo device, it is reasonable to assume that the photoSCR has operating characteristics very similar to the ones we have already discussed for the other devices. Figure 14-42 bears this out as we look at the LASCR's spectral response characteristics. The peak response once again

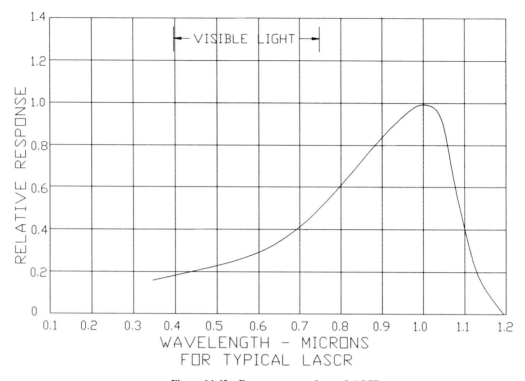

Figure 14-42   Response curve for an LASCR.

occurs in the infrared portion of the spectrum, in this case the peak response occurring at 1 μm.

The response time for a typical LASCR varies with voltage and current that is being switched; however, a response time of less than 0.5 μs can be expected.

14-7.4 Practical Applications

The LASCR may be thought of as a light-activated relay that is the solid-state equivalent of an electromechanical relay being controlled by a photocell and the relay allowing current to travel in only one direction. The LASCR has the capability of handling much larger voltages and currents than do many of the other light-sensitive devices that we have discussed. However, one of its drawbacks is a somewhat slower response time than those of the photodiode and phototransistor. Figure 14-43 shows a latching type of relay circuit, where a light triggers the LASCR, causing the load to become energized with the voltage supply. Removal of the light source does not affect the relay; it remains latched. The only way to shut off the LASCR is to interrupt the circuit by opening switch SW_1. This behavior is very similar to that of a SCR circuit.

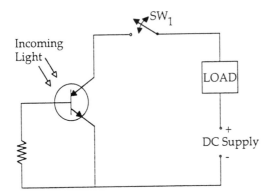

Figure 14-43 Latching relay circuit using an LASCR.

REVIEW QUESTIONS

14-1. Explain the difference between the *lumen* and the *candela*.

14-2. How can one determine velocity and acceleration from data determined from a photocell that was used to gather data for a plot showing time versus position for a particular machine member.

14-3. Show how force may be measured by using a photocell.

14-4. Explain the self-generating characteristics of a photovoltaic cell.

14-5. How are thermistors and photoresistors similar?

14-6. How does the photodiode have a photo-multiplying mode?

14-7. What are the three diode junction devices? List their characteristics.

14-8. What is an opto-isolator? Explain its function.

14-9. What is an LASCR? How are they used? Draw a simple control circuit showing how an LASCR controls a device.

14-10. Show how data can be transported through fiber optical cable using a light source at one end and a light receiver at the other. Design a simple light receiver using a photoFET for the detecting circuit.

PROBLEMS

14-11. A photovoltaic light meter produces a current reading that is proportional to the illuminance shining on it. If the light meter produces a reading of $5\mu A$ at a distance of 2 m from a light source, what would be the light meter's reading when the meter is held at a distance of 7 m?

14-12. A 5-cd light source is placed at one end of a 5-m stick. A 2-cd light source is placed at the opposite end. If a light meter is placed between the two sources, find the location along the 5-m stick where the light meter will read the same from both sources.

14-13. Using eq. (14-4) in Section 14-3.5 and Figure 14-14, design a circuit around the 555 timer IC that will produce a 2700-Hz signal for a photoresistor having a 750-Ω resistance value. Use commonly available values for C. Choose your final value for R that is closest to a common 5% tolerance resistor after making your calculations.

14-14. The general mathematical expression that describes the response time for any device is $y = 1 - e^{-t/\tau}$, where e is the natural log, 2.718, t is time in seconds, and τ is a constant; assume a value of 1 for τ. Plot a graph on x-y coordinate paper using this equation, plotting t as the abscissa and y as the ordinate. Do this very neatly. On your completed graph determine the approximate spot on the curve where it levels off horizontally. Drop vertically from this point to the x-axis and make a mark. Now divide the horizontal distance between this mark and the origin of your plot into six equal spaces, each space representing one time constant. Prove that after one time constant has elapsed on the x-axis, the y-axis value is a little over 63% of the total height of your curve.

REFERENCES

Allocca, John A., and Allen Stuart, *Transducers: Theory and Applications.* Reston, VA: Reston, 1984.

Fink, Donald G., *Electronics Engineers' Handbook.* New York: McGraw-Hill, 1975.

Photodiodes, 1989–90 Catalog. Hamamatsu Corporation (360 Foothill Rd., P.O. Box 6910, Bridgewater, NJ 08807–0910).

15

THE PHOTOEMISSIVE SENSOR

CHAPTER OBJECTIVES

1. To review the theory of photoemission.
2. To study the vacuum photodiode.
3. To study the image converter tube.
4. To understand how the gas-filled diode works.
5. To study the photomultiplier.
6. To understand the image intensifier tube and how it works.
7. To study the CCD.

15-1 INTRODUCTION

The expression *photoemissive* is a rather descriptive term, as we will soon learn. The photoemissive sensor warrants a separate chapter of explanation because of the method that these sensors use to make themselves sensitive to light. Photoemission is one of the earliest methods of light detection used in the development of light-sensitive devices. Because of the amount of development work done on these sensors over many years, much is known about the behavior of these devices. We begin our discussion with the necessary theory.

15-2 THEORY OF PHOTOEMISSION

The basic premise of photoemission is based on the phenomenon of a particle of light, a photon, colliding with, and imparting its energy of collision to, an electron associated with an atom. This collision causes that electron to become separated from its parent atom and to be propelled off into another direction. This is somewhat similar to what occurs in a game of billiards. But let us now discuss in more scientific terms exactly what actually goes on here in this process.

The energy content, E_{photon}, of an electromagnetic wavelength of light that is transferred when light falls on an object is easily determined by the following equation, which was first mentioned in Chapter 1 and is repeated here:

$$E_{photon} = hf \quad \text{[see eq. (1-15)]}$$

where E_{photon} = energy content of photon (J)

h = Planck's constant, 6.62×10^{-34} J-s

f = photon frequency (1/s)

Equation (1-15) tells us that this energy transfer is done in energy "packets" whose energy contents are proportional to the light's frequency. This equation also implies that since blue light is higher in frequency than red light, blue light has a higher energy content.

Having now established the energy capabilities of light, let's now find out how much energy is required to remove electrons from certain materials. The reason we want to do this is evident in Figure 15-1. We plan to bombard the surface of a material, the material comprising our light sensor, with light energy. Hopefully, in this process we will manage to dislodge electrons from the surface, so that, for a given input of light, we can create a flow of electrons as our sensor's output. A flow of output electrons will then constitute a current whose magnitude of flow will be ideally related to the sensor's incident light.

We now have to somehow determine the "bonding" energy of the electrons within our material. If we can exceed this energy level, referred to as that material's *work function*, we are then assured of being able to produce our output flow of current. The *maximum* bonding energy or work function level we can tolerate and still have a current flow will be an energy level equal to the value E_{photon}. In other words, E_{photon} must equal E_{wf} (energy of work function) or be greater. If E_{photon} is greater than E_{wf}, the "leftover" energy is imparted to the dislodged electron in the collision process in the form of kinetic energy. That is,

$$E_{photon} - E_{wf} = \frac{mv^2}{2} \tag{15-1}$$

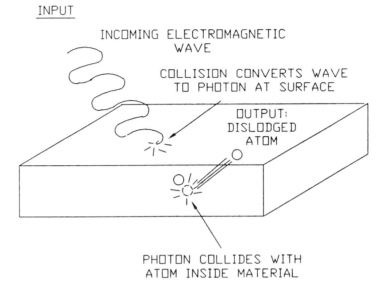

INPUT

INCOMING ELECTROMAGNETIC
WAVE

COLLISION CONVERTS WAVE
TO PHOTON AT SURFACE

OUTPUT:
DISLODGED
ATOM

PHOTON COLLIDES WITH
ATOM INSIDE MATERIAL

Figure 15-1   Incoming electromagnetic wave becomes converted into a photon at the surface of a light-sensitive material, the collision causing an electron within the material's surface to become displaced in the process.

where E_{photon} = energy of photon (J)

E_{wf} = energy of work function (J)

m = mass of electron (9.10×10^{-31} g)

v = velocity of impacted electron (km/s)

Or we can rewrite eq. (15-1) as

$$hf - E_{wf} = \frac{mv^2}{2} \qquad (15\text{-}2)$$

The values for E_{wf} are often stated in terms of volts for different materials rather than given in units of energy (see Table 15-1). Since the energy expended in moving one electron through a potential difference of v is equal to Q times v, where Q is the electron charge in coulombs, it is reasonable to assume that by taking any value of working function voltage from Table 15-1 and multiplying that value by 1.6×10^{-19} C/electron, the charge on one electron, we will obtain E_{wf} expressed in joules. That is,

$$E_{wf} = v_{wf} Q \qquad (15\text{-}3)$$

where E_{wf} = energy of work function (J)

v_{wf} = working function voltage (V, from Table 5-1)

Q = charge per electron (1.6×10^{-19} C/electron)

TABLE 15-1 WORKING FUNCTIONS FOR SOME MATERIALS (VOLTS)

Aluminum	3.38	Mercury	4.50
Barium	1.73	Nickel	4.96
Bismuth	4.17	Platinum	5.36
Cadmium	4.00	Potassium	1.60
Cobalt	4.21	Rubidium	4.52
Copper	4.46	Silicon	4.20
Germanium	4.50	Silver	4.44
Gold	4.46	Tantalum	4.10
Iron	4.40	Tin	4.09
Lead	3.94	Tungsten	4.38
Magnesium	3.63	Zinc	3.78
Manganese	4.14		

Let's look at an example problem that now demonstrates all that has been discussed so far concerning photoemission.

Example 15-1

Determine the frequency and wavelength of light needed to *just overcome* the work function of a surface made of barium so that one photon of this light will just dislodge one electron.

Solution: From Table 15-1 we find the work function for barium to be 1.73 V. We multiply this value by 1.6×10^{-19} C to obtain a work function energy level, E_{wf}, of 2.77×10^{-19} J [eq. (15-3)]. We now apply eq. (15-2), remembering that the kinetic energy component of this equation is zero since we are just dislodging an electron from the barium surface, therefore making the electron's velocity zero. We now solve for frequency, f.

$$hf = 2.77 \times 10^{-19}$$

$$f = \frac{2.77 \times 10^{-19}}{h}$$

$$= \frac{2.77 \times 10^{-19} \text{ J}}{6.62 \times 10^{-34} \text{ J-s}}$$

$$= 0.418 \times 10^{15} \text{ s}^{-1} \text{ or Hz}$$

To determine the light's wavelength, we use eq. (1-13):

$$c = \lambda f$$

Transposing, we get

$$\lambda = \frac{c}{f}$$

$$= \frac{3 \times 10^{8} \text{ m/s}}{0.418 \times 10^{15} \text{ s}^{-1}}$$

$$= 0.717 \times 10^{-6} \text{ m} \quad \text{or} \quad 0.717 \text{ } \mu\text{m}$$

This places the color of the light wave into the red portion of the spectrum.

What if the incoming photon in Example 15-1 had more than enough energy to dislodge an electron from the barium surface? What would be this electron's velocity? The following example explores that situation.

Example 15-2

The barium surface in Example 15-1 is now exposed to yellow light whose wavelength is 0.58 μm. Determine the velocity of the electrons dislodged.

Solution:
According to eq. (15-2),

$$\frac{mv^2}{2} = hf - E_{wf}$$

We must first determine f. Since

$$f = \frac{c}{\lambda} \quad \text{[from eq. (1-13)]} \tag{15-4}$$

we can then substitute eq. (15-4) into eq. (15-2) and solve for v:

$$\frac{mv^2}{2} = h\frac{c}{\lambda} - E_{wf}$$

$$v^2 = \frac{2}{m}\left(h\frac{c}{\lambda} - E_{wf}\right)$$

$$v = \sqrt{\frac{2}{m}\left(h\frac{c}{\lambda} - E_{wf}\right)} \tag{15-5}$$

Placing the known values into eq. (15-5), we get

$$v = \sqrt{\frac{2}{9.1 \times 10^{-34} \text{ kg}}\left[\frac{(6.62 \times 10^{-34} \text{ J-s})(3 \times 10^8 \text{ m/s})}{0.58 \times 10^{-6} \text{ m}} - 2.77 \times 10^{-19} \text{ J}\right]}$$

$$= \sqrt{1.437 \times 10^{14} \text{ m}^2/\text{s}^2}$$

$$= 1.2 \times 10^7 \text{ m/s}$$

It is interesting to note that the electron's velocity in Example 15-2 is traveling at approximately $\frac{1}{25}$ the speed of light.

15-3 COMMONLY SENSED MEASURANDS

Because we are dealing with yet another light-sensing method whose only difference from the photoconductive method described in Chapter 14 is the sensing mechanism used, the measurands that can be sensed by the photoemitter are the same as those mentioned in Section 14-3.1: position, motion, sound, temperature, flow, and force (pressure).

15-4 TYPES OF SENSORS PRESENTLY AVAILABLE

The types of photoemissive sensors that are presently being used both in industry and research fall under the following three broad categories: (1) the vacuum photodiode, (2) the gas-filled photodiode, and (3) the photomultiplier tube. There are other types of photoemissive sensors that are very complex in construction and have very esoteric designs, but we will cover only those that have readily been accepted by industry and are in common use today. There is one other type of photomultiplier tube that does deserve discussion here despite its complexity. This is a type of photomultiplier tube called an *image intensifier*. Because of the large amounts of research that have been done to perfect this device, and because of its value in security and research, its function is discussed later in this chapter in our discussion of photomultipliers.

Concerning construction, the three groups above all have one feature in common; they all are contained by a glass envelope called a *phototube*. Inside this tube is contained the photo-sensitive surface and photoelectron-gathering elements, all of which are protected from outside contaminants. We begin our discussion with the first category of photoemitters from our list.

15-5 VACUUM PHOTODIODE

The term *photodiode* results from the fact that the device is constructed from only two internal, electrically active components, an *anode* and a *cathode*. The name originated in the days of the vacuum tube, when tubes were identified by the number of internal components used in the amplification or rectification process. In the case of the vacuum phototube, the term *diode* does not refer to its mode of operation; that is, it does not rectify alternating current in the classic sense of a diode.

15-5.1 How the Vacuum Photodiode Operates

Figure 15-2 shows a simplified schematic of a vacuum photodiode. Light enters this unit through a transparent window and strikes the phototube's cathode. At this point the electromagnetic wave becomes converted to a photon, which, in turn, dislodges an electron from the cathode's surface. The dislodged photoelectron becomes attracted to the anode of the phototube because of a positive voltage placed on the anode relative to the phototube's cathode. The anode then transports all the photoelectrons to the exterior of the phototube to the signal-processing circuits located elsewhere. This current is usually then amplified so that it can perform a useful task.

15-5.2 Operating Characteristics

Spectral response

Because the cathode material is a factor in determining the vacuum photodiode's response frequency, and because such a wide variety of materials can be used for the cathode (each material having its own particular work function), there are

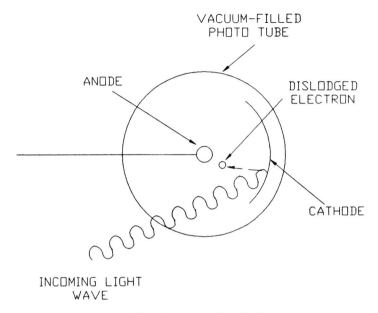

Figure 15-2 How the vacuum photodiode operates.

numerous tubes manufactured to suit a wide range of response needs. Figure 15-3 shows typical response curves for a wide variety of phototubes. It is obvious from these curves that peak responses are available from the tubes from well outside the range of visible light.

As was the case for certain photoconductors explained in Chapter 14, vacuum phototubes also exhibit a dark current characteristic when completely shielded from radiation. The origin of this current can be traced to *thermionic emission* taking place at the phototube's cathode. This emission results from the more positive voltage at the anode literally pulling off electrons from the cathode's surface because of slight warming of this surface due to ambient (above absolute zero) temperature conditions. Artificial cooling of the cathode can significantly reduce this emission and thus increase the phototube's sensitivity for very weak light sensing.

Another typical characteristic of the vacuum phototube is its relatively low current output for a given light input to the cathode. Typical anode current values are tens of microamperes per lumen of light. (Refer to Section 14-2 for a review of the lumen.) Compared to other photo-sensitive devices, this output amount is quite low. Because of this low output the phototube's signal-to-noise ratio is rather low.

15-5.3 Typical Construction

Vacuum phototubes are usually constructed from a glass envelope mounted on a plastic or phenolic base similar to the old 8-pin octal-tube sockets that were popular several years ago. The glass envelope contains the anode and cathode. The cathode

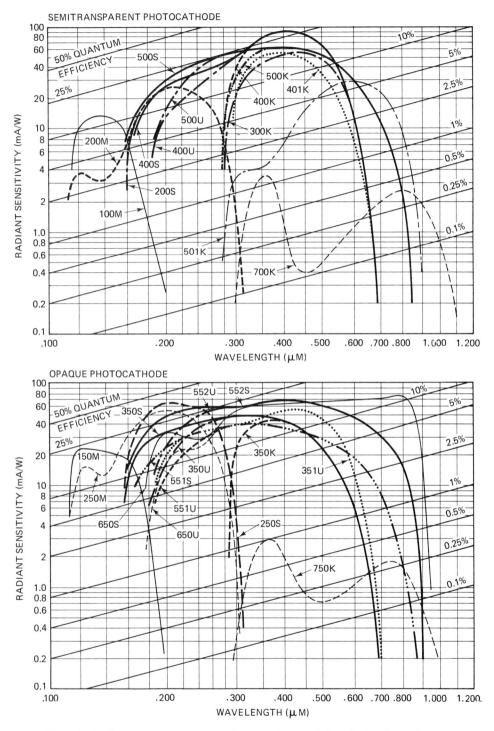

Figure 15-3 Frequency response curves for several types of photodiode tubes. (Courtesy of Hamamatsu Corp., Bridgewater, NJ.)

is so constructed that light must enter from the side of the tube to activate the anode rather than entering through the tube's top end. This configuration, called the *side-on* configuration, is used in most vacuum phototubes. Typically, the phototube is about 3 in. high by approximately 2 in. in diameter.

The cathode of the vacuum phototube is coated with a metal whose work function has been specially chosen for its response to a particular range of desired light frequencies. In addition, the cathode is usually curved, creating a focal point that terminates at the anode, which is the collection point for electrons leaving the cathode.

The phototube's glass envelope is evacuated to remove as many obstacles as possible from the paths of the fast-moving photoelectrons leaving the cathode. Any collisions with other particles along the way only deflects the electrons or slows their forward velocities and therefore eventually reduces the overall current flow within the anode.

Not only do the cathode materials determine the spectral response for the phototube, but so does the transparent window admitting the incident light. The makeup of this window can be formulated to act as a filter to pass only small selected portions of the light spectrum. For instance, ordinary glass blocks ultraviolet radiation and freely admits infrared radiation, whereas certain glasslike quartz materials operate in just the opposite manner. They transmit ultraviolet radiation but block out the infrared.

Because an anode voltage is required for the proper operation of this device, external power sources are required. Typical anode voltages may be as low as 90 V dc to as high as several thousands of volts dc, depending on the unit's design and purpose.

15-5.4 Practical Applications

Probably the earliest use for the vacuum photodiode was in the motion picture industry in the late 1920s. This was when the first "talkies" were produced. The vacuum phototube was used in the motion picture projector to convert the sound track recorded alongside the photoframes into sound. The phototube was well suited because of the high-temperature environment created by the arc lamps inside the projector. It was very rugged and performed remarkably well for its day. Unfortunately, because of its characteristically low output of current for a given light input, it lacked sensitivity and modulated frequency response. The vacuum phototube was later replaced by a more sensitive phototube, the gas-filled tube. But later, development of more compact and lower-voltage-operating photocells gradually replaced phototube devices completely for this particular application.

A typical vacuum phototube circuit is shown in Figure 15-4. The phototube's output appears across load resistor, R_L, which is in series with the tube's anode voltage supply, E_s. Typical supply voltages are in the upper tens or low hundreds of volts, whereas typical current flows that are experienced in this circuit are in the hundreds of microamperes to perhaps as high as 1 mA or so. Any variations in the

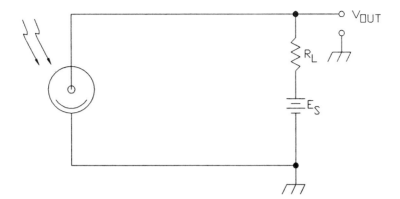

Figure 15-4   Simplified vacuum phototube circuit.

anode-to-cathode current because of light-intensity variations will cause a proportional variation of voltage across the load resistor. Typically, this output voltage is then sent to another amplifier stage for further processing, eventually to drive an actuator such as a relay, motor, or other actuator.

15-6 IMAGE CONVERTER TUBES

Image converter tubes are a special adaptation of the vacuum photodiode tube. These special devices were developed during World War II to enable ground troops to see the enemy at night. This was possible through a unique optical conversion process involving a substance acting as the cathode that was sensitized to infrared radiation. Although this in itself was not unusual, the construction of the phototube *was* unusual.

Figure 15-5 shows a schematic of this device. The entire tube is constructed of glass. The front of the glass is coated with a layer of cesium. This material possesses the desired ability to convert infrared radiation into photoelectronic radiation. A telescopic objective lens and image-inverting optical system is used to project an erect infrared image of the target onto the cathode's surface. The target is "illuminated" with an infrared spotlight which is, of course, invisible to the target. At the converter tube's cathode, electrons are ejected from the back side of the cathode and accelerated to the rear of the tube by means of a very high positive anode voltage, usually operated between 3000 and 6000 V dc. The velocities of these electrons are in direct proportion to the intensity of the infrared image formed on the cathode's front side.

At the back end of the phototube is the anode. This is a glass surface coated with zinc sulfide containing an embedded wire gridwork. The grid is charged with the extremely high voltage (but at a very low current for personal safety) which attracts the dislodged photoelectrons streaming from the tube's front end. When

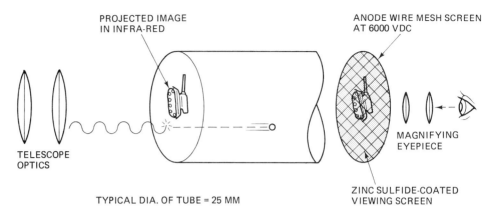

Figure 15-5   Image-converter tube.

these electrons strike the anode they cause the zinc sulfide surface to fluoresce in proportion to their accelerated velocities. As a result, a reconstituted visual image appears on this screen and can be inspected in detail under the high magnification of an eyepiece located at the rear of the converter system.

This type of converter tube system can also be designed to be sensitive to the ultraviolet end of the visible spectrum. Converter tubes having this capability are used by scientists for determining the fluorescent characteristics of minerals and other materials when exposed to ultraviolet radiation.

15-7 GAS-FILLED PHOTODIODE

The gas-filled photodiode was developed to correct a serious deficiency in the vacuum photodiode—low sensitivity. The theory of operation of the gas-filled photodiode is explained next.

15-7.1 How the Photodiode Works

In the case of the vacuum photodiode, recall that every photon that struck the surface of the cathode caused the release of one electron toward the tube's anode. This one-for-one relationship did nothing as far as signal amplification was concerned. On the other hand, the phototube's sensitivity could be greatly enhanced if somehow it were possible to release more electrons for each photon colliding with the cathode's surface. This is exactly the principle behind how the gas-filled phototube works.

In place of the usual vacuum in a phototube a gas is installed that is easily ionized. That is, the electrons that comprise the outer shell of this gas's molecules are easily removed or stripped from the nucleus. When the initial electron that is dislodged from the cathode strikes a molecule of this gas, one or more electrons are released as a result of the original incoming photon. Because of this ricocheting effect, considerable anode current amplification is attained.

The anode voltages used in the gas-filled photodiodes are generally lower than those used in vacuum photodiodes. The reason for this is to prevent the ionization of the gas due strictly to the high anode voltage. Should ionization happen, the gas acts as a dead short between the cathode and anode, allowing the conductance of large amounts of uncontrolled current flow to take place.

15-7.2 Operating Characteristics

Virtually all the operating characteristics of the vacuum phototube are retained in the gas-filled phototube, with one or two notable exceptions. One exception is the signal-to-noise ratio. Because the anode currents are considerably higher in the gas-filled tube, typically five to six times as high per lumen of light exposure, the signal-to-noise ratios are also correspondingly higher. Luminous sensitivity values ranging from 100 μA/lm to as much as 300 μA/lm are common.

Another notable exception is the sensitivity to frequency modulation. The vacuum photodiode has a sensitivity to only very high modulation frequencies in the audio range. Low frequencies are not detected as well by the phototube. On the other hand, the gas-filled phototube has a low-frequency response characteristic which allows it to function well at the low end of the audio band. Because of this, the vacuum-tube phototubes were eventually replaced by gas-filled models on all motion picture projectors because of their superior sound reproduction capabilities.

One serious drawback to the gas-filled tube, however, is the fact that unlike its vacuum phototube counterpart, it is a nonlinear device. It also has a significantly slower response time than that of the vacuum phototube.

15-7.3 Typical Construction

The only difference in construction between the gas-filled and the vacuum phototube is the gas used in the former device. Otherwise, physically, it is very difficult to tell one from the other when placed side by side.

15-7.4 Practical Applications

As mentioned in Section 15-7.2, one of the primary applications of the gas-filled photodiode was in the motion picture industry. However, even these photodiodes were eventually replaced with more modern solid-state light detection systems such as the photovoltaic cell (described in Chapter 16) or the photoconductive cell (Chapter 14).

15-8 PHOTOMULTIPLIER TUBE

Another type of photoemissive device is the *photomultiplier tube* (PMT). While the gas-filled photodiode permits weaker signal detection, there is still a need to obtain even weaker signal-detecting capabilities. The way the photodiode is modified to increase its sensitivity is described in the following sections.

15-8.1 How the Tube Works

Just like the vacuum photodiode, the photomultiplier tube incorporates the cathode emitter and anode collector. However, the photomultiplier has an added feature, an electron multiplier section, located between the cathode and the anode (Figure 15-6). This section consists of additional plates called *dynodes*, made from low-work-function material similar to that of the cathode. However, these additional plates all have more positive anode-like voltages, with each adjacent plate having a somewhat higher voltage than the plate before it. Photomultiplication with this system works like this: A photon strikes the cathode of a photomultiplier tube as shown in Figure 15-7. (The dynodes shown in this figure are somewhat more

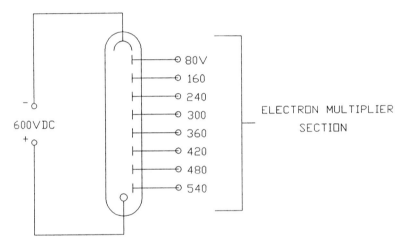

Figure 15-6 Dynodes (anodes) in a photomultiplier tube.

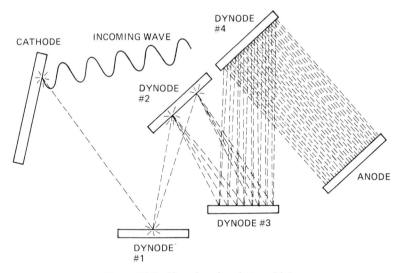

Figure 15-7 How dynodes photomultiply.

representative of the phototube's actual construction than what is shown in Figure 15-6). When exposed to light, photon dislodges the usual electrons from the cathode. These electrons are immediately attracted to the first dynode plate. Upon colliding with this first dynode several additional electrons are dislodged for every incident electron; this additional episode or generation of electrons is called *secondary emission*. This new batch of electrons is then attracted to the next dynode. Each colliding electron dislodges several more electrons, which are attracted to the next dynode in line, and so on. What began as a single electron therefore ends with a deluge of electrons entering the tube's anode. This increase in flow is in direct proportion to the light intensity entering the tube.

15-8.2 Operating Characteristics

Photomultiplier gain

The gain of a photomultiplier tube is dependent on the number of dynodes it has and on the magnitude of the voltage between its anode and cathode. The relationship between gain and dynode and anode voltage is quite nonlinear and can be represented approximately by the equation

$$G = K\left(\frac{V}{n + 1}\right)^{0.75n} \tag{15-6}$$

where G = overall gain of multiplier tube

$\quad\quad K$ = a constant furnished by the manufacturer

$\quad\quad V$ = anode-to-cathode voltage (V)

$\quad\quad n$ = number of dynodes used in photomultiplier's construction

Inspecting eq. (15-3), we see that for very small changes in anode-to-cathode voltage, rather large variations in the photomultiplier's gain are produced. Let's look at an example problem.

Example 15-3

Assume that a particular photomultiplier tube contains six dynodes. The anode voltage is 500 V. Determine the percentage change in gain due to a 10-V increase in anode voltage.

Solution: Substituting the initial values above into eq. (15-3), we calculate a G_1 value:

$$G_1 = K\left(\frac{500}{6 + 1}\right)^{0.75 \times 6}$$

$$= K{\cdot}220 \times 10^6$$

We now perform the same calculation using the 510-V value.

$$G_2 = K\left(\frac{500}{6 + 1}\right)^{0.75 \times 6}$$

$$= K{\cdot}241 \times 10^6$$

Comparing the two gains, we get

$$\frac{G_2 - G_1}{G_1} \times 100 \quad \text{or} \quad \frac{K \cdot 241 \times 10^6 - K \cdot 220 \times 10^6}{K \cdot 220 \times 10^6} \times 100 = 9.5\%$$

A 2% change in anode voltage produces a 9.5% change in gain in the photomultiplier.

We conclude from Example 15-3 that because the gain changes so markedly due to a smaller percentage change in anode voltage, it is essential that the photo-tube's power supply dc voltage be very closely regulated. Even the smallest voltage fluctuations will manifest themselves in the form of larger fluctuations in tube gain. Typical dc supply voltages for the photomultiplier tube range from 1000 to as high as 3000 V. A typical current amplification factor is 1×10^6.

Sensitivity

The transfer function of a photomultiplier tube's cathode is, in reality, its sensitivity. In other words, taking the ratio of its output, measured in microamperes, versus its input, light, measured in lumens, produces its sensitivity figure, expressed in the units $\mu A/lm$. (The color of light used in making this determination is from a pure tungsten white-light source.) The type of cathode material will determine the magnitude of this figure. Table 15-2 lists some of the most used substances for cathodes.

TABLE 15-2 PROPERTIES OF TYPICAL PHOTOCATHODE MATERIALS

Cathode material	White-light sensitivity ($\mu A/lm$)
Rb_3Sb	25
K_3Sb	60
Cs_3Sb	300
$Na_2KSb + Cs_3Sb$	450
$Si + Cs_2O$	500
$InAsP + Cs_2O$	1200
$GaAs + Cs_2O$	2062

Spectral response

Determining the theoretical spectral response of a photomultiplier tube can be frustrating. In addition to the individual spectral response of the cathode (Figure 15-8 shows the spectral response curves for the cathode materials listed in Table 15-2), we also have the response characteristics of each dynode to contend with. Therefore, when the spectral response curve for the entire phototube is actually measured, its curve looks generally as shown in Figure 15-9. The curve showing the peak response occurs in the infrared portion of the light spectrum. However, through the application of filters and using different cathode materials, it is possible to have the peak occur in virtually any portion of the light spectrum. We discuss the types of filters and cathode construction materials used in more detail below.

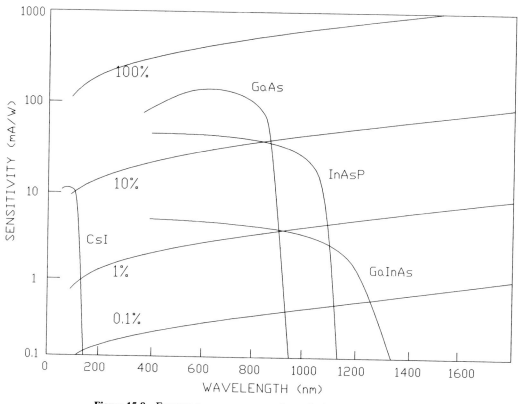

Figure 15-8   Frequency response curves for cathode materials listed in Table 15-2.

Magnetic field susceptibility

Because the photomultiplier contains lengthy paths of traveling electrons going from one dynode to another, the otherwise straight-line paths can be seriously distorted through the introduction of magnetic fields. Any electronic device containing electron paths between a cathode and an anode has this potential problem. In the case of the photomultiplier tube, however, this becomes especially critical because of the very high amplification factors involved. The slightest path deviation is exaggerated, creating intolerable distortion of the signal arriving at the anode. Special design construction steps must be taken to lessen this problem. These steps are discussed in the following section.

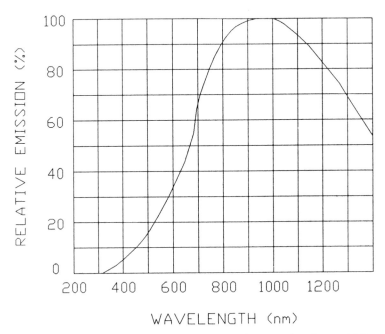

Figure 15-9 Typical frequency response curve for an entire photomultiplier assembly.

15-8.3 Typical Construction

There are two basic construction types used in photomultiplier tube designs: *end-on* design and *side-window* design, both of which are illustrated in Figure 15-10. End-on design uses a semiopaque cathode that is incorporated into the glass end of the phototube. This is done by fusing the cathode to the underside of the tube's glass end or by using an adhesive. Unfortunately, this type of construction has one serious drawback: the head-on construction is more susceptible to stray magnetic fields (see the discussion above) than is the alternative construction method. However, end-on construction does create better distortion-free signals at the anode, due to fewer tight turns in the paths needed in this type of construction. (This will be better understood in the discussion below concerning side-window construction.) Because of the straighter paths employed, the anode's electron collection efficiency is better.

In side-window design the incoming radiation enters the phototube through the side rather than through the top. Because of this, the phototube can be made much more compact. The dynodes may be assembled in a circular, spiraling fashion that conserves space within the glass tube's envelope. This tends to shorten the electrons' path of travel between dynodes, and consequently, reduces the problem of stray magnetic field interference. Unfortunately, because of the sharp turning and twisting of the electron paths, the likelihood of signal distortion becomes somewhat greater.

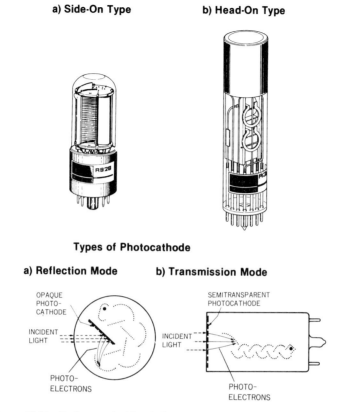

Figure 15-10 End-on and side-window photomultiplier designs. (Courtesy of Hamamatsu Corp., Bridgewater, NJ.)

Four basic dynode arrangements are used in the construction of photomultiplier tubes. Figure 15-11 illustrates these arrangements. Each type has its own construction and operating characteristics.

Venetian blind type

This is probably the simplest type of dynode construction (Figure 15-12). It allows for large total dynode surface areas, resulting in larger current-handling capabilities and better uniform signal density across the anode surface. Used only in end-on-type tubes. This type has the poorest response time of the four types.

Box-and-grid type

The dynode construction in this type is comprised of cylindrical quarter-sections arranged in cascading fashion as shown in Figure 15-13. This is a popular design found in many end-on phototubes.

(a) Circular-Cage Type

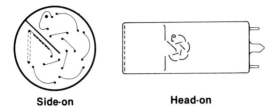

Side-on   **Head-on**

(b) Box-and-Grid Type

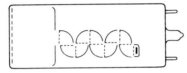

(c) Linear Focused Type

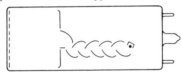

(d) Venetian Blind Type

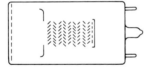

Figure 15-11 The four types of dynode construction. (Courtesy of Hamamatsu Corp., Bridgewater, NJ.)

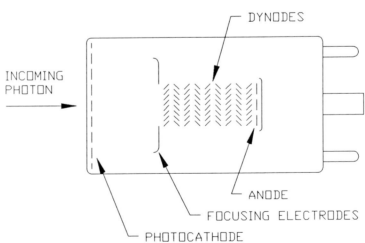

Figure 15-12 Construction detail of the venetian blind dynode. (Courtesy of Hamamatsu Corp., Bridgewater, NJ.)

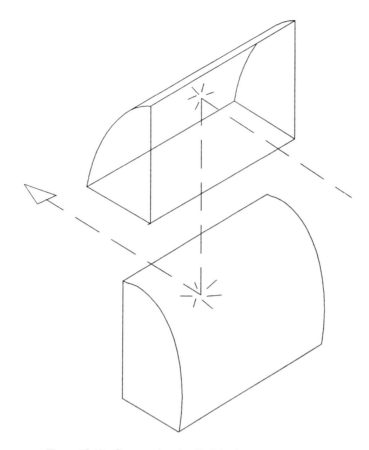

Figure 15-13 Construction detail of the box-and-grid dynode.

Circular cage type

This type requires the least volume of the four styles, resulting in very compact construction. This type is found in both end-on and side-window tubes. The side-window configuration results in the most compact design, as evident in Figure 15-14.

Linear type

The linear type (Figure 15-15) has the fastest response time of the four types. However, this type also requires a larger housing than the other types. Used only in end-on tubes.

15-8.4 Typical Design Specifications

Typical catalog specification sheets for head-on and side-on photomultiplier tubes are given in Figures 15-16 and 15-17, respectively.

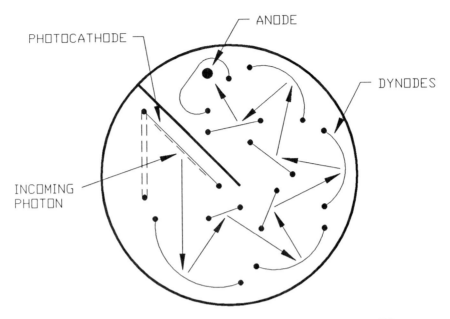

Figure 15-14   Construction detail of the circular-cage dynode. (Courtesy of Hamamatsu Corp., Bridgewater, NJ.)

15-8.5 Practical Applications

Because of their susceptibility to stray magnetic fields, most photomultiplier tubes
are shielded with a grounded cover. This is especially true for linear tubes having
a higher exposure risk than the other tube types.

Photomultipliers are used today for the detection of extremely weak light
sources. One application is in nuclear physics, where light radiation is often the
by-product of a nuclear reaction. Photomultiplier tubes are used to count and record
the occurrence of single photon emissions from certain reactions. Additional amplification of the photomultiplier's output current is often needed to activate a
recorder's marking pen or to move an actuator of some sort, such as a servomotor.

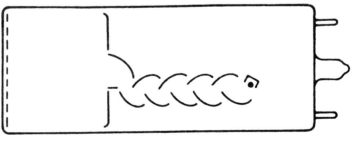

Figure 15-15   Construction detail of the linear dynode.

3/8 inch (10 mm) Dia. Types

Type No.	Remarks	Spectral Response B Curve Code	Range (nm)	Peak Wavelength (nm)	C Photocathode Material	D Window Material	E Outline No.	F Dynode Structure / No. of Stages	G Socket	Maximum Ratings H Anode to Cathode Voltage (Vdc)	I Average Anode Current (mA)	Cathode Sensitivity Luminous J Min. (µA/lm)	Typ. (µA/lm)
R1893	Solar blind response, Cs-Te photocathode, synthetic silica	200S	160~320	210	Cs-Te	Q	1	L/8	E678-11N	1500	0.01	—	—
R1635	Visible response, bialkali photocathode	400K	300~650	420	Bi	K	1	L/8	E678-11N	1500	0.03	60	90
R1635-02	For photon counting, HA coating with magnetic shield				Bi	K	1	L/8	E678-11N	1500	0.03	60	90
R1894	S-20 response, multialkali photocathode	500K (S-20)	300~850	420	M	K	1	L/8	E678-11N	1500	0.03	80	120
*R2496	Low TTS, high speed time response	400S	160~650	420	Bi	Q	2	L/8	E678-11N	1500	0.03	60	90

1/2 inch (13 mm) Dia. Types

Type No.	Remarks	Spectral Response B Curve Code	Range (nm)	Peak Wavelength (nm)	C Photocathode Material	D Window Material	E Outline No.	F Dynode Structure / No. of Stages	G Socket	Maximum Ratings H Anode to Cathode Voltage (Vdc)	I Average Anode Current (mA)	Cathode Sensitivity Luminous J Min. (µA/lm)	Typ. (µA/lm)
R1081	For VUV detection, MgF2 window, temporary base	100M	115~200	140	Cs-I	MF	5	L/10	E678-12A	2250	0.01	—	—
R1080	Variant of R1081 with Cs-Te photocathode	200M	115~320	210	Cs-Te	MF	5	L/10	E678-12A	1250	0.01	—	—
R759	Solar blind response, Cs-Te photocathode, synthetic silica	200S	160~320	210	Cs-Te	Q	3	L/10	E678-13A	1250	0.01	—	—
R647	10-stage, bialkali photocathode				Bi	K	3	L/10	E678-13A	1250	0.1	40	80
R647-01	Selected for scintillation counting	400K	300~650	420	Bi	K	3	L/10	E678-13A	1250	0.1	60	90
R647-04	For photon counting, HA Coating with magnetic shield				Bi	K	4	L/10	E678-13A	1250	0.1	60	90
R760	Variant of R647 with synthetic silica window	400S	160~650	420	Bi	Q	3	L/10	E678-13A	1250	0.1	40	90
R1591-01	High temp. bialkali photocathode, ruggedized type	401K	300~650	375	H Bi	K	3	L/10	E678-13E	1800	0.1	—	40
R1463	High gain variant of R761				M	U	3	L/10	E678-13A	1250	0.03	80	120
R1463-01	For photon counting, HA Coating with magnetic shield	500U	185~850	420	M	U	4	L/10	E678-13A	1250	0.03	80	120

Figure 15-16 Typical specifications for head-on photomultiplier tubes. (Courtesy of Hamamatsu Corp., Bridgewater, NJ.)

(at 25°C)

Cathode Sensitivity			Anode to Cathode Supply Voltage (Vdc)	Anode Characteristics									Notes	Type No.
Blue Typ. (μA/lm-b)	Red/White Ratio Typ.	Radiant Typ. (mA/W)		Anode Sensitivity			Current Amplification Typ.	Anode Dark Current (after 30 min.)		Time Response				
				Luminous Min. (A/lm)	Luminous Typ. (A/lm)	Radiant Typ. (A/W)		Typ. (nA)	Max. (nA)	Rise Time Typ. (ns)	Electron Transit Time Typ. (ns)			
—	—	12[b]	1250 ⑤	—	—	3600[b]	3.0×10^5	0.5	2.5	0.8	7.8			R1893
10.5	—	82	1250 ⑤	30	90	8.2×10^4	1.0×10^6	5.0	50	0.8	7.8	Silica window type (R2055) available.		R1635
10.5	—	82	1250 ⑤	50	200	1.8×10^5	2.2×10^6	100[c] cps	400[c] cps	0.8	7.8			R1635-02
—	0.2	51	1250 ⑤	10	50	2.1×10^4	4.2×10^5	2.0	20	0.8	7.8			R1894
10.5	—	82	1250 ⑧	30	90	8.2×10^4	1.0×10^6	5.0	50	0.7	7.8			R2496*
—	—	9.8[a]	2000 ⑪	—	—	980[a]	1.0×10^5	0.03	0.05	1.8	18			R1081
—	—	20[b]	1000 ⑪	—	—	$2.0\times10^{4\,b}$	1.0×10^6	0.3	0.5	2.5	24			R1080
—	—	20[b]	1000 ⑪	—	—	$1.0\times10^{4\,b}$	5.0×10^5	0.3	1.0	2.5	24			R759
9.5	—	75	1000 ⑪	30	80	7.5×10^4	1.0×10^6	5.0	15	2.5	24	UV glass window type (R960) available.		R647
10.5	—	80	1000 ⑪	30	90	8.0×10^4	1.0×10^6	1.0	2.0	2.5	24			R647-01
10.5	—	80	1000 ⑪	70	200	1.8×10^5	2.2×10^6	80[c] cps	400[c] cps	2.5	24			R647-04
10.5	—	80	1000 ⑪	10	90	8.0×10^4	1.0×10^6	5.0	15	2.5	24			R760
6.0	—	50	1500 ⑪	10	20	2.5×10^4	5.0×10^5	0.5	10	2.0	20	Flexible lead type (R1591) available.		R1591-01
—	0.2	51	1000 ⑪	30	120	5.1×10^4	1.0×10^6	4.0	20	2.5	24	HV resistant type (R761) and silica window type (R2007) available.		R1463
—	0.2	51	1000 ⑪	30	120	5.1×10^4	1.0×10^6	900[c] cps	1000[c] cps	2.5	24			R1463-01

Figure 15-17 Typical specificatons for side-on photomultiplier tubes. (Courtesy of Hamamatsu Corp., Bridgewater, NJ.)

Astronomers use photomultipliers to record the light intensities of distant stars. The photomultiplier tube is mounted at the focal point of a telescope's objective lens or mirror where the telescope's light-gathering power is most concentrated. In this particular application, the telescope is used much like a huge telephoto lens, with the photomultiplier replacing the observer's eyepiece. Figure 15-18 shows such a telescope with an installed photomultiplier ready for use. With an arrangement like this it is possible to measure a star's light intensity even though it may not be visible to the naked eye, and it is hundreds or perhaps thousands of light years away from the earth (1 light year is the equivalent of 5.9×10^{12} miles).

15-9 IMAGE INTENSIFIER TUBE

Image intensifier systems are used for the purpose of enhancing the vision of an object through the amplification of available light that is illuminating that object. A common application of this type of system is for night vision. In the following sections we discuss how this is done. Also, we discuss the solid-state equivalent to the image intensifier tube, the CCD.

Figure 15-18   Telescope with installed photomultiplier system.

15-9.1 How the Tube Works

Essentially, there are two types of image-enhancement systems. Both systems use the familiar cathode emitter and the anode found in all photoemission tubes. However, in place of the secondary emission dynodes found in photomultipliers, a system of plates is used to concentrate or focus the photoelectrons as they stream back toward the anode. There are two methods of activating these focusing plates, thereby creating the two types of image intensifiers just mentioned.

Electrostatic-focus image intensifier

In Figure 15-19 we see the schematic of an *electrostatic-focus* image intensifier tube. The initial image to be intensified is projected onto a flat surface comprised of many thousands of fiber optic bundles. These bundles transport the image back to the cathode, where it is converted to photoelectrons. At this point the electrons are attracted and then repelled into a narrow beam by high electrostatic voltages (the term *electrostatic* merely indicates that the voltages are constant, similar to how they are on the dynodes in photomultiplier tubes) from nearby electrodes constructed on the inside walls of the intensifier tube. An *anode cone* electrode serves as a concentrator, directing the electrons into a *microchannel plate array*. This is a device that consists of an array of long glass tubes, each tube being similar to a fiber optics rod except that each tube is hollow. The inside wall of each tube is a cathodic secondary emitter, so that as an electron cascades down the tube, many more electrons are freed from the sidewalls. A high voltage potential is applied across the length of the tube to attract the secondary emissions. What may have started with a single photoelectron at the front end of each tube may end in several thousand electrons at the opposite end. These emissions then strike a phosphor screen where a high-

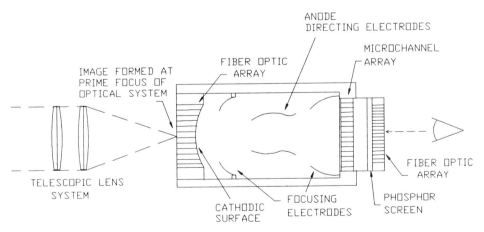

Figure 15-19   Image intensifier tube with electrostatic focusing.

resolution image appears. Often, the phosphor imaging screen is followed by another fiber optic array which directs this image to another intensifier stage, where the process of intensification is begun again. In Figure 15-20 we see a three-stage image intensifier where each successive stage supplies the next stage with a more intense image of the object being viewed.

Magnetic-focus image intensifier

The second type of image intensifier, the *magnetic-focus* image intensifier, uses a different method for propelling or attracting emitted electrons from the photocathode surfaces of the tube: Magnetic guides are used rather than high-voltage focusing. These are either permanent magnets or electromagnets in the form of toroids that surround the tube. Through magnetic attraction and deflection it is possible to direct the cathodic emissions to their proper image-receiving areas.

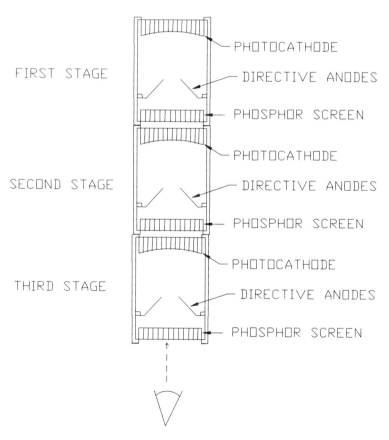

Figure 15-20   Three-stage electrostatic-focus intensifier.

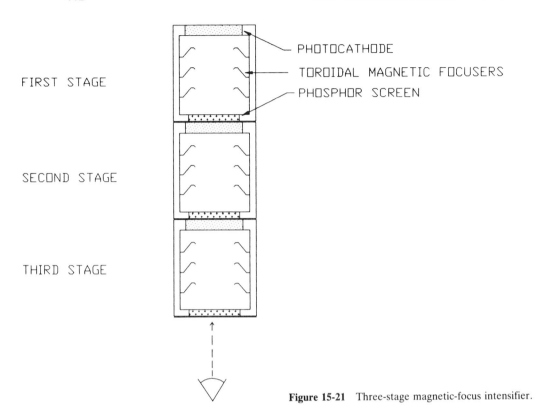

Figure 15-21   Three-stage magnetic-focus intensifier.

Figure 15-21 shows a three-stage intensifier that operates on the same staging principle as that of the electrostatic-focus device. As in the preceding case, the phosphor screen of the previous stage is adjacent to the photocathode of the next imaging stage, and so on. While the magnetic-focus system is characterized by high-resolution images, this system suffers from a lack of portability due to the weight and bulk created by the toroidal magnetic focusing systems.

15-9.2 Operating Characteristics

Both types of image intensifiers are noted for producing high-resolution images of the object being viewed. However, the magnetic-focusing unit is especially noted for its very low image distortion.

The ability of an image intensifier to brighten the image of an object is referred to as the intensifier's *luminous gain*. This quantity is the ratio of the luminous flux output of the image intensifier's viewing screen to the luminous flux input to the intensifier's first photocathode. The luminous gain is the intensifier's transfer function. Typical values for electrostatic-focus devices are 30,000 to 35,000. Typical values for magnetic-focus devices are 75 to as high as 200,000.

15-9.3 Typical Construction

In electrostatic-focus intensifiers the cathode's light-sensitive area typically has a diameter from about 18 mm to as large as 40 mm. The phosphor screen viewing area is typically the same area, that is, 18 to 40 mm in diameter.

The physical size and weight of electrostatic intensifiers make them very desirable for portable use. Figure 15-22 shows a typical unit for night-vision applications. However, they do require high-voltage power supplies that are usually contained in a separate power supply and battery packet when transported by personnel. Supply voltages range from 2.5 to 24 V dc. These voltages are then converted to the much higher focusing voltages required by the intensifier unit.

The much heavier magnetic-focus imaging units do not require high-voltage power packs, but because of their additional bulk, are usually operated at fixed sites.

The primary photocathodes of the magnetic-focus units typically have a diameter somewhat larger than those of the electrostatic-focus units. Diameters are around 40 mm. The same dimensions are found for the viewing screens also, that is, around 40 mm or so.

Multiple-stage units of the magnetic-focus imaging devices are usually built inside one continuous glass envelope and are furnished in two, three, or four stages.

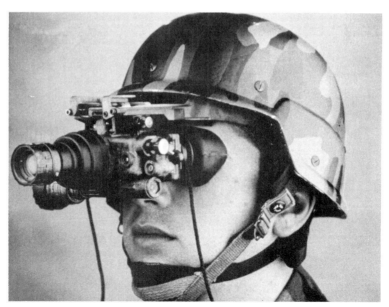

Figure 15-22 Image intensifier unit used for night vision. (Courtesy of ITT Electro-Optical Products, Roanoke, VA.)

A distinguishing feature of this type of unit is that, unlike the other unit, it usually does not contain fiber optics.

15-9.4 Practical Applications

Of the two image intensifer types, the electrostatic-focus type is presently enjoying greater use. This is probably due to the military's and many law enforcement agencies' employment of this system for surveillance applications.

The electrostatic-focus intensifier is best noted for its use as a night-vision surveillance instrument. This is undoubtedly because of its lightweight, compact design. One such application is in the use of night-vision goggles that can be worn by personnel while performing such functions as driving a vehicle or discharging a firearm.

The magnetic-focus intensifier is finding more and more use as a research tool. In addition to using photomultipliers for light-measuring duties, the astronomer is also beginning to use the magnetic-focus intensifier for deep-space exploration with the telescope. In place of lengthy time-exposure photography, the astronomer is now using the image intensifier to shorten the exposure times and yet still record any subtle hidden detail in distant galaxies and other stellar structures. Also, another device is rapidly evolving that may even take the place of the image intensifier. This is the *charge-coupled device* (CCD), an all-solid-state device whose operating characteristics are similar to those of the image intensifier units.

15-10 CHARGE-COUPLED-DEVICE COLD CAMERA

It can be argued that the charge-coupled device is not really a photoemissive device and does not belong in a discussion of photoemitter devices. However, because of its unique application and performance in low-light-level photography and because it behaves so much like the photomultiplier cell, it is presented here as a matter of interest.

The CCD is an integrated circuit that behaves somewhat like a capacitor in that it acts as a storage area for charge. However, its construction is radically different from that of a capacitor. The CCD is a dynamic memory chip that has been made light sensitive. Each memory cell is made up of a conducting plate (see Figure 15-23) that is separated from a semiconductor substrate by means of an insulator. If positive voltage is applied to each conducting plate, a depletion region capable of holding electrons is formed. This region develops immediately below the substrate's surface opposite each conducting plate. All that is needed now is a source of electrons to fill these depletion areas in order to form a binary "1." An absence of electrons is a binary "0."

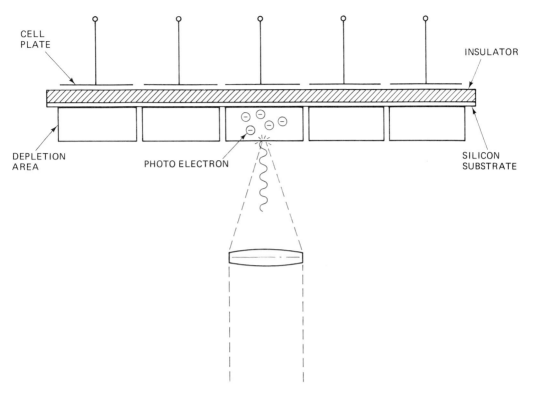

Figure 15-23 How the charge-coupled device works.

A transparent window is constructed over the depletion formation area of the CCD. The electrons needed to fill each formed depletion area to form a charge of binary "1" is furnished by photoelectrons from a lens system. An optical image is projected onto this region, where dark patterns result in no photoelectrons being injected into the depletion regions, and light patterns produce an injection of photoelectrons. The result is an image that has been transformed into a binary mosaic or *bit map* comprised of "1's" and "0's." What remains is to scan each depletion area to ascertain whether or not a charge is present. This information can then be reconstituted into an image using a computer. The computer can even be used to enhance the image if desired.

The sensitivity of a CCD can be greatly enhanced by cooling the CCD to very low temperatures. This encourages the production of photoelectrons even for extremely weak lighting conditions. As a result of this characteristic, the *CCD cold camera* was developed. This work was pioneered by astronomers who wanted to record extremely weak light sources with their telescope while using very short exposure times. This technique has produced spectacular photographic results that are beginning to rival the detail and sensitivity presently obtained with ordinary photographic plates.

REVIEW QUESTIONS

15-1. Why do different metals have different work functions? How does the work function of a metal affect the spectral response of a photoemissive device?

15-2. Explain the origin of a photoemitter's dark current.

15-3. How does the artificial cooling of a photoemitter reduce thermionic emission?

15-4. What is the purpose of a dynode in a photoemission tube?

15-5. Explain the purpose of the gas used in gas-filled photo diodes.

15-6. What differences are there in the operating characteristics of the gas-filled photodiode and the vacuum photodiode?

15-7. Explain the differences in construction between end-on design and side-on design used in photomultipliers.

15-8. Explain why a linear-type photomultiplier tube has a higher susceptibility to stray magnetic fields than do the other types of photomultiplier construction.

15-9. Explain the function of a microchannel plate array.

15-10. What is the constructional difference between the electrostatic-focus image converter tube and the magnetic-focus image converter tube?

PROBLEMS

15-11. Convert a light source's wavelength of 7800 Å to its wavelength.

15-12. Calculate the energy content, in joules, of a photon whose wavelength is 0.88 μm.

15-13. What is the work function, expressed in joules, of the element gold?

15-14. Calculate the kinetic energy (if any) of an electron that has been displaced from a cobalt surface due to irradiation by an electromagnetic beam having a wavelength of 0.13 μm.

15-15. Calculate the wavelength of light needed to just overcome a cadmium surface so that a photon will just dislodge an electron. What is the frequency of this light?

15-16. A photomultiplier tube contains four dynodes. If the anode-to-cathode voltage is 1600 V dc and the K value is 0.023, determine the tube's gain.

15-17. Plot a curve showing how the gain of a photomultiplier is affected by adding more dynodes. Assume a constant anode-to-cathode voltage of 1500 V dc and a constant K value of 0.03. Select dynode numbers of 3, 4, 5, 6, 7, and 8.

15-18. Plot a curve showing how the gain of a photomultiplier is affected by the anode-to-cathode voltage. Use a dynode value of 6 and a K value of 0.035. Plot voltages of 500, 750, 1000, 1250, 1500, and 1750 V dc. Compare your results to those obtained in Problem 15-17. Which has the greater influence on gain, the dynode number or the anode voltage?

REFERENCES

FINK, DONALD G., *Electronics Engineers' Handbook*. New York: McGraw-Hill, 1975.

MILLER, GARY M., *Modern Electronic Communication*. Englewood Cliffs, NJ: Prentice Hall, 1988.

Photomultiplier Tubes. Hamamatsu Corporation (360 Foothill Rd., P.O. Box 6910, Bridgewater, NJ 08807-0910).

TAUB, HERBERT, *Digital Circuits and Microprocessors*. New York: McGraw-Hill, 1982.

16

PHOTOVOLTAIC SENSORS

CHAPTER OBJECTIVES:

1. To understand the operation of the selenium photo cell.
2. To study the silicon photovoltaic cell.
3. To study the germanium cell.
4. To understand how the gallium–arsenide phosphor cell and the indium arsenide and antimonide cells work.

16-1 INTRODUCTION

The photovoltaic sensor is a photo-sensitive device that generates a dc voltage whose magnitude varies with the intensity of light illuminating the device. One kind of photovoltaic cell that has received considerable publicity in recent years is the *solar cell*. The solar cell is considered a photovoltaic device because it has been specifically designed to generate electrical power as a result from being exposed to the sun's light. While the solar cell was not designed or intended to be used specifically for industrial applications or in instrumentation for processing data, the device is still discussed in this chapter because of its close relationship to the other photovoltaic devices. Also, the concept of powering instruments using the sun's energy is definitely a reality. This has been proven by the many space probes and communications satellites launched by various industrial nations over the past several years. Therefore, on this basis it is justifiable to include a discussion of solar cells with the other photovoltaic devices.

There are several types of photovoltaic cells presently being manufactured, the type's name being based on the materials used in making the cells' light-sensitive surfaces. The names of these cell types are (1) the selenium cell, (2) the silicon cell, (3) the germanium cell, (4) the gallium arsenide phosphor (GaAsP) cell, (5) the indium arsenide (InAs) cell, and (6) the indium antimonide (InSb) cell. With the exception of the selenium cell, all these cell types behave like a solid-state p-n junction diode. This behavior is described in the next section.

16-2 SELENIUM CELL

The element selenium was discovered in 1817 in Sweden. It was soon discovered that it had very interesting properties, one of which pertained to its electrical conductivity. It seems that its conductivity varied according to the amount of light exposed to it, although no practical use was made of this phenomenon until almost a century later. It was also noted that selenium allowed an electrical current to travel through it in one direction only. Practical use was made of this characteristic in the 1930s; the first ac power supply rectifiers were made of this material.

As light strikes the surface of selenium, its surface and the electromagnetic energy are converted to photons or photoelectrons, and the electrons within the selenium become displaced away from the surface (i.e., away from the light source). Because of the "rectifying" nature of selenium, the selenium electrons can *only* move backward.

Figure 16-1 shows a photovoltaic cell comprised of a base layer of iron covered by a layer of selenium. Attached to the iron base is an electrode that becomes the photocell's positive electrode. The selenium layer is covered with a very thin layer of cadmium oxide and cadmium (the cadmium acting as a conductor), both layers being so thin that light can pass through. These two layers form the photocell's viewing "window." Surrounding this window is a ring made of silver or gold that is in contact with the cadmium layer. This electrode acts as a negative electrode.

When light enters the cadmium window the photons travel through this very thin material to the junction of the cadmium and cadmium oxide layers. At this point the photons displace those electrons occurring naturally at the junction and because of the orientation of the selenium layer (the selenium acting as a rectifying diode oriented to oppose electron flow back to the iron electrode) have no choice but to flow into the cadmium and into the silver or gold ring electrode. Meanwhile, the iron backing supplies more electrons to fill the holes vacated by the cadmium-bound electrons, thereby developing a positive charge in the process. This process is illustrated in Figure 16-1. It is interesting to note that this action of electron displacement at the cadmium–cadmium oxide junction and the electron replacement from the iron base can go on for just so long for a given light intensity. This process continues until a counter "back pressure" of electrons is developed at the open-circuited negative electrode. When this happens, the displacement–replacement

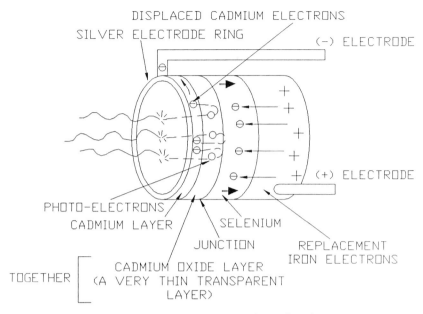

Figure 16-1 How the cadmium photocell works.

process stops. If it is to continue, either the light's intensity has to be increased or a load must be placed across the photocell's electrodes to allow partial or full circulation of the electrons.

This type of photovoltaic cell construction, where metals are "layered" to form photo-sensitive surfaces and areas of rectification, is no longer used in industry. Instead, this type of cell has been replaced with the more modern junction type of construction, which is described in the following sections.

16-3 SILICON PHOTOVOLTAIC CELL

The silicon photovoltaic cell consists of an n-type silicon wafer that has been doped with arsenic. Another silicon layer is treated with boron to form a p-type barrier, which becomes the light-entering surface for the entire cell structure. The construction is illustrated in Figure 16-2. At the p-n junction a permanent electric field is developed similar to the fields developed in semiconducting devices that have pn junctions. This electric field serves to function as a charge separator, directing the positive charges to the top surface's electrode or terminal and the negative charges or photoelectrons to the negative terminal at the cell's base. Because of this charge separation, a potential is developed across the two terminals that is proportional in amount to the light's intensity falling on the cell's surface.

There are several subtypes or categories of silicon photovoltaic cells. Some of these are (1) the low-capacitance planar diffusion type, (2) the pnn$^+$ type, (3) the pin type, and (4) the inverted layer type. A short description of each type follows.

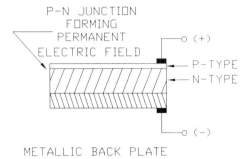

Figure 16-2 Construction of the silicon photovoltaic cell.

16-3.1 Low-Capacitance Planar Diffusion Type

A very high resistance, extremely pure n-type material is used in place of the usual n-type material. Also, an extremely thin p layer is used, both modifications resulting in a photovoltaic cell having an extremely low response time and a spectral response that favors the ultraviolet end of the visible spectrum (Figure 16-3).

16-3.2 PNN $^+$ Type

In this type of cell a very low resistance, thick-layered n $^+$-type material is used in addition to the usual p and n materials, to cause an increase in ultraviolet sensitivity to occur within the photocell.

16-3.3 PIN

The pin-type cell is a modification of the low-capacitance planar diffusion type. An additional extra high-resistance layer, called an i-layer (the i stands for *intrinsic*) is placed between the p and n $^+$ layers (Figure 16-4). This further reduces the cell's

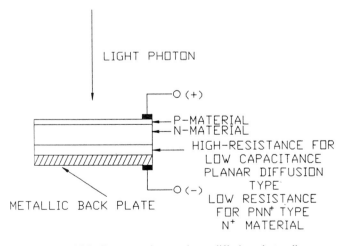

Figure 16-3 Low-capacitance planar diffusion photocell.

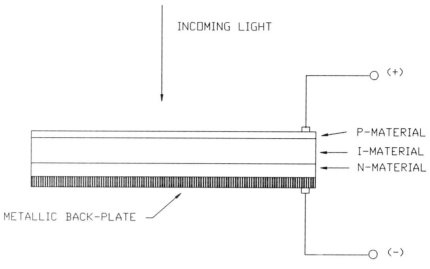

Figure 16-4   The pin photocell.

response time when used with a reverse bias circuit (see the discussion in Section 16-7 on the reverse biasing of photovoltaic cells).

16-3.4 Inverted-Layer Type

In the inverted layer type of cell, the p-layer becomes a substrate on which the cell itself is constructed. A very thin n-layer is placed on top of this substrate, which causes the cell to become very sensitive to ultraviolet radiation (Figure 16-5).

16-4 GERMANIUM CELL

Operation of the germanium cell is virtually identical to that of the silicon cell. Germanium is used instead of silicon for the ingredients used in making the p and n wafers in the cell's construction. The only major difference resulting from this change in construction material is in the cell's spectral response. The germanium cell has a spectral response peak near 1600 nm compared to the typical 800-nm region of the silicon cell (see the discussion of spectral response below).

16-5 GALLIUM–ARSENIDE PHOSPHOR CELL

The gallium–arsenide phosphor (or GaAsP) cell is constructed much like the silicon and germanium cells. Again, the only difference in performance is in the resulting spectral response characteristics.

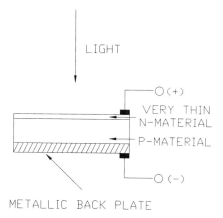

LIGHT

○ (+)

VERY THIN
N-MATERIAL

P-MATERIAL

○ (−)

METALLIC BACK PLATE

Figure 16-5 Inverted-layer photocell.

16-6 INDIUM ARSENIDE AND INDIUM ANTIMONIDE CELLS

Indium arsenide (InAs) cells have spectral response curves near 3200 nm, while indium antimonide (InSb) cells have spectral responses near 6800 nm. As a consequence, both cell types are well suited for use in infrared applications.

16-7 DIODE BEHAVIOR OF THE PHOTOVOLTAIC CELL

To understand the behavior of a photovoltaic cell, it is first necessary to understand the characteristics of the ordinary p-n junction diode. We develop this understanding by first looking at the current–voltage relationship that exists in this device. The current flow is related to the biasing voltage that is applied across the diode by means of the expression

$$I = I_s(e^{qv/\sigma kT} - 1) \tag{16-1}$$

where I = forward current flow through diode and therefore through load resistor (A)

I_s = reverse saturation current, or leakage current; at room temperature for silicon diode, use 1×10^{-8} A; for germanium diode, use 1×10^{-6}

q = one electron charge, 1.602×10^{-19} C

V = bias voltage across diode (V)

σ = curve shape factor; for Ge diode, $\sigma = 1$, for Si diode, $\sigma = 2$; in both cases, true only for very small values of I, typically less than *approximately* 50 mA or so; beyond this value, use $\sigma = 1$ for both diode types

k = Boltzmann constant, 1.38×10^{-23} J/K

T = diode temperature (K)

Figure 16-6 shows typical *I–V* (current–voltage) curves for both the silicon and germanium diodes. These curves represent the graphs obtained with eq. (16-1). Note the temperature variable in this equation. Temperature has an effect on both the forward and reverse current conductance, especially in the case of the germanium diode. The following example will help in becoming further acquainted with eq. (16-1) and the *I–V* characteristics described in Figure 16-6.

Example 16-1

A germanium diode operating at room temperature (25°C) operates with a 0.35-V forward bias. Calculate the theoretical current flowing through the diode.

Solution:
In eq. (16-1), the exponent in that expression has the following value:

$$\frac{qV}{\sigma kT} = \frac{1.602 \times 10^{-19})(0.25)}{(1)(1.381 \times 10^{-23})(273 + 25)}$$

$$= 9.73$$

Therefore,

$$I = I_s(e^{qV/\sigma kT} - 1) = (1 \times 10^{-\sigma})(e^{9.73} - 1)$$

$$= 0.0168 \text{ A}$$

$$= 16.8 \text{ mA}$$

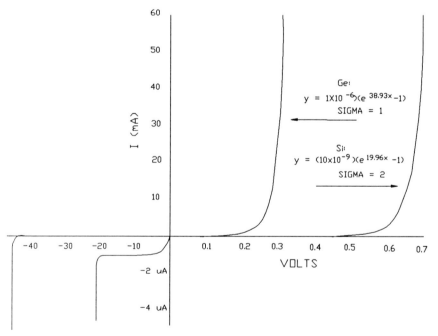

Figure 16-6 Typical *I–V* curves for the silicon and germanium diodes.

This value is confirmed by Figure 16-6. It should be stressed at this point that eq. (16-1) and its curves in Figure 16-1 are theoretical only. However, the measured I–V curves for many silicon and germanium diodes will be found to be rather close to these theoretical results.

Most photovoltaic cells behave in a fashion similar to the pn diode; the photocell's I–V characteristics are quite similar. The one major difference in a photovoltaic cell's behavior is that its internal diode configuration's bias voltage is derived from light striking the cell rather than from an external voltage source. The equivalent circuit for this type of cell is seen in Figure 16-7. In this equivalent circuit we see the photocell replaced by an ordinary diode and a constant-current source. This current source, I_{CS}, is the result of photon energy being converted to photoelectrons at a constant rate of conversion for a given illumination. This current will vary proportionally with the illumination. At this same time, the reverse current (resulting from an applied reverse bias voltage) of the cell will vary proportionally with this illumination, as will be demonstrated in Figure 16-9. (It is important at this point to note that the reverse current must not be confused with the leakage or saturation current associated with this cell. The leakage current can be measured only under zero illumination conditions.)

As a matter of fact, I_{CS} may be thought of as a variable function of I_s by means of a proportionality factor or constant which we call β. That is, β remains constant for a given level of illumination. We can now write the expression

$$I_{CS} = \beta I_s \tag{16-2}$$

where I_{CS} = constant current source

β = proportionality constant

I_s = reverse saturation or leakage current

Example 16-2

Determine a β value for a silicon photovoltaic cell whose short-circuited current at its output was measured to be 100 mA.

Solution: Referring to Figure 16-7, we see that when R_L is short-circuited, the constant current supplied by the source must be the same current as that measured at the cell's

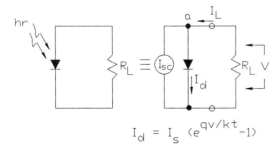

$$I_d = I_s \, (e^{qv/kt} - 1)$$

Figure 16-7 Equivalent circuit for a photovoltaic diode.

short-circuited output. That is, $I_{CS} = 100$ mA. Since we are dealing with a silicon cell, a typical value for I_s is 1×10^{-8} A. Therefore,

$$\beta = \frac{I_{CS}}{I_s} = \frac{0.100 \text{ A}}{1 \times 10^{-8} \text{ A}}$$

$$= 1 \times 10^7$$

Equation (16-2) is put to further use in our discussion of the cell's open-circuit voltage.

To understand further the characteristics of a photovoltaic cell, we now observe what happens when we first create an open circuit at the output of this cell and determine its open-circuited output voltage. Next, we observe what happens when we short-circuit the cell's output in an effort to determine its short-circuit current.

16-7.1 Determining the Open-Circuit Voltage

Referring again to Figure 16-9 and its equivalent circuit, if we remove R_L so that we now have an open circuit, V becomes V_o for the open-circuit voltage across the diode. Kirchhoff's current law states that *all current flowing into a node must equal all current flowing out of that same node*. Therefore, at node a,

$$I_{CS} = I_L + I_d \tag{16-3}$$

Because we now have an open--circuit condition, $I_L = 0$. Therefore,

$$I_d = I_{CS} = I_s(e^{qV_o/\sigma kT} - 1)$$

Then

$$\frac{I_{CS}}{I_s} = e^{qV_o/\sigma kT} - 1$$

or

$$e^{qV_o/\sigma kT} = \frac{I_{CS}}{I_s} + 1$$

Solving for V_o gives

$$\ln e^{qV_o/\sigma kT} = \ln\left(\frac{I_{CS}}{I_s} + 1\right)$$

$$\frac{qV_o}{\sigma kT} = \ln\left(\frac{I_{CS}}{I_s} + 1\right)$$

$$V_o = \frac{\sigma kT}{q} \ln\left(\frac{I_{CS}}{I_s} + 1\right)$$

and since $I_{CS} = \beta I_s$ [from eq. (16-2)],

$$V_o = \frac{\sigma kT}{q} \ln\left(\frac{\beta I_s}{I_s} + 1\right)$$

$$= \frac{\sigma kT}{q} \ln(\beta + 1) \qquad (16\text{-}4)$$

Equation (16-4) allows us now to determine a value for V_o assuming that we know the value of the proportionality constant β for a given illumination.

Example 16-3

Using the value of β found in Example 16-2, determine V_o for the cell described in that example. Assume an operating temperature of 25°C.

Solution:

In Example 16-2, β was found to be 1×10^7. We now place this value into eq. (16-4) along with $\sigma = 1$. We use this value for σ because we are interested in finding the open-circuit voltage, V_o, which occurs along the straight, almost vertical portion of the I–V curve [see Figure 16-10 and also the variable discussion for eq. (16-1)]. Therefore (note that the units are dropped for solution clarity),

$$V_o = \frac{\sigma kT}{q} \ln(\beta + 1) = \frac{(1)(1.38 \times 10^{-23})(298)}{1.602 \times 10^{-19}} \ln(1 \times 10^7 + 1)$$

$$= (0.026)(16.12)$$

$$= 0.42 \text{ V}$$

We can now conclude two very important facts from eq. (16-4): *(1) The open-circuit voltage, V_o, varies logarithmically with the illumination, and (2) the temperature no longer has the exponential affect it had when a load resistor was present. It now has only a linear effect.*

16-7.2 Determining the Short-Circuit Current

Substituting eq. (16-1) into eq. (16-3) [noting that I in eq. (16-1) is equivalent to I_d in eq. (16-3)], we get

$$I_{CS} = I_L + I_s(e^{qV/\sigma kT} - 1)$$

Because of the cell's now short-circuited output, $V = 0$. Therefore,

$$I_{CS} = I_L + 0 - I_s$$

Because the diode in our equivalent circuit is now short-circuited, the saturation current, $I_s = 0$. We now let $I_L = I_{SH}$, our short-circuit current. Therefore,

$$I_{SH} = I_{CS} \qquad (16\text{-}5)$$

The fact that the two currents are shown to be equivalent supports what was ascertained earlier in Example 16-2.

Equation (16-5) illustrates a very important characteristic of the photovoltaic cell. Under short-circuited conditions, the cell's output current varies *proportionally* with the level of illumination. This is because I_{CS} varies proportionally with illumination level, a statement that was made earlier.

We have now explored two extreme operating conditions of the photovoltaic cell, the open-circuited condition and the short-circuited condition; these are illustrated in Figure 16-8. A more complete picture may be seen by showing how illumination varies both I_{SH} and V_o; this is illustrated in Figure 16-9. For a linear increase in illumination we obtain the linearally occurring crossover points for the various I_{SH} values while obtaining nonlinear crossover points for the V_o values. Because our cell will be operating with a load resistance somewhere between zero and infinity, we must now be able to determine a value for the load resistance that produces an acceptable current linearity but with a reasonable voltage variation.

As we have already seen, when the amount of illumination striking a photovoltaic cell is varied, a family of *I–V* curves are generated as shown in Figure 16-9. Looking at this figure, if we extended a straight line outward from the origin in the figure's third quadrant, the slope of this line would represent the conductance or inverse of the load resistance applied across the cell's output. This line is called a *load line*. An example is illustrated in Figure 16-9. Because of the nature of these *I–V* curves, it would be advantageous to choose a load line that is close to the *V*-axis in the first quadrant, where maximum voltage changes occur for changes in illumination. At the same time, however, it would be equally advantageous to locate near the *I*-axis in the fourth quadrant, where linear changes in current occur. The optimum or compromise location is usually determined through experimentation. There may be cases of applications where only linear output current responses are desired. If this is the case, very low resistance values for R_L must be selected. If a

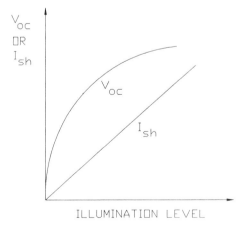

Figure 16-8 Comparison of the open-circuit voltage and short-circuit current with illumination.

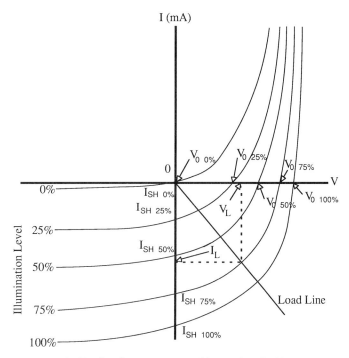

Figure 16-9 Family of curves generated by varying the illumination.

voltage output is needed, such as in the case where the cell's output is sent to a voltage amplifier, relatively high resistance values must be chosen.

In cases where maximum power must be obtained, such as in the case of utilizing the photovoltaic cell as a solar cell, determination of an optimum R_L is not as straightforward. One would assume that this optimum value could simply be obtained by using the calculation $R_L = V_o \div I_{SH}$, since we would be using the maximum obtained voltage with the most current, which would, in turn, give us the most power (since power in this case would be $V_o \times I_{SH}$. Unfortunately, this would not be possible because the photovoltaic cell is not a constant-voltage source. Its output voltage would have to be something less than V_o since R_L would have a resistance something less than infinite. Therefore, to determine the optimum R_L for maximum power transfer, a complex mathematical proof must be utilized. This proof, however, is beyond the desired scope of this book. A much simpler mathematical rule may be applied instead. This rule has been proven empirically and it states the following for determining R_L:

$$R_L = 0.8 \frac{V_o}{I_{SH}} \qquad (16\text{-}6)$$

Example 16-4

Using the V_o and I_{SH} figures calculated in Examples 16-2 and 16-3, determine the optimum load resistor to be used in conjunction with the photovoltaic cell that we have been working with in the earlier examples.

Solution:

From Example 16-2 we ascertained that I_{CS} was 100 mA. Later, we concluded that I_{SH} was also 100 mA, since these two currents must equal each other under short-circuit conditions. In Example 16-3 we calculated V_o to be 0.42 V. Placing these two values into eq. (16-6), we get

$$R_L = 0.8\frac{V_o}{I_{SH}} = 0.8\frac{0.42 \text{ V}}{0.1 \text{ A}}$$

$$= 3.4 \ \Omega$$

16-8 SPECTRAL RESPONSE CHARACTERISTICS

The spectral response curves for many of the photovoltaic cells that we have discussed thus far are graphed in Figure 16-10.

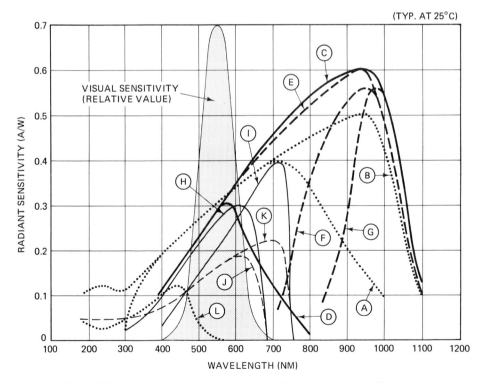

Figure 16-10 Spectral response curves for several photovoltaic cells. (Courtesy of Hamamatsu Corp., Bridgewater, NJ.)

TYPE	FEATURES	SPECTRAL RESPONSE CHARACTERISTICS			TYPE NO.	LISTED PAGE
		RANGE	PEAK WAVELENGTH	MARK		
SILICON PHOTODIODES	ULTRAVIOLET TO VISIBLE LIGHT, FOR PRECISION PHOTOMETRY	190 ~ 1000	720	Ⓐ	S1226, S1227 SERIES	14–15
	ULTRAVIOLET TO INFRARED, FOR PRECISION PHOTOMETRY	190 ~ 1100	960	Ⓑ	S1336, S1337 SERIES	16–17
	VISIBLE LIGHT TO INFRARED FOR PRECISION PHOTOMETRY	320 ~ 1100	960	Ⓒ	S2386, S2387 SERIES	18–19
	VISIBLE LIGHT, FOR GENERAL PURPOSE PHOTOMETRY	320 ~ 730	560	Ⓓ	S1087, S1133, S2833 ETC.	20–21
	VISIBLE LIGHT TO INFRARED, FOR GENERAL PURPOSE PHOTOMETRY	320 ~ 1100	960	–	S1087-01, S1133-01, etc.	
PIN SILICON PHOTODIODES	HIGH-SPEED RESPONSE FOR OPTICAL COMMUNICATION OPTICAL FIBER DATA LINK, ETC.	320 ~ 1000	800	–	S2839, S2840	22 ~ 25
		320 ~ 1060	900	–	S2216, S1721, ETC.	
		320 ~ 1100	960	Ⓔ	S1190, S1223, ETC.	
	VISIBLE LIGHT CUTOFF TYPE	700 ~ 100	960	Ⓕ	S2506	
		840 ~ 1100	980	Ⓖ	S2506-01	
	LARGE SENSITIVE AREA, HIGH ULTRAVIOLET SENSITIVITY	190 ~ 1060	960	Ⓑ	S1723-05, ETC.	
GaAsP PHOTODIODES (DIFFUSION TYPE)	FOR VISIBLE LIGHT	300 ~ 680	640	Ⓗ	G1115, G1116, G1117, ETC.	26–27
	EXTENDED RED SENSITIVITY	400 ~ 760	710	Ⓘ	G1735, G1736, G1737, ETC.	
GaAsP PHOTODIODES (SCHOTTKY TYPE)	ULTRAVIOLET TO VISIBLE LIGHT	190 ~ 680	610	Ⓙ	G1125-02, G1126-02, ETC.	28–29
	EXTENDED RED SENSITIVITY	190 ~ 760	710	Ⓚ	G1745, G1746, G1747	
GaP PHOTODIODES	ULTRAVIOLET TO GREEN LIGHT	190 ~ 550	440	Ⓛ	G1961, G1962, G1963	30–31
SILICON AVALANCHE PHOTODIODES (APD)	HIGH-SPEED RESPONSE AND HIGH GAIN	400 ~ 1000	800	–	S2381, S2382, S2383, S2384 S2385	32–33

Figure 16-10   (cont.)

16-9 SOLAR CELLS

Solar cells are specially designed photovoltaic cells. They have been specifically designed to respond to the sun's yellow-white color and to withstand the sun's intense heat.

The maximum energy received by the earth from the sun per square area of the earth's surface is 1 kW. This value is realized only under the best conditions. The goal of any solar photovoltaic cell design is to convert as much of this 1 kW/m² figure to electrical energy as possible. Presently, the best *practical* conversion efficiency is around 20% when used with solar concentration systems. Figure 16-11 shows the attainable efficiencies for many of the presently fabricated photovoltaic cells being produced today. Notice that while there are a number of cell types that appear to be capable of efficiencies greater than 20% in this graph, their practical implementation is doubtful at this point.

The single-crystal silicon solar cell is perhaps the most popular cell in use today. It is the best understood, least expensive, and easiest of all cell types to manufacture.

As with all solar voltaic cells, the solar conversion efficiency increases with an

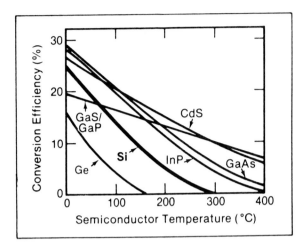

Figure 16-11 Solar cell conversion efficiencies. (From Kenneth Zweibel, *Basic Photovoltaic Principles and Methods,* copyright 1984, p. 33. Reprinted by permission of Van Nostrand Reinhold Co., New York.)

increase in solar illumination concentration. Solar illumination is increased through the use of lenses or mirrors or other reflective or refractive means (Figure 16-12). Unfortunately, the increase in concentration also increases the cell's temperature. With silicon this is especially critical since its dc resistance increases with temperature, causing a decrease in its conversion efficiency. As a result, it is necessary to keep the cell cooled by some means. Inverted layer cells have some of the highest conversion efficiencies found among silicon cells. These efficiencies are typically around 18% or so.

Figure 16-12 Typical photovoltaic solar cell array. (Courtesy of Design News Magazine, Sept. 21st, 1987, "Solar Energy Begins to see the Light, pg. 24.)

REVIEW QUESTIONS

16-1. Explain the function of selenium in a photovoltaic cell.

16-2. Explain why a photovoltaic cell with an open circuit cannot "build up" voltage amounts to infinitely high levels for a particular light-level input to the cell.

16-3. What are the output voltage characteristics for a photovoltaic cell having an open circuit versus one having a heavily loaded circuit output?

16-4. What is the purpose of the i layer in a pin cell?

16-5. What is the basic difference in performance when comparing a germanium photovoltaic cell to a silicon photovoltaic cell?

PROBLEMS

16-6. If the short-circuit current of a silicon photovoltaic cell were found to be 85 mA, and its dark current (i.e., no illumination current) were found to be 6.3 μA, determine the value of β.

16-7. Determine the open-circuit voltage that would be expected from a photovoltaic cell whose β value is 1.2×10^7 and whose operating temperature is 35°C. Assume that $\sigma = 1$ for this solution.

16-8. For a given illumination amount, how would you prove that the open-circuit voltage varies proportionally with the photovoltaic cell's temperature?

16-9. The I–V curves in Figure 16-13 are for a photovoltaic cell under the illumination conditions indicated. Calculate values for V_o and I_{SH} for 50% illumination conditions. Assume a temperature of 25°C.

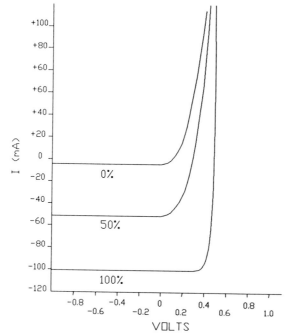

Figure 16-13 I–V curves for Problem 16-9.

16-10. In Problem 16-9, determine the optimum load resistance, R_L, for this cell. If you wanted a more linear output current response from this cell, how would you vary R_L?

REFERENCES

BOYLESTAD, ROBERT, and LOUIS NASHELSKY, *Electronic Devices and Circuit Theory*, 3rd ed. Englewood Cliffs, NJ: Prentice Hall, 1982.

FINK, DONALD G., *Electronic Engineers' Handbook*. New York: McGraw-Hill, 1975.

JOHNSON, CURTIS D., *Process Control Instrumentation Technology*. New York: Wiley, 1988.

ROSE, ROBERT W., LAWRENCE SHEPARD, and JOHN WULFF, *The Structure and Properties of Materials*, Vol. IV, *Electronic Properties*. New York: Wiley, 1966.

SOLAR ENERGY RESEARCH INSTITUTE, *Basic Photovoltaic Principles and Methods*. New York: Van Nostrand Reinhold, 1984.

SZE, S. M., *Physics of Semiconductor Devices*. New York: Wiley, 1981.

YANG, EDWARD S., *Fundamentals of Semiconductor Devices*. New York: McGraw-Hill, 1978.

17

CHEMICAL AND NUCLEAR IONIZATION SENSORS

CHAPTER OBJECTIVES

1. To determine what comprises a chemical ionization sensor.
2. To analyze the conductive electrolytic transducer.
3. To understand the potential-electrolytic transducer.
4. To study ionization transducers.

17-1 INTRODUCTION

Of all the sensor types discussed in this book, the chemical or nuclear sensor probably evokes the least understanding from its general description name. Despite this identity crisis, both the chemical and nuclear ionization sensors are vital sensor types in industry. To understand their operation fully, however, one must have a basic understanding of the nuclear chemistry involved in making these sensors work. In the following sections we discuss this chemistry and describe several very important transducers that utilize the principles involved.

17-2 WHAT COMPRISES A CHEMICAL IONIZATION SENSOR?

A chemical ionization sensor is a sensor comprised of a chemical or group of chemicals that when exposed to the measurand, become ionized. Ionization provides an electrical current that varies in proportion to the intensity or concentration of the measurand being sensed. The measurand is usually a chemical element, compound, or mixture.

17-3 COMMONLY SENSED MEASURANDS

The measurands that are sensed using this type of sensor are chemical constituents rather than the usual measurands we find from the list of general measurands. (This is the list described in Section 1-5 that we have abided by in all earlier chapters. Instead, these are specific substances having associated with them either units of *concentration* or electrical conductivity or units of *pH*. (There are other ion activity quantities that can be determined with chemical ionizing sensors but we will study only the ones just mentioned.) If the former is the case, typical units would be "parts per thousand," "parts per million" (ppm), and "percent concentration." The unit of electrical conductivity is the *mho*, although the *siemens* (which is the same as the mho) is gradually replacing the mho in usage. Actually, conductivity is expressed in units of conductance/length, for reasons that we explain in Section 17-5.1. Units of pH are expressed in decimal values ranging from 0 through 14.

There is one general measurand that can be determined using the chemical ionization sensor. This is the measurand length, associated with level measurements, described in Section 17-5.4.

17-4 TYPES OF TRANSDUCERS PRESENTLY AVAILABLE

A variety of transducers are used as chemical ionization devices. Those discussed in this book are the following: (1) electrolytic (two types), and (2) ionization (not to be confused with the general category name *chemical ionizer*, which refers to *all* the types listed here). We begin our discussion with the two types of electrolytic transducers.

17-5 CONDUCTIVE ELECTROLYTIC TRANSDUCER

The *conductive electrolytic transducer* is a transducer that depends on the electrical conductivity of an electrolytic solution between its electrodes for its proper operation. Its theory of operation is discussed below.

17-5.1 Theory of Operation

Liquids that contain compounds contain whole molecules of these compounds. These same liquids also contain incomplete molecules called *disassociated ions* or simply *ions*. These are fragmented molecular parts that have either a positive or a negative electrical charge. They are fragmented as a result of the ionic bonding force that holds the molecule together becoming broken or disrupted within the molecule. This disruption may be caused by high temperature, a chemical reaction of some sort, or through the application of an electrical current or polarizing voltage. Whatever the cause, the individual ions are free to drift about within the solution among any other unaffected electrically neutral whole molecules, until the disruption is removed. At that time the ions may reunite to form once again into whole electrically neutral molecules.

The following is an example of what was just discussed. If ordinary table salt (NaCl) is dissolved in water, we now have *for the most part*, a liquid solution comprised of electrically neutral NaCl molecules. We must keep in mind that sodium chloride is comprised of one ion of sodium that has one positive charge, and one ion of chloride having one negative charge (Figure 17-1). When the two ions are brought together, they form the electrically neutral molecule NaCl. Now, depending on the solution's temperature, this will have a bearing on the degree of disassociation among the ionic bonds of the NaCl molecule. The higher its temperature, the more likely it is to find separated ions of chloride and sodium. Temperature acts as a catalyst to encourage ionic separation among otherwise neutral molecules.

The operation of an electrolytic sensor depends on the electrical conductance of an electrolytic liquid, as pointed out in an earlier paragraph. This conductance can be correlated with the concentration of the chemical compound or element whose disassociated ions are creating the magnitude of the conductivity path. The path's magnitude is then measured between a set of electrodes submerged in the liquid (Figure 17-2). In reality an ammeter measures the current furnished by an attached voltage source. This meter can easily be calibrated in units of concentration rather than in amperes.

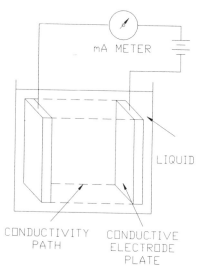

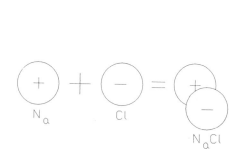

Figure 17-1 Forming the sodium chloride molecule.

Figure 17-2 Measuring the conductivity of electrolytes.

How a liquid's conductivity is specified

Recalling that a material's resistivity is defined and calculated from the equation $R = \rho(l/A)$ [eq. (7-3)], we have the *inverse* of this equation, which represents the *conductance*, G, which is

$$G = \delta \frac{A}{l} \qquad (17\text{-}1)$$

where G = conductance [siemens (S)]

   δ = conductivity (S/cm)

   A = cross-sectional area of electrodes (cm^2)

   l = length between electrodes (cm)

The unit siemens/cm (S/cm) is too large for most conductivity values. Consequently, the units mS/cm and μS/cm (corresponding to their reciprocal counterparts, the kΩ and MΩ) are generally used instead. Frequently, in the design of conductivity electrodes, the term *cell constant* is used to describe a particular design. An electrode pair's cell constant is defined as l/A.

Example 17-1

According to the manufacturer, a set of conductivity probes has a cell constant of 1.5 cm/cm^2. Determine the conductivity, in microsiemens, of a 2% solution of sodium hydroxide (NaOH).

Solution: Using the definition of cell constant, which states that the cell constant, K_c, is defined as l/A, eq. (17-1) may be rewritten as

$$G = \frac{\delta}{K_c} \qquad (17\text{-}2)$$

where K_c is the cell constant (cm/cm^2). The conductivity value, δ, for a 2% solution of NaOH is (from Table 7-1) 1×10^5 μS/cm. Solving now for G, we get

$$G = \frac{1 \times 10^5 \ \mu\text{S/cm}}{1.5 \ \text{cm/cm}^2}$$

$$= 0.67 \times 10^5 \ \mu\text{S}$$

17-5.2 Operating Characteristics and Physical Description

Figure 7-3 shows the results of running a conductivity test with a set of probes made from 3.0-mm-diameter stainless-steel conductors that are 13 mm in length and spaced 7.0 mm apart. The probes used in generating the data in Figure 17-3 are shown in Figure 17-4. They were fully immersed in the salt solution at the solution's maximum depth when the data were recorded. As long as the same set of probes is used in future unknown saltwater mixtures, the curve in Figure 17-3 can be used in determining the percentage of salt concentration. (It is assumed that the solution's

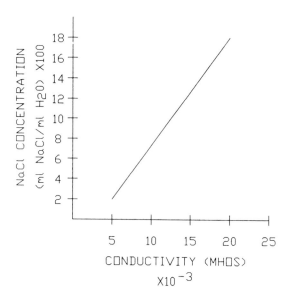

Figure 17-3   Conductivity curve.

concentration is the same throughout its volume. Adequate mixing must be done to the solution to assure that this is the case.) It can be seen from these results that conductivity varies inversely with the salt's concentration. This relationship is true, however, only for relatively weak solutions of saltwater. Higher concentrations produce an exponential relationship.

Standards governing the dimensions for a conductivity probe set do exist. The standard states that the conductivity probes must be made from conductive plates

Figure 17-4   Conductivity probes.

that are 1 cm × 1 cm and are fully submersed in the test liquid and separated by a distance of 1 cm. The obvious advantage to us of using this standard is that we are able to compare one set of data with that of another without regard to the probe dimensions, provided, of course, that both probe sets were manufactured according to those standard dimensions. As a result of this standard, we now have a means of producing a voltage or current reading based on conductivity that varies with concentration.

17-5.3 Practical Applications

Using the conductive electrolytic sensor to check for sodium chloride concentrations is not an accurate method if other salts are present. Many other salts also produce electrolytic solutions and would therefore fool the conductivity circuit into thinking that only the presence of sodium chloride was producing the reading. As a result, the conductive electrolytic sensor must be calibrated for a specific salt only. Once that is done, it becomes a feasible checking device for monitoring concentrations for that particular electrolytic-producing substance. Table 17-1 shows the conductivity/resistivity ranges for several other common materials.

Figure 17-5 illustrates a process control system that employs a conductive electrolytic transducer for controlling a salt's concentration in a water–salt mixture. The conductivity sensor replaces one of the resistors in a Wheatstone bridge. The bridge's output is amplified to drive a servomotor that operates the flow valve for the salt supply and also the inlet flow valve for the mixing vat's water supply. The Wheatstone bridge circuit is nulled at the desired concentration. If the vat's concentration is lower than that desired, the bridge's null is upset, producing a voltage that causes the servomotor to open the salt supply a proportional amount and to close the inlet fresh water supply a proportional amount. If the concentration is too high, the servomotor "sees" a reversed polarity and causes both control valves to reverse their travel, thereby causing a decrease in concentration.

Another application of the conductive electrolytic sensor is in level detecting. In this particular application two conductors are immersed in a container of salt solution whose level or depth must be monitored (Figure 17-6). By taking conductivity measurements at various intervals, it is possible to construct a curve similar to the one in Figure 17-7. With this curve now serving as a calibration curve for the salt concentration indicated on the graph, it is possible to know precisely the liquid depth within the container after measuring the conductivity.

Because many liquids exhibit a measurable conductivity of some value for a given set of electrodes, it is possible to use the depth-measuring scheme just described to control the depth in a container automatically. Figure 17-8 illustrates this process, which is somewhat similar to the one illustrated in Figure 17-5. Again, we see the Wheatstone bridge being used as a nulling circuit. The bridge is nulled at the desired level by adjusting R_1. Assuming that a continuous dumping of the liquid in the vat or container is taking place (because of bottling, etc.), whenever the bridge experiences an unbalancing of the center-leg current, this current is amplified as in the previous case and drives a servomotor that controls the valve on a makeup supply

Table 17-1

Resistivity in Ohm-cm	100M	10M	1M	0.1M	10K	1K	100	10	1
Conductivity in uS/cm	0.01	0.1	1	10	100	1K	10K	0.1M	1M
Pure Water		▭							
Demineralized Water			▭						
Condensed Water Vapor			▭▭▭	▭▭▭	▭				
Natural Fresh Water				▭▭	▭▭	▭			
Cooling Tower Water				▭▭	▭▭	▭▭	▭		
Ocean Water							▭		
Water from Great Lakes				▭▭	▭▭	▭			
NaOH Solution < 50% Concentrate								▭	▭
HCl Solution < 50% Concentrate									▭

pipe. It is assumed in this circuit that the servomotor is "torque-loaded." That is, when the signal is reduced, the servomotor reverses its rotation due to a mechanical spring mechanism and continues to rotate until the countertorque generated by the reduced current is equal to the spring-produced torque. However, since the current will continue to diminish until the Wheatstone bridge is once again nulled, the valve will continue to close until it is at its fully shut position, the original position at the null condition.

17-6 POTENTIAL-ELECTROLYTIC TRANSDUCER

The *potential-electrolytic transducer* is a far more complex device than the conductive electrolytic type. It is also used for an entirely different application as far as solution detection is concerned. The potential-electrolytic transducer is comprised of certain

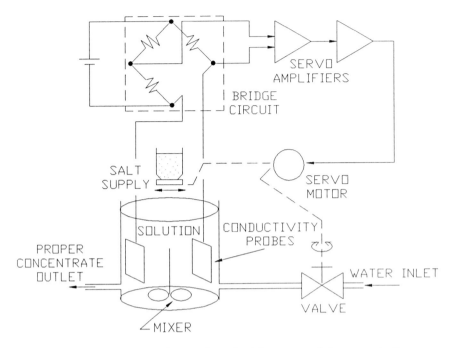

Figure 17-5   Process control system for maintaining proper solution concentration.

chemicals that when exposed to a particular electrolytic substance, produce a potential difference that is proportional either to the pH of the material being detected or to the ion activity or oxidizing/reducing properties of the solution. In our discussions here, we are concerned only with the pH determination of solutions. The ion activity or oxidizing/reducing properties of solution's applications are rather com-

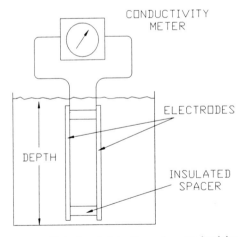

Figure 17-6   Level detecting using conductivity probes.

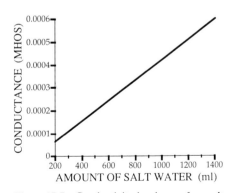

Figure 17-7   Conductivity-level curve for a salt solution.

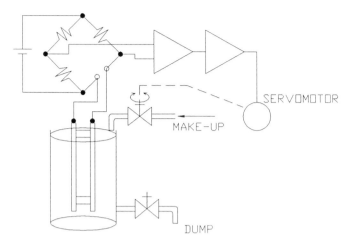

Figure 17-8   Automatic liquid depth controller system.

plex and beyond the scope of this book. (See the references at the end of the chapter for further information on this subject.)

In the case of the pH of a solution, this quantity is a direct indication as to whether the solution is acidic or alkaline in its chemical makeup. Figure 17-9 shows a pH scale listing values for a few common liquids. Note that this scale extends from zero to 14 (or on some scales, to 15) with water (which has a pH of 7) at its midpoint. Water is considered neither a base nor an acid; it is chemically neutral.

17-6.1 Theory of Operation

What is pH?

The "p" in the term "pH" comes from the French word *puissance*, meaning "power." The "H" is the chemical symbol for the element hydrogen. The value of pH is defined by the following equation:

$$pH = -\log_{10}(H^+) \qquad (17\text{-}3)$$

where pH = pH number

H^+ = hydrogen ion concentration (see Table 17-2)

Table 17-2 lists the hydrogen ion concentration found in one *mole per liter* of a solution at 25°C. (A mole is the weight of a molecule expressed in grams, which is equal to the atomic weight of each of the atoms comprising the molecule, all added together.) As an example, for a solution having a pH of 11, that solution would have 1×10^{-11} mole of hydrogen ions; and according to the definition of a mole, there would be 1×10^{-11} gram of hydrogen ions in 1 liter of solution.

To determine the pH value for a given solution, it is necessary to use two electrodes. One of these electrodes, the *glass electrode* or *pH-sensing electrode*,

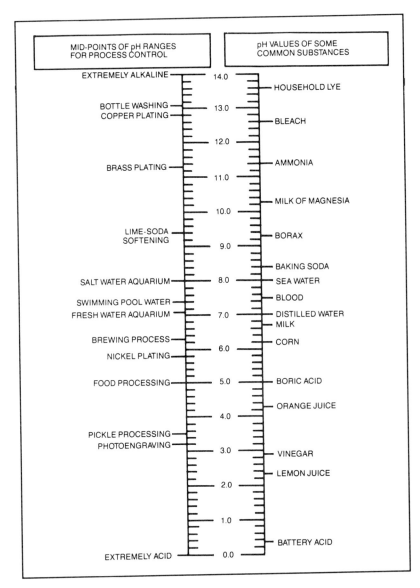

Figure 17-9 pH scale for some common substances. (REPRODUCED WITH THE PERMISSION OF OMEGA ENGINEERING, INC.)

generates a voltage much like that of a wet cell battery, where the liquid electrolyte could be an acid or a base. The other electrode, the *reference electrode*, also acts as a battery but acts as a special reference voltage type of battery, whose output remains the same regardless of the electrolyte it is placed in.

Both electrodes have sensing bulbs that are manufactured from materials that act as membranes. In the case of the pH electrode, this material may be glass; in

TABLE 17-2 HYDROGEN ION CONCENTRATION FOR ANY SOLUTIONS HAVING THE pH SHOWN

pH	H^+	
0	10^{0}	1
1	10^{-1}	0.1
2	10^{-2}	0.01
3	10^{-3}	0.001
4	10^{-4}	0.0001
5	10^{-5}	0.00001
6	10^{-6}	0.000001
7	10^{-7}	0.0000001
8	10^{-8}	0.00000001
9	10^{-9}	0.000000001
10	10^{-10}	0.0000000001
11	10^{-11}	0.00000000001
12	10^{-12}	0.000000000001
13	10^{-13}	0.0000000000001
14	10^{-14}	0.00000000000001

the case of the reference electrode, this material may be a porous ceramic or a quartz or asbestos fiber. Figure 17-10 shows a schematic representation of both electrodes.

When both electrodes are immersed in a sample solution whose pH is to be determined, the combined voltages generated by both probes are processed by the

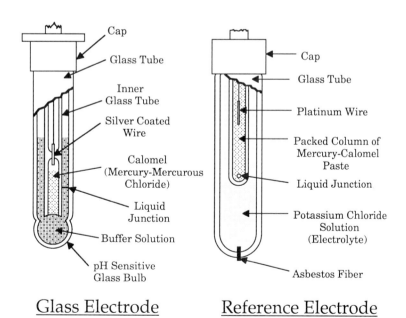

Glass Electrode   Reference Electrode

Figure 17-10   Glass and reference electrodes. (Courtesy of Honeywell Inc.)

pH analyzer. There must be a completed circuit between these two electrodes. This is obtained by an electrolytic material that "leaks" through the reference electrode over to the glass electrode, thus creating the circuit path that is needed. The ions in the glass electrode are exchanged for ions in the sample and filling solution, thereby completing this electrical circuit. The potentials developed by both electrodes are combined at the pH amplifier shown in Figure 17-11. The amplifier interprets the combined voltages and displays the results in the form of the sample solution's pH.

The potassium chloride solution seen inside the reference electrode of Figure 17-11 is the material that slowly seeps through the electrode's bulb, forming the electrical circuit path. This material is sometimes called the *filling solution*. It is very critical to proper operation of the pH measuring system that the reference electrode's filling solution "leak" adequately. Consequently, proper maintenance and supervision of this electrode is needed to obtain reliable pH results.

17-6.2 Operational Characteristics

Temperature has a profound effect on the ionic activity inside an electrolytic solution. To state this in another way, temperature has a measurable affect on a solution's pH value. Although it is not desirable to compensate for this effect, since it is

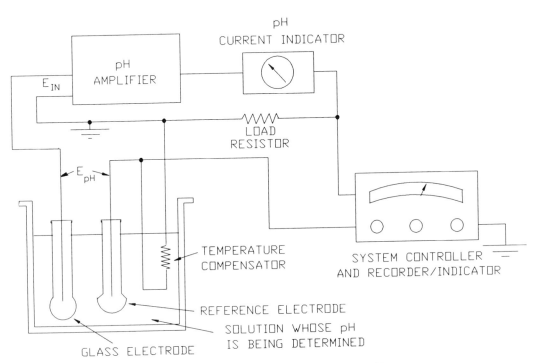

Figure 17-11 Measuring a solution's pH.

assumed that the true, existing pH of a substance at the existing temperature is what is wanted, it *is* necessary to maintain the pH electrode at a constant temperature. The voltage output of the pH electrode is significantly affected by temperature. The amount is illustrated in the graph depicted in Figure 17-12. It is interesting to note that only at a pH of 7 is there no temperature effect. The largest effects are experienced for the highest and lowest pH values.

It was implied earlier that one of the problems with using reference electrodes is that an improper leakage or diffusion rate of filling material will create erroneous pH readings. Quite often a very long stabilization time or rapid change of pH reading is an indication that the reference electrode is faulty. A very slow continuous drifting of a pH reading is usually an indication that the reference electrode is plugged or dirty.

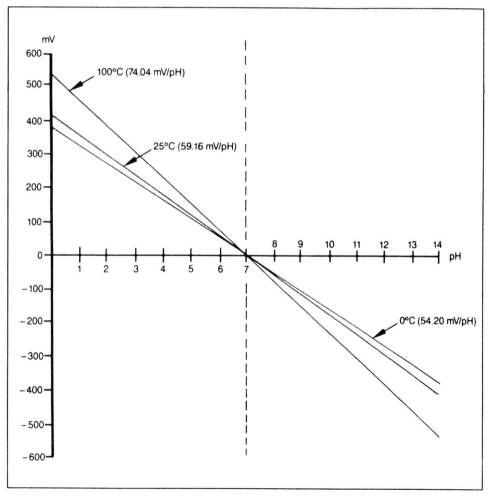

Figure 17-12 Effect of termperature on a pH electrode's output. (REPRODUCED WITH THE PERMISSION OF OMEGA ENGINEERING, INC.)

Because ionic exchanges take place at the pH electrode's glass surface, it is essential that this glass remain clean at all times. Therefore, the electrodes are usually stored immersed in slightly acidic solutions so that the glass is protected from atmospheric contaminants. Also, if any glass impurities are present, these are drawn out into the acidic solution to be replaced by hydrogen ions. This also serves to increase the probe's sensitivity to pH readings.

17-6.3 Typical Construction

Figure 17-13 shows a typical pH electrode constructed with a glass body. Figure 17-14 are typical reference electrodes. These have been designed to accept refill solutions to make up for the loss of the filling solution that escapes through the electrode's specially designed body membrane material. The membrane material used in the electrodes' construction pictured in Figure 17-14 is glass.

Often the pH probe and reference probe are housed in one electrode, called a *combination probe*. This type of construction increases the ease of usage and reduces the chances of breakage. Figure 17-15 shows examples of this type of electrode construction.

Quite often an entire pH controller system can be purchased that includes the necessary electrodes. This system has the advantage of being able to furnish a control signal for operating a final control element, the operation being based on a desired pH value. Such a system is shown in Figure 17-16.

17-6.4 Practical Applications

A pH analyzer system lends itself to process control applications very readily. In Figure 17-17 we see a mixing operation where a chemical is added to a liquid whose pH is being monitored by a pH controller. The controller operates a flow control valve that controls the amount of liquid being added to the mixing tank. Ultimately, then, the pH value determines the setting of the chemical's flow control valve.

17-6.5 Typical Design Specifications

Figure 17-18 shows the design specifications for a potential-electrolytic transducer.

17-7 IONIZATION TRANSDUCERS

The ionization sensor is used primarily for the detection of atomic radiation. In very general terms, atomic radiation (or nuclear radiation—the two names are often used interchangeably) is the radiation that is emitted from atomic nuclei under certain conditions. This radiation may consist of charged particles, uncharged particles, and electromagnetic radiation. We now define these terms.

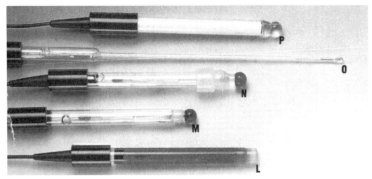

COMBINATION ELECTRODES

	Part No. BNC	Part No. U.S. Std.	Application	Length (mm)	Diameter (mm)	pH Range	Temp. °C
L	PHE-2171	PHE-2171-U	Flat surface for measurement of meats or agar plates	110	12	0-12	0-100
M	PHE-2161	PHE-2161-U	Soil pH measurements, directions included	110	12	0-13	0-100
N	PHE-2191	PHE-2191-U	High junction flow rate for difficult samples (slurries, emulsions, high-purity water)	130	14	0-13	0-110
N	PHE-2791	PHE-2791-U	High junction flowrate with Calomel reference	130	14	0-13	−5 to 60
N	PHE-2291	PHE-2291-U	High junction flowrate with double junction	130	14	0-13	0-110
O	PHE-2754	PHE-2754-U	Extra long 3.5 mm diameter for NMR tubes and small lab glassware	183	3.5	0-12	0-60

Figure 17-13 Typical pH electrodes. (REPRODUCED WITH THE PERMISSION OF OMEGA ENGINEERING, INC.)

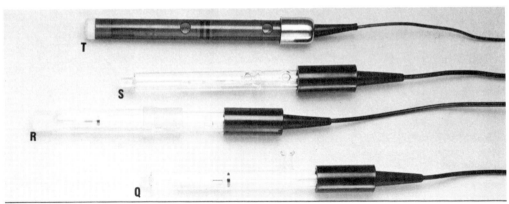

REFERENCE HALF CELLS

Q	PHE-3711	Calomel junction for protein substances and high-purity water	110	12	—	−5 to 60
Q	PHE-3111	General purpose reference half cell	110	12	—	−5 to 105
R	PHE-3291	High junction flow rate for low conductivity or dirty solutions	115	15	—	−5 to 105
R	PHE-3791	High junction flow rate with Calomel double junction	115	15	—	−5 to 60
S	PHE-3211	Double junction reference half cell with 10% KNO₃ salt bridge	110	12	—	−5 to 105
T	PHE-3217	Teflon® double junction reference, for difficult or dirty solutions	127	12	—	0 to 100

Figure 17-14 Reference electrodes. (REPRODUCED WITH THE PERMISSION OF OMEGA ENGINEERING, INC.)

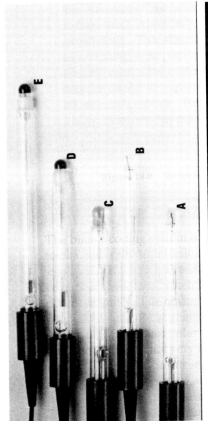

	Part No. BNC	Part No. U.S. Std.	Application/Features	Length (mm) Insertion	Dia. (mm) Insertion	pH Range	Temp. °C
A	PHE-2121	PHE-2121-U	General Purpose Laboratory	55	8	0-12	0-100
A	PHE-2721	PHE-2721-U	General Laboratory with Calomel Reference	55	8	0-12	0-60
B	PHE-2122	PHE-2122-U	General Laboratory with removeable bulb guard	55	8	0-12	0-100
B	PHE-2722	PHE-2722-U	Calomel Reference for measurement of protein samples-removeable bulb guard	55	8	0-12	0-60
C	PHE-2111	PHE-2111-U	General Purpose with large spherical bulb for fast response	110	12	0-12	0-100
C	PHE-2711	PHE-2711-U	Calomel Reference with large spherical bulb for fast response	110	12	0-12	0-60
D	PHE-2114	PHE-2114-U	General Purpose large bulb	110	12	0-13	0-100
D	PHE-2714	PHE-2714-U	Calomel Reference large bulb, can be used for high-purity water	110	12	0-13	−5 to 60
D	PHE-2414	PHE-2414-U	Large bulb sealed double junction	110	12	0-13	0-110
D	PHE-2314	PHE-2314-U	General Purpose large bulb sealed	110	12	0-13	0-110
D	PHE-2214	PHE-2214-U	Sealed inner refillable outer junction	110	12	0-13	0-110
D	PHE-2319	PHE-2319-U	High pH sealed	110	12	0-14	5-110
E	PHE-2116	PHE-2116-U	Rugged bulb design	140	12	0-13	5-110

Figure 17-15 pH combination electrodes. (REPRODUCED WITH THE PERMISSION OF OMEGA ENGINEERING, INC.)

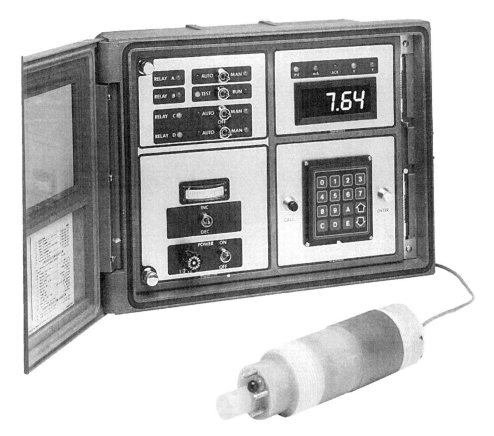

Figure 17-16   pH controller system. (REPRODUCED WITH THE PERMISSION OF OMEGA ENGINEERING, INC.)

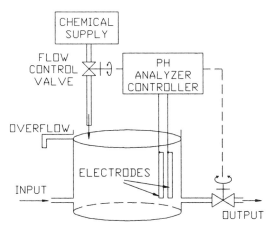

Figure 17-17   Simplified pH analyzer controller system for maintaining a solution's pH.

Application	Insertion Length (mm)	Diameter (mm)	pH Range	Temp. °C
General Purpose	110	12	0-12	– 5 to 80
General Purpose for Samples Requiring Double Junction	110	12	0-12	– 5 to 80
For use with sulfides or other contaminants found in plating baths and process streams	110	12	0-12	– 5 to 80
General Purpose with removeable bulb guard	110	12	0-12	– 5 to 80
Double junction design for use with interfering ions such as zinc, copper or sulfide.	110	12	0-12	– 5 to 80
Test Tubes	180	6.5	0-12	– 5 to 80
Test Tubes for samples requiring double junction	180	6.5	0-12	– 5 to 80
Extra long Test Tubes (detachable style shown)	300	6	0-13	0-100
Measurement of flat, moist surfaces such as meat or paper	110	12	0-12	– 5 to 80
Measurement of flat surfaces for samples requiring double junction	110	12	0-12	– 5 to 80
Extra rugged puncture tip for meats, cheeses, fruits, leather	55	8	0-10	– 5 to 100
Extra rugged puncture tip (Calomel)	55	8	0-10	– 5 to 60
Extra rugged puncture tip for samples requiring double junction	55	8	0-10	– 5 to 100
Rugged puncture tip for meats, cheeses, fruits, leather	55	8	0-13	0-100
Economical with removeable guard and Teflon® junction	110	12	0-13	0-100
Economical with double Teflon® junction	110	12	0-12	0-80
Economy	90	12.5	0-12	0-80
Economy	89	13	0-12	0-80
General Purpose-ORP	110	12	± 5000 mV	– 5 to 80
Dbl. junction for interfering ions such as zinc, copper or sulfide-ORP	110	12	± 5000 mV	– 5 to 80
For use with sulfides or other contaminants found in process streams-ORP	110	12	± 5000 mV	– 5 to 80

Figure 17-18 Design specifications for a typical pH glass electrode. (REPRODUCED WITH THE PERMISSION OF OMEGA ENGINEERING, INC.)

Nuclear radiation terminology

Charged particles consist of (1) *alpha particles*, which are helium-atom nuclei constructed from two neutrons and two protons and possessing a double positive charge value; (2) *beta particles*, which are negative electrons or positive electrons (called *positrons*); and (3) *protons*, which are positively charged particles. These are often referred to as elementary particles since they represent the very building block of matter.

Uncharged particles are represented by the *neutron*, which like the proton, is an elementary particle. As a matter of fact, it has the same mass as the proton.

Electromagnetic radiation consists of (1) *x-rays* and (2) *gamma rays*.

Types of ionization detectors

The types of ionization detectors that are discussed here in detail are the following: (1) the ionization chamber, (2) the proportional counter, (3) the Geiger–Mueller tube, (4) the semiconductor, and (5) the scintillation.

17-7.1 Ionization Chamber

Theory of operation

Nuclear radiation is characterized by its very high energy content or *quanta*. The ionization chamber, which is basically a sealed metallic container filled with an easily ionized gas such as neon or argon (refer to Figure 17-19), takes advantage of this fact by allowing the radiation that is to be sensed to enter through an entranceway or window located at one end or side of the chamber. Because of the radiation's high initial energy, the atomic particles brush by the atoms of the contained gas, causing some of the atoms' outer electrons to become stripped off, thereby ionizing the gas and creating an ionized pair of particles. The particle pair is made up of the positively charged atoms (because of having lost one or more electrons) and the negatively charged electrons themselves that have been stripped from their parent atoms. These ionized particles are then attracted to a high-voltage anode or a cathode, depending on the particle's charge. This attraction flow now constitutes a flow of current.

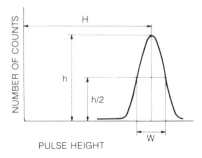

Figure 17-19 How the ionization chamber works. (Courtesy of Hamamatsu Corp., Bridgewater, NJ.)

The chamber itself is constructed from a high-density metal that is opaque to the radiation being detected. This opacity will vary, of course, depending on the radiation that is desired to be detected. Also, the chamber gas used inside the chamber is selective. Because different gases require different levels of ionization energy to become ionized, the gas chosen for a particular chamber design must be compatible for the energy-level capabilities of the atomic particles to be detected.

The foregoing principles of operation of the ionization chamber are the same basic principles as those used in the operation of the proportional counter and Geiger–Mueller tube. We discuss these devices next.

17-7.2 Proportional Counter

Theory of operation

The proportional counter differs from the ionization chamber only in the chamber gas that is used and the higher anode voltages that are present. The purpose of the proportional counter is to encourage an *avalanche effect* within the chamber. This effect is created by the incoming radiation entering the counter's ionization chamber and striking or stripping off the first electron from the chamber-gas atom. Because the anode voltage in the counter is now much higher than that normally experienced in an ionization chamber, the free electron has a much higher forward velocity and therefore a greater kinetic energy than it would have normally. This electron now has the capability of ionizing another atom, thereby creating another high-energy electron capable of ionizing. The process repeats itself many times before eventually dampening out. The gases used for this purpose are usually argon, neon, krypton, or xenon.

The avalanching effect is, in reality, an amplification process that can produce very high values of gain. The amount of anode voltage needed to cause this *gas amplification* or *gas gain* process to take place varies widely with the gas being used and the initiating radiation particle that is being detected. In general, therefore, for a given avalanching gas, the anode voltage determines the amount of gas gain. Gas amplifications in excess of 10,000 are possible, but it is usually recommended that gains above that value be avoided to assure stable operation of the detector. The amount of ionization that takes place within the ionization chamber is directly proportional to the energy content of the radiation being detected.

Operating characteristics

The gas gain, G_{gas}, for a proportional counter tube is determined from the following relationship:

$$\ln G_{gas} = K_1 \frac{V}{\ln(b/a)} \cdot \ln\left[\frac{V}{Pa\ \ln(b/a)} - \ln K_2\right] \qquad b > a \qquad (17\text{-}4)$$

where G_{gas} = gas gain (sometimes called the gas ratio)

K_1, K_2 = gas constants (torr/V)

a = radius of anode in chamber (in., mm, cm, etc.)

b = radius of cathode (or chamber, if cathode) (in., mm, cm, etc.)

P = chamber gas pressure (torr)

Note: One torr = 1 mm Hg, which is a pressure head measurement used primarily for measuring very small pressure values usually associated with vacuums.

Example 17-2

A proportional counter has the following dimensions and operating parameters: anode diameter = 15.0 mm, chamber diameter = 25.4 mm, chamber vacuum pressure = 12.8 torr, operating anode-to-cathode voltage = 1100 V dc. The chamber is the cathode in this particular proportional counter design. Determine the gas gain. Values for K_1 and K_2 are 0.0021 torr/V and 2.4×10^6 torr/V, respectively.

Solution: We place all the known values into eq. (17-4), first noting that radius a = 15.0 mm/2 and radius b = 25.4 mm/2:

$$\ln G_{gas} = K_1 \frac{V}{\ln(b/a)} \cdot \ln\left[\frac{V}{Pa \ln(b/a)} - \ln K_2\right]$$

$$= (0.0021)\frac{1100}{\ln(12.7/7.5)} \cdot \ln\left[\frac{1100}{(12.8)(7.5)\ln(12.7/7.5)} - \ln 2.4 \times 10^6\right]$$

$$= \frac{2.31}{0.527} \cdot \ln(21.74 - 14.69)$$

$$= (4.383)(1.95)$$

$$= 8.55$$

Then

$$e^{\ln G} = e^{8.55}$$

$$G = 5167$$

Equation (17-4) shows that the gas gain is an exponential function of the anode voltage. That is, for a given change in voltage there will be a much larger change in gas gain. Consequently, it is very critical to utilize a precisely regulated voltage supply for this type of application.

It is interesting to note how the particle counter's gas gain varies with the type of "fill" gas used in the chamber. Figure 17-20 shows this relationship. This figure also shows which gas is best suited for a particular anode voltage being used. Figure 17-21 shows the relationship that exists between xenon gas used as a fill gas and various gas pressures. It can readily be seen that the lower chamber pressures produce the higher gas gains.

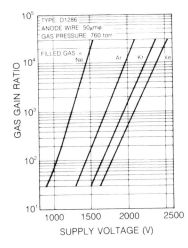

Figure 17-20 Relationship between type of fill gas and gas gain. (Courtesy of Hamamatsu Corp., Bridgewater, NJ.)

Energy resolution

The amplitude of the pulse signal generated by a proportional counter is determined by the energy content of the incident photon (nuclear particle). Unfortunately, the pulse's height may also vary, due to a variation in the gas gain ratio created by, say, a poorly regulated power supply condition. The *pulse height resolution*, sometimes called the *energy resolution* of a particle counter, refers to the counter's ability to differentiate between the energy content represented by the pulse generated by a nuclear particle and the variation of that pulse height due to a fluctuating gas gain created by instrument instability. This quantity is somewhat analogous to using a

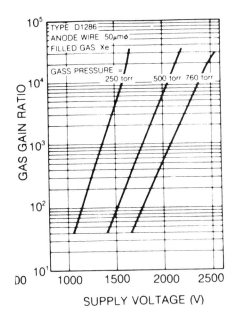

Figure 17-21 Relatonship between fill gas pressure and gas gain. (Courtesy of Hamamatsu Corp., Bridgewater, NJ.)

signal-to-noise ratio to rate the signal reception performance of an instrument, although the units of measurement involved are entirely different.

To determine the energy resolution of a counter, the following method is used. First, a distribution curve is constructed based on numerous counts that have been recorded for a particular nuclear particle. Second, this information is plotted against the pulse heights received from those counts. A curve similar to the one pictured in Figure 17-22 is typical of this effort. Finally, the following calculation is made based on the information obtained from the distribution curve:

$$E_{\text{res}} = \frac{W}{H} \times 100 \qquad (17\text{-}5)$$

where E_{res} = energy resolution (%)

W = width of distribution curve measured at half-height (see Figure 17-22)

H = pulse mean value (see Figure 17-22)

Note: *Energy resolution* and *pulse height resolution* mean the same thing.

If the percentage energy resolution is high, eq. (17-5) indicates that the counter, to a much lesser degree, is unable to distinguish between the energy of incoming particles and "noise" due to a decreased gas gain ratio. Lower percentage values imply that the counter is more capable of making the distinction between a particle's energy level and the pulse height that has been effected by decreasing the gas gain.

Typical design specifications

Typical design specifications are shown in Figures 17-23 and 17-24.

Practical applications

The proportional counter tube has found many uses in industry for applications that involve radiation. In medicine the proportional counter is used for the detection and analysis of x-rays. It is also used in applications where the proportional counter

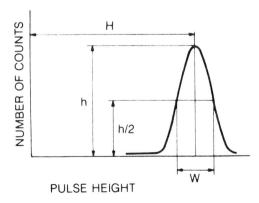

Figure 17-22 Distribution curve of pulse heights versus number of particle counts. (Courtesy of Hamamatsu Corp., Bridgewater, NJ.)

Head-On Types

Type No.	Outline No.	Window Material	Window Thickness (μm)	Window Effective Area (mm)	Cathode Material	GAS Filling	Gas Pressure (torr)
D988	❶	Beryllium	100	20 dia.	SUS304	Ⓐ Ar + CO_2	350

Side-On Types

Type No.	Outline No.	Window Material	Window Thickness (μm)	Window Effective Area (mm)	Cathode Material	GAS Filling	Gas Pressure (torr)
D1215	❷	Beryllium	50	20 x 20	SUS304	Ⓑ Xe + CO_2	500
D1286	❸	Beryllium	50	20 x 20	SUS304	Ⓒ Ne + CO_2	500
D1873	❹	Beryllium	25	3 x 30	Aluminum	Ⓒ Ne + CO_2	400
D1875	❺	Beryllium	150	10 x 30	Aluminum	Ⓓ Xe + CO_2	250
D2040	❻	Beryllium	300	82.5 x 25	SUS304	Ⓔ Ar + CO_2	760
D2278	❼	Beryllium	200	25 x 50	SUS304	Ⓓ Xe + CO_2	760
D2279	❽	Beryllium	200	25 x 50	SUS304	Ⓓ Xe + CO_2	760
D2756	❾	Beryllium	200	14 x 14	SUS304	Ⓑ Xe + CO_2	1520

Ⓐ Kr gas is available.
Ⓑ Ne, Ar or Kr gas is also available.
Ⓒ Ar, Kr or Xe gas is also available.
Ⓓ Ar or Kr gas is also available.
Ⓔ Kr or Xe gas is also available.
Ⓕ Measured with ^{55}Fe source (5.9 keV).

Figure 17-23 Specification sheet for proportional counter tubes. (Courtesy of Hamamatsu Corp., Bridgewater, NJ.)

Path Length (mm)	Maximum Length (mm)	Maximum Diameter (mm)	Operating Voltage Range (Vdc)	Recommended Operating Voltage (Vdc)	Typical Energy Resolution (%)	Capacitance (pF)	Temperature Range		Type No.
							Storage (°C)	Operating (°C)	
100	130	25	1200~1400	1300	21	1.5	−40~+90	−20~+60	D988
30	86	38	1700~1900	1800	17	2.5	−40~+90	−20~+60	D1215
30	120	38	1200~1400	1300	19	2.5	−40~+90	−20~+60	D1286
30	94	94 x 46 x 35	1100~1300	1200	21	2.0	−40~+90	−20~+60	D1873
30	94	94 x 46 x 35	1400~1600	1500	20	2.0	−40~+90	−20~+60	D1875
47	250	50.7	1700~1900	1800	17	3.0	−40~+90	−20~+60	D2040
47	209	50.7	1900~2150	2000	17	5.0	−40~+90	−20~+60	D2278
47	200	50.7	1900~2150	2000	17	2.5	−40~+90	−20~+60	D2279
20	82	24	2100~2350	2200	17	2.5	−40~+90	−20~+60	D2756

Figure 17-24 Additional specifications for proportional counter tubes. (Courtesy of Hamamatsu Corp., Bridgewater, NJ.)

becomes a densiometer. This is where a measured amount of radiation is sent through a material whose density and thickness are monitored by a counter tube. The intensity of the radiation passing through the material is recorded and then correlated with the material's thickness or density. From this information any future measurements pertaining to thickness or density can then be determined.

In the chemical analysis field the proportional counter is used in an instrument called a *sulfur content analyzer* (Figure 17-25). This device is used for determining the sulfur content in certain materials and depends on the principle that when x-rays bombard a sample, the energy content of the secondary emissions identifies the elements within that sample. The proportional counter is used for making the energy content determinations.

17-7.3 Geiger–Mueller Tube

Theory of operation

The Geiger–Müller tube, better known as the *Geiger counter* (Figure 17-26), is a proportional counter with two minor but very important modifications. The first modification relates to the anode voltage used inside the ionization chamber. This voltage is considerably higher than that used in the proportional counter system. Second, a second gas called a *quenching gas* is added to the chamber's ionization gas. The purpose of this gas is to shut down the discharge effect that would normally take place over a lengthy period of time throughout the ionization chamber as a result

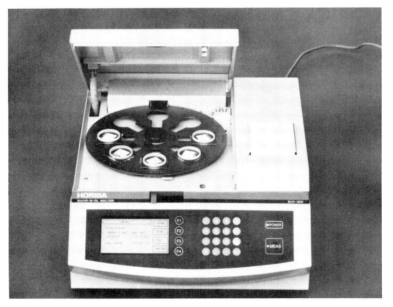

Figure 17-25   Sulfur analyzer. (From Dale R. Patrick and Stephan Fardo, *Industrial Process Control Systems,* copyright 1985, p. 151. Reprinted by permission of Prentice Hall, Englewood Cliffs, NJ.)

Figure 17-26   Geiger counter. (Courtesy of Edmund Scientific Corporation, Barrington, N.J.)

of the very high anode voltage. With the combination of the very high anode voltage and the quenching gas, an incoming radiation particle produces a short conduction pulse having a very short rise time (typically 1 μs or less) and a pulse duration of about 5 to 10 μs before quenching takes place. These sharply defined pulses, when amplified, are the characteristic "clicking" sounds that are heard when the Geiger counter is exposed to radioactive materials. Again, because of the ionization characteristics of the chamber's primary gas, its quenching gas, and the high anode voltage, the pulse amplitudes are independent of the radiation's energy content or type.

Operating characteristics

The relationship that exists between the number of pulses or cycles per second indicated by a Geiger counter and the dosage rate causing the cycles per second is shown graphically in Figure 17-27.

   The anode voltages used in the Geiger counter are typically less than 1000 V dc. Voltages in the 500-V range are quite common. Higher anode voltages will increase the counting tube's sensitivity somewhat but not dramatically. Also, the speed of response, or *dead time*, to an incoming particle is not affected appreciably, as shown by the dead-time curves in Figure 17-28.

Design and physical hardware specifications

Typical design specifications for Geiger counter tubes are shown in Figure 17-29. Physical hardware specifications are shown in Figure 17-30.

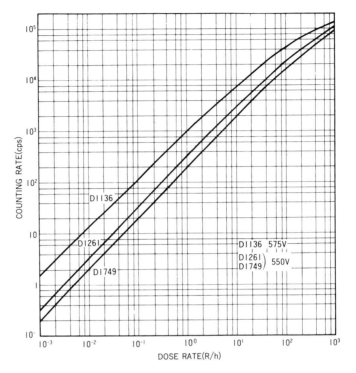

Figure 17-27 Typical relationship between particle counts and dosage rates. (Courtesy of Hamamatsu Corp., Bridgewater, NJ.)

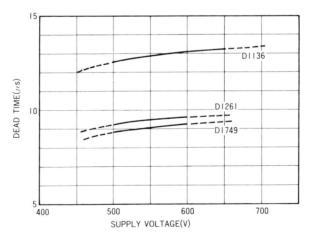

Figure 17-28 Dead-time curve for a Geiger counter. (Courtesy of Hamamatsu Corp., Bridgewater, NJ.)

SPECIFICATIONS

General

Parameters	D3372	D3517	D3553	Unit
Cathode Material	28% Cr, 72% Fe			—
Thickness	80 to 100			mg/cm²
Effective Length	16	5	8	mm
Fill Gas	He, Ne, Halogen			—
Capacitance of Anode to Cathode	3	2.5	2.5	pF

Maximum Ratings

Parameters	D3372	D3517	D3553	Unit
Anode Resistance		2.2		MΩ (Min.)
Applied Voltage	650	600	600	V
Temperature — Operation		−40 to +50		°C
Temperature — Storage		−40 to +75		°C

Characteristics (at 25°C)

Parameters	D3372	D3517	D3553	Unit
Starting Voltage	380	400	400	V (Max.)
Plateau Voltage	500 to 650	500 to 600	500 to 600	V
Operating Voltage	Arbitrary within plateau			—
Plateau Slope	0.15	0.3	0.3	%/V (Max.)
Background[A]	2	1	1	cpm (Max.)
Dead Time at 600V	20	15	15	μs (Max.)
Life Expectancy[B]	5×10^{10}	1×10^{10}	1×10^{10}	counts
Equivalent Tube	Philips 18509 Hamamatsu D1136	Hamamatsu D1749	Philips 18529 Hamamatsu D1261	—

A: Shielded with 50mm Pb and 3mm Al at 575V (D3372), at 550V (D3553, D3517).

B: Count rate 4500 cps (D3372), 3200 cps (D3553, D3517) at 25°C.

Figure 17-29 Specification sheet for Geiger–Mueller tube. (Courtesy of Hamamatsu Corp., Bridgewater, NJ.)

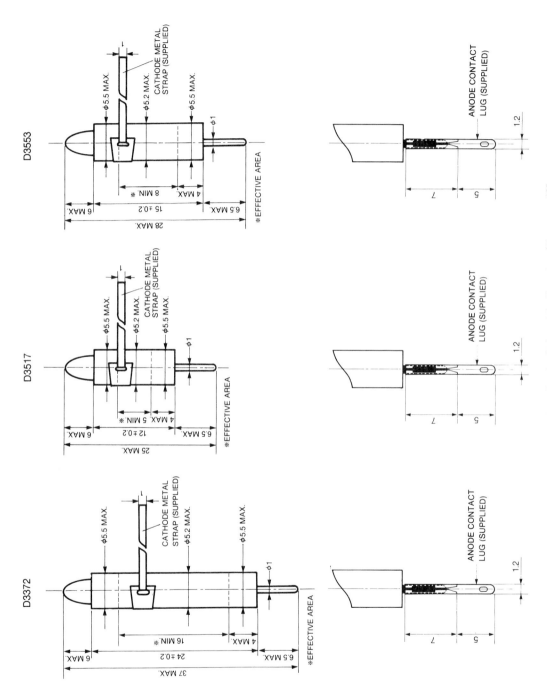

Figure 17-30 Hardware specifications for the Geiger–Mueller tube. (Courtesy of Hamamatsu Corp., Bridgewater, NJ.)

Practical applications

Figure 17-31 shows a simplified circuit utilizing a Geiger counter tube. Basically, the main tube circuit is surprisingly simple and straightforward. The output circuitry comes from the cathode rather than the anode, in order to reduce high-voltage hazards in the design layout. The 2.2-MΩ resistor in series with the high-voltage supply merely acts as a current-limiting resistor in case of accidental shorting within the counter tube itself. Otherwise, because of the very small current flow through that resistor, the voltage drop is almost negligible under normal operating conditions. The tube's output is then sent to a pulse-shaping circuit to enable the pulses to be counted by a frequency counter. If a certain counting rate is exceeded, an alarm is sounded indicating that a particular radiation level has also been exceeded.

17-7.4 Semiconductor Ionization Detector

Theory of operation

Virtually any solid-state device is affected by particle radiation. It has been known for some time that instruments and equipment containing semiconductor circuitry fail to operate properly when exposed to intense radiation from nuclear sources. In many cases these effects are detrimental and permanent to the solid-state devices themselves. However, when designed properly, a semiconductor device may act as a radiation sensor, reacting in a very reliable and safe manner.

The theory of operation of these specially designed semiconductors is quite similar to the photodetection processes discussed in Chapters 14 and 16. Figure 17-32 shows the p-n junction of a semiconductor being exposed to a radiation particle. As the particle collides or brushes by the atoms contained within the junction material, the atoms become ionized, along with donating a dislodged electron in the process. Because of the usual charge separation at the pn junction, an electric field is present. This field acts as a further charge separator, causing the newly ionized positive charges to veer in one direction when encountering this field, and the negative

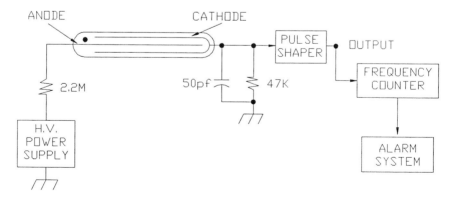

Figure 17-31   Circuit for radiation detection and annunciating.

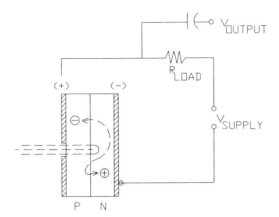

Figure 17-32 How the semiconductor ionization detector works.

charges (the dislodged electrons) to swerve in the opposite direction. A thin conducting film is deposited onto the back of the semiconducting material to form a backplane and a cathode; this surface acts as a collection area for the ionized positive charges. The front or entrance side of the detector is coated with a conductor to serve as an anode; it collects the dislodged electrons. Together, these two surfaces complete a circuit whose current represents the intensity of the nuclear radiation entering the detector. The detector's output is in the form of pulses, much like that of the Geiger–Mueller tube, each pulse representing one ionizing event created by one nuclear particle.

17-7.5 Scintillation Detector

Theory of operation

Scintillation detectors are photomultiplier tubes that have been modified to detect nuclear radiation. The modification is fairly simple. The photomultiplier's light-sensitive window is covered with a special chemical coating called a *phosphor* that emits light when it is struck by a nuclear particle. One material is sodium iodide that has been doped or activated with the element thallium. The surface is then installed into a light-tight area directly over the front end of a photomultiplier tube. The construction layout of a scintillator tube is shown in Figure 17-33. When a nuclear particle collides with the sodium iodide surface, a faint light is emitted momentarily. The photomultiplier tube then amplifies the light pulse many times, to a usable, detectable value. Like the proportional counter, the amplitudes of the pulses measured by the photomultiplier contain the energy-value information of each light photon–converted particle.

Operating characteristics

Like the proportional counter, the quantity *pulse height resolution* is used to rank the scintillator's ability to differentiate between actual data measured and anomalies generated by the device itself. Figure 17-34 shows the approximate energy distribu-

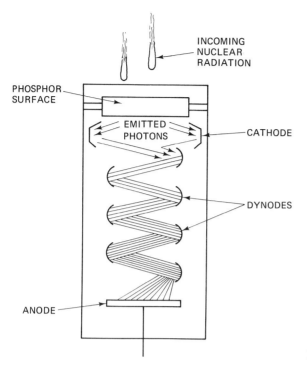

Figure 17-33   Scintillation detector.

**Pulse Height Distribution
(Energy Spectrum)**

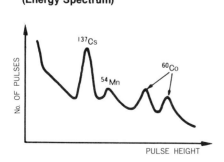

Definition of Pulse Height Resolution

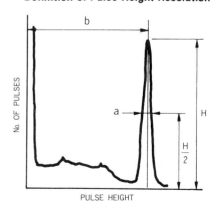

Pulse Height Resolution $= \dfrac{a}{b} \times 100$ (%)

Figure 17-34   Energy distribution of atomic particles and scintillator's pulse height resolution. (Courtesy of Hamamatsu Corp., Bridgewater, NJ.)

Spectral Response of PMT and Spectral Emission of NaI (Tl) Scintillator

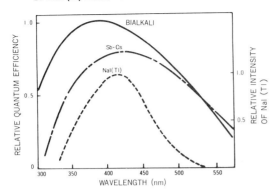

Figure 17-35 Frequency response for a sodium iodide surface in a scintillator. (Courtesy of Hamamatsu Corp., Bridgewater, NJ.)

tion of several atomic particles, showing their relative pulse heights versus the number of pulses, along with the definition of pulse height.

Figure 17-35 shows the spectral emission capability of a thallium-activated sodium iodide detector-emitter surface. To make an effective scintillator detector, this emission characteristic must match closely the receptive spectral response of the photomultiplier's cathode. Fortunately, bialkali cathodes are the most popular cathode types used, and their spectral response curves are very similar to that of the sodium iodide phosphor.

Design specifications

The design specifications for a photomultiplier tube that is to be used as a scintillator tube are shown in Figures 17-36 and 17-37. The model chosen has been marked with an arrow on the specification sheet. The manufacturer chose this particular model because of its bialkali cathode and its high anode sensitivity.

17-7.6 Summarizing Relationship between Anode Voltage and Output of Gas-Filled Detectors

Having discussed the various nuclear particle detectors it becomes obvious that one of the basic differences that separates one device from the other is the magnitude of anode voltage that is used. This is what determines the behavior of the chamber gas that is used; it also determines the magnitude of the pulse that is generated. Figure 17-38 summarizes the relationships between three of the four gas-filled detection devices, their anode voltages, and the resultant pulse heights. The shape of the curve seen in Figure 17-38 is due to the ionization characteristics of the gas being used. The zone to the left of the ionization chamber range is called the *recombination range* because it is in this area where the anode voltage does not yet have a strong enough influence to prevent the ionized particles from recombining. The two curves E_1 and E_2 are for two different incident photon energy levels. The Geiger counter operates in the E range, as shown in the curve.

Type No.	Remarks	Spectral Response Curve Code	Range (nm)	Peak Wavelength (nm)	Photo-cathode Material	Window Material	Out-line No.	Dynode Structure No. of Stages	Socket	Maximum Ratings Anode to Cathode Voltage (Vdc)	Average Anode Current (mA)	Cathode Sensitivity Luminous Min. (μA/lm)	Typ. (μA/lm)
3/8 inch (10 mm) Dia. Types													
R1893	Solar blind response, Cs-Te photocathode, synthetic silica	200S	160~320	210	Cs-Te	Q	①	L/8	E678-11N	1500	0.01	—	—
R1635	Visible response, bialkali photocathode	400K	300~650	420	Bi	K	①	L/8	E678-11N	1500	0.03	60	90
R1635-02	For photon counting, HA coating with magnetic shield				Bi	K	①	L/8	E678-11N	1500	0.03	60	90
R1894	S-20 response, multialkali photocathode	500K (S-20)	300~850	420	M	K	①	L/8	E678-11N	1500	0.03	80	120
*R2496	Low TTS, high speed time response	400S	160~650	420	Bi	Q	②	L/8	E678-11N	1500	0.03	60	90
1/2 inch (13 mm) Dia. Types													
R1081	For VUV detection, MgF2 window, temporary base	100M	115~200	140	Cs-I	MF	⑤	L/10	E678-12A	2250	0.01	—	—
R1080	Variant of R1081 with Cs-Te photocathode	200M	115~320	210	Cs-Te	MF	⑤	L/10	E678-12A	1250	0.01	—	—
R759	Solar blind response, Cs-Te photocathode, synthetic silica	200S	160~320	210	Cs-Te	Q	③	L/10	E678-13A	1250	0.01	—	—
R647	10-stage, bialkali photocathode				Bi	K	③	L/10	E678-13A	1250	0.1	40	80
R647-01	Selected for scintillation counting	400K	300~650	420	Bi	K	③	L/10	E678-13A	1250	0.1	60	90
R647-04	For photon counting, HA Coating with magnetic shield				Bi	K	④	L/10	E678-13A	1250	0.1	60	90
R760	Variant of R647 with synthetic silica window	400S	160~650	420	Bi	Q	③	L/10	E678-13A	1250	0.1	40	90
R1591-01	High temp. bialkali photocathode, ruggedized type	401K	300~650	375	H Bi	K	③	L/10	E678-13E	1800	0.1	—	40
R1463	High gain variant of R761	500U	185~850	420	M	U	③	L/10	E678-13A	1250	0.03	80	120
R1463-01	For photon counting, HA Coating with magnetic shield				M	U	④	L/10	E678-13A	1250	0.03	80	120

Ⓐ * : Newly listed in this catalog.
Ⓑ Typical spectral response characteristics are shown on pages 76 and 77.
Ⓒ Photocathode materials
   Bi : Bialkali
   M : Multialkali
   H Bi : High temperature bialkali
Ⓓ Window materials
   Q : Synthetic silica
   K : Borosilicate glass
   MF : MgF2
   U : UV glass
Ⓔ Basing diagram symbols are explained on page 17.
Ⓕ Dynode Structure
   L : Linear focused
Ⓖ A socket will be supplied with a tube.
Ⓗ The maximum ambient temperature range is −80 to +50°C except high temperature bialkali photocathode types which withstand up to +175°C. When using tubes with glass base at −30°C or below, see precautions on page 74.
Ⓙ Averaged over any interval of 30 seconds maximum.
Ⓚ At the wavelength of peak response.

Ⓛ Voltage distribution ratios used to measure characteristics are shown on page 62.
Ⓜ Anode characteristics are measured with the supply voltage and the voltage distribution ratio specified by Note Ⓛ.
a : At 122 nm
b : At 254 nm
c : Dark counts per second (cps)

Figure 17-36 Specification sheet for photomultiplier tube. (Courtesy of Hamamatsu Corp., Bridgewater, NJ.)

Cathode Sensitivity			Anode to Cathode Supply Voltage (Vdc)	Anode Characteristics								Notes	Type No.
Blue Typ. (μA/lm-b)	Red/White Ratio Typ.	Radiant Typ. (mA/W)		Anode Sensitivity			Current Amplification Typ.	Anode Dark Current (after 30 min.)		Time Response			
				Luminous		Radiant Typ. (A/W)		Typ. (nA)	Max. (nA)	Rise Time Typ. (ns)	Electron Transit Time Typ. (ns)		
				Min. (A/lm)	Typ. (A/lm)								
—	—	12^b	1250 ⑤	—	—	3600^b	3.0×10^5	0.5	2.5	0.8	7.8		R1893
10.5	—	82	1250 ⑤	30	90	8.2×10^4	1.0×10^6	5.0	50	0.8	7.8	Silica window type (R2055) available.	R1635
10.5	—	82	1250 ⑤	50	200	1.8×10^5	2.2×10^6	100^c cps	400^c cps	0.8	7.8		R1635-02
—	0.2	51	1250 ⑤	10	50	2.1×10^4	4.2×10^5	2.0	20	0.8	7.8		R1894
10.5	—	82	1250 ⑧	30	90	8.2×10^4	1.0×10^6	5.0	50	0.7	7.8		R2496*
—	—	9.8^a	2000 ⓕ	—	—	980^a	1.0×10^5	0.03	0.05	1.8	18		R1081
—	—	20^b	1000 ⓕ	—	—	$2.0 \times 10^{4\,b}$	1.0×10^6	0.3	0.5	2.5	24		R1080
—	—	20^b	1000 ⓕ	—	—	$1.0 \times 10^{4\,b}$	5.0×10^5	0.3	1.0	2.5	24	UV glass window type (R960) available.	R759
9.5	—	75	1000 ⓕ	30	80	7.5×10^4	1.0×10^6	5.0	15	2.5	24		R647
10.5	—	80	1000 ⓕ	30	90	8.0×10^4	1.0×10^6	1.0	2.0	2.5	24		R647-01
10.5	—	80	1000 ⓕ	70	200	1.8×10^5	2.2×10^6	80^c cps	400^c cps	2.5	24		R647-04
10.5	—	80	1000 ⓕ	10	90	8.0×10^4	1.0×10^6	5.0	15	2.5	24		R760
6.0	—	50	1500 ⓕ	10	20	2.5×10^4	5.0×10^5	0.5	10	2.0	20	Flexible lead type (R1591) available.	R1591-01
—	0.2	51	1000 ⓕ	30	120	5.1×10^4	1.0×10^6	4.0	20	2.5	24	HV resistant type (R761) and silica window type (R2007) available.	R1463
—	0.2	51	1000 ⓕ	30	120	5.1×10^4	1.0×10^6	900^c cps	1000^c cps	2.5	24		R1463-01

Figure 17-37 Additional specifications for photomultiplier tube. (Courtesy of Hamamatsu Corp., Bridgewater, NJ.)

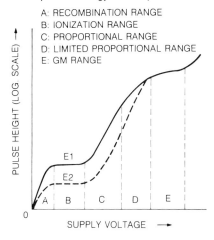

Operating Range of Gas-Filled Detectors
(Incident photon energy E1 > E2)

A: RECOMBINATION RANGE
B: IONIZATION RANGE
C: PROPORTIONAL RANGE
D: LIMITED PROPORTIONAL RANGE
E: GM RANGE

PULSE HEIGHT (LOG SCALE)

E1
E2
A   B   C   D   E

0

SUPPLY VOLTAGE ⟶

Figure 17-38 Summary of relationship between anode voltage and pulse heights for the three gas-filled detection devices shown. (Courtesy of Hamamatsu Corp., Bridgewater, NJ.)

REVIEW QUESTIONS

17-1. Define what comprises a chemical ionization sensor. What measurands are detected by this type of sensor?

17-2. Name the two types of transducers used for ionizing sensors. What distinguishes one from the other?

17-3. What is the theory of operation of the conductive electrolytic transducer?

17-4. What is the inverse of conductance? What standards govern the measurement of conductance in a solution?

17-5. What is a practical application of a conductance sensor other than the ones mentioned in the text?

17-6. Explain how the glass electrode and the reference electrode work in a pH-measuring system.

17-7. What is a combination probe?

17-8. What particles comprise the radiation detected by ionization transducers?

17-9. Explain how the ionization chamber in an ionization detector works.

17-10. Explain how gas amplification takes place inside an ionization chamber.

PROBLEMS

17-11. The sensing plates of a set of electrodes for a conductivity tester have the following dimensions: height 6.51 cm and width 2.5 cm. The separation between electrodes is 10 mm. If the solution whose conductivity is being tested is 20% hydrochloric acid (HCl), determine its conductance in siemens. What is the electrodes' cell constant?

17-12. A conductivity tester is to be used as a liquid-level detector for a container of fresh water. The two electrodes are separated by a distance of 5 mm and extend the full depth of the container, which is 5.0 m. Each electrode is 30 mm wide. Plot a curve made from at least six different level readings so that the resultant curve could be used as a calibration curve for this particular level-measuring system. Do all plotting on x-y coordinate graph paper.

17-13. A proportional counter contains an anode having a diameter of 17 mm and a cathode having a diameter of 22.3 mm. If $K_1 = 0.0025$ torr/V and $K_2 = 3.05 \times 10^6$ torr/V, and the counter's chamber has a measured vacuum pressure of 16 mm Hg, find the counter's gain. The counter's anode operates from a 750-V dc power supply.

17-14. Determine how much the gain varies in the proportional counter described in Problem 17-13 if the power supply voltage varies by $\pm 3\%$.

REFERENCES

NORTON, HARRY N., *Sensor and Analyzer Handbook*. Englewood Cliffs, NJ: Prentice Hall, 1982.

The pH and Conductivity Handbook. Omega Engineering (P.O. Box 4047, Stamford, CT 06907-0047).

Photomultiplier Tubes. Hamamatsu Corporation (360 Foothill Road, P.O. Box 6910, Bridge-water, NJ 08807-0910).

18

============================

ENCODERS

CHAPTER OBJECTIVES

1. To understand how the linear and rotary encoder functions.
2. To study their construction techniques.
3. To study the potentiometric encoder.
4. To understand the differences between incremental and absolute encoding.
5. To analyze the problems associated with digital and analog encoders.

18-1 INTRODUCTION

Now that many of the existing sensors being used today in industry and in research have been discussed, it is time to look at a special implementation of some of the sensors. A few of these sensors, specifically displacement-type sensors, lend themselves very well to being installed in special enclosures specifically designed for the purpose of sensing displacement, velocity, or acceleration. These enclosures, together with their sensors and contained electronics, are called *linear* and *rotary encoders*.

   The output from one of these encoders is typically in the form of an electrical signal that has either been transformed into a coded digital signal or is in the form of an analog voltage or current. Regardless of the form, the signal is a proportional representative of the measurand.

The encoder's housing is a specially designed containment comprised of a mechanical sliding or rotating shaft extending from the housing itself. The shaft's opposite end is attached to an internal mechanism within the housing, which may be part of the actual sensing device. These shafts are designed to be in direct physical contact with the measurand, so that relative movement takes place between the shaft and the encoder's built-in sensor. Some of the better known and most commonly used encoders are listed below.

18-2 TYPES OF ENCODERS (BOTH ROTARY AND LINEAR) PRESENTLY AVAILABLE

There are several encoder types in use in industry at the present time. The following is a list of some types that have enjoyed a large degree of widespread use: (1) optical encoders, (2) potentiometric encoders, (3) brush contact encoders, (4) magnetic encoders, and (5) synchromechanisms: synchros and resolvers. Most of these devices have already been discussed as to their theory of operation and operational characteristics. Optical sensors were discussed in Chapters 14, 15, and 16; potentiometric devices were discussed in Chapter 7; synchromechanisms were discussed in Section 10-4. However, we are now interested in the characteristics of linear and rotary encoders themselves, how they are constructed, and specifically, the various methods used for generating digital and analog output signals.

18-3 LINEAR ENCODER CONSTRUCTION TECHNIQUES

Linear encoders are comprised of an encoder housing with a shaft that extends from one end of the housing and is constructed so that it can move linearly into and out of the housing freely and unrestricted. A typical linear encoder design is illustrated in Figure 18-1. The shaft itself is usually made from hardened steel with a highly polished surface which surface is often chrome-plated. The shaft rides on a lubricated bearing surface built into the housing, the bearing being either of the roller type or of a bronzed sleeve construction. The bearing construction is often accompanied by "shaft wipers" built into the bearing housing to prevent foreign particles from entering the encoder housing and also to keep the shaft wiped clean of any abrasive materials that might otherwise scratch the shaft's polished surfaces. Scratches may eventually wear away the bearing surfaces and ultimately lead to the bearings' destruction. It is also important that these bearings be able to withstand appreciable side loading to accommodate minor misalignments between the encoder shaft and any mechanical attachments made to it. Side loading is further prevented by an internal guiderail or guide-slide design, which helps to maintain the encoder's shaft centerline alignment and concentricity with the housing's bearing assembly. In addition to serving as a guide plate, this member serves as an attachment point for the encoder's sensing device, requiring relative movement for motion detection.

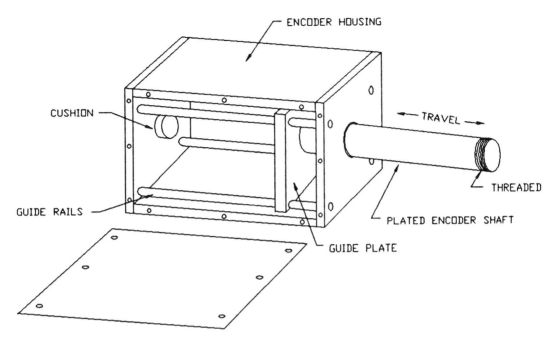

Figure 18-1 Linear encoder housing construction.

At the shaft's extreme ends of travel a cushion made of soft rubberlike material is installed to absorb any impact in case the shaft becomes fully extended or retracted. In some instances a spring-loaded pad is used in place of or in addition to the rubberlike pad for absorbing this impact. Otherwise, any sudden impacts could shatter the portion of the sensor that is attached to the sliding plate.

18-4 OPTICAL ENCODER

Figure 18-2 shows how the guide plate is used for a mounting surface for a photodetection and light-source system. Other encoder types can certainly be used here, but we use the photodetection system only as an example. A ruled-glass plate comprised of many finely spaced opaque lines is shown mounted between the light source and photocell. The mounting is such that a light shines between the ruled markings onto the photocell. As the guide plate is moved by the encoder shaft, the photocell detects the alternate light and dark variations of the light source created by the lines on the glass as they pass between the light and photocell. The photocell translates these variations into a pulsating signal whose frequency is directly determined by the number of lines per second of travel passing beneath the photocell. Because the line spacing is known precisely, the amount of shaft movement is determined merely by multiplying the photocell's frequency by this spacing.

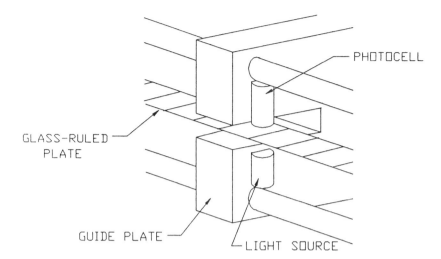

Figure 18-2 How a guide plate is modified for photocell, light source, and ruled-glass plate.

As already mentioned, the system just described is but one of many types of encoder assemblies that are constructed for linear encoder usage. Other types are discussed later. But now we look at another type of encoder, the rotary encoder.

18-4.1 Rotary Digital Encoders

A rotary encoder is comprised of an encoder housing and a rotating shaft extending from one end of the housing. Internally, the shaft is attached to one portion of the sensor, a ruled-glass disk, while the opposite end is attached to a rotating driving member (see Figure 18-3). As in the case of the linear encoder, the front-housing shaft bearings may be either roller or bronzed sleeves. However, in addition to the bearings mentioned, a thrust bearing is needed at one end of the encoder shaft to accept the axial loading that often takes place in a driven rotating member situation. Also, care must be taken not to exceed the side-thrust limitations of the front-housing bearings due to radially applied forces created by the driving member.

For our rotary encoder example we again select the photodetection system. An encoder wheel is attached to the rotating shaft and is made to pass between a photocell and a light source, as illustrated in Figure 18-3. The wheel is made of the same glass material as that used in the linear optical encoder and contains the same ruled opaque lines for producing pulses in the photocell circuitry. As in the case of the linear encoder, by counting the pulses and knowing the angular spacing between each line, the total angular rotation of the encoder shaft is easily obtained.

Rotary encoders utilizing sensing methods other than optical encoding are constructed somewhat similarly to the one just described. Other rotary types are described in later sections.

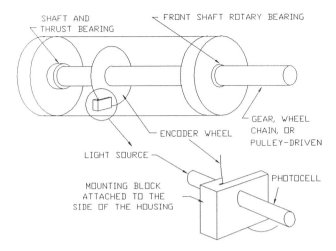

SHAFT AND
THRUST BEARING

FRONT SHAFT ROTARY BEARING

GEAR, WHEEL
CHAIN, OR
PULLEY-DRIVEN

ENCODER WHEEL

LIGHT SOURCE

PHOTOCELL

MOUNTING BLOCK
ATTACHED TO THE
SIDE OF THE HOUSING

Figure 18-3   Rotary encoder equipped with photocell, light source, and ruled-glass plate.

18-4.2 Analog Linear and Rotary Encoders

Analog linear encoders are made up primarily of only one type of sensing element, the variable resistor. The rotary encoder, on the other hand, may be made up of a variable resistor or of a reluctance type of sensor called a synchromechanism (discussed in Section 10-4.3).

18-5 POTENTIOMETRIC ENCODER

In place of the glass slide in the linear encoder or rotating glass disk in the optical encoder, a resistive element is mounted instead. Figure 18-4 shows typical installations for both the linear encoder and rotary encoder. In both cases a ramp-type curve is generated (Figure 18-5), indicating the amount of rotation of the shafts. In addition, the direction of travel of both shafts is ascertainable by the increasing or decreasing magnitude values of the output signals. This very simple characteristic has encouraged the manufacture of relatively inexpensive analog encoders and as a consequence has led to their quite widespread use.

18-6 PROBLEMS WITH ANALOG ENCODING SYSTEMS

Much of the electrical noise that exists today (ignition noise in radio and television receivers, electrical static produced by lightning discharges, commutator noise from electric motors, etc.) is analog in nature. Because of this characteristic, it becomes

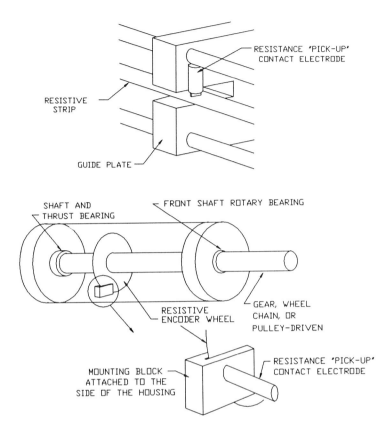

Figure 18-4 Linear and rotary variable resistance encoders.

difficult to filter out undesirable noise superimposed on analog data so that it does not become processed and confused with the desirable data. Filtering analog-type noise signals from digital signals is not as difficult, due to the radically different signal characteristics of the two signal types.

Figure 18-6 is an example showing how digital data can be separated from analog noise. The noisy signal is fed to a circuit called a *Schmitt trigger*, a circuit that has an adjustable threshold level capability. That is, only a signal having a particular threshold voltage level (L_2 in Figure 18-6) will cause the circuit to generate an output voltage signal of a particular level. When the input signal plus its noise *drops* below a certain preset level (L_1 in Figure 18-6), the trigger also drops its output signal to a lower level. The result is an output signal that is a fairly faithful reproduction of the input signal as seen from the input level of the Schmitt trigger. Of course, the trigger cannot distinguish between noise and data if either occur at either threshold L_1 or L_2. For the filtering system to work well, it is important that both L_1 and L_2 be set for values above the highest peak of existing noise.

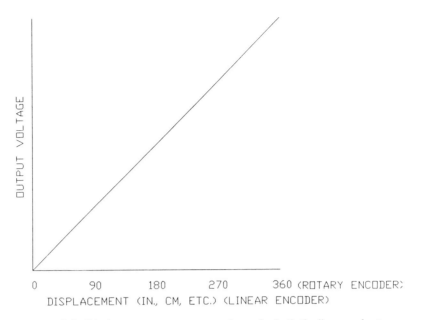

Figure 18-5 Displacement versus output voltages for both the linear and rotary encoders.

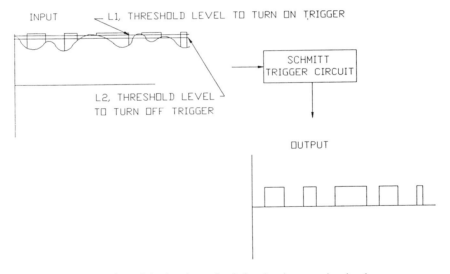

Figure 18-6 Using a Schmitt trigger circuit for cleaning up noisy signals.

18-7 PROBLEMS WITH DIGITAL ENCODING SYSTEMS

Digital encoding schemes are not without their problems also. The major problem with pulse-generating encoders is that it is difficult to determine the vector quantity of the data. For instance, in the case of the linear digital encoder, how would you determine if the linear shaft's velocity were negative or positive if all you had to go on were the number of pulses generated per unit time? In other words, how would you know if the shaft were moving into or out of the encoder? To complicate this determination further, what if the shaft changed its direction of travel momentarily during the time of recording its velocity? There is no provision for the subtraction of pulses with the system as presently described. This type of system, the system used in the units depicted in Figures 18-2 and 18-3, is called *incremental encoding*. With this type of system there is no way to determine the "direction" or vectoring of the data. Only the encoder's increments of motion are available, and these are good only for determining the data's magnitude.

18-8 ABSOLUTE ENCODING

Encoding systems that allow us to determine the data's direction or vector relative to a reference point or zero point utilize encoders called *absolute encoders*. There are two kinds of systems or methods being used at present to generate this type of information: (1) the phasing method, and (2) the binary coding method.

18-8.1 Phasing Method

There are several circuit designs that allow you to determine in which direction the encoder shaft is traveling. One method involves the usage of two photodetectors, as shown in Figure 18-7. The ruled eclipsing lines on the moving encoder strip or disk have been purposely slanted as shown. This is to create a progression of photocell eclipses rather than simultaneous eclipse of both cells. For the direction shown, photocell P_1 will be eclipsed before photocell P_2. Therefore, there will be an instant when P_1 will be eclipsed but not P_2. Then both P_1 and P_2 will be eclipsed, P_1 will be exposed while P_2 is still eclipsed, and finally, both P_1 and P_2 will be exposed. This sequence of events can be summarized as in the truth table in Figure 18-7, assuming that a binary 1 means an exposed cell and a binary 0 means a covered or eclipsed cell. As long as the order of binary counting is maintained as shown, you can infer correctly that the encoder's shaft is moving as shown. However, any reversal in the counting sequence is an indication that the shaft is moving in the opposite direction. The digital circuitry needed for detecting the direction of this count is a relatively easy circuit to design and can be installed neatly inside the encoder housing itself. (These circuits will not be presented here but may be found in any good textbook dealing with digital electronics.) In essence, this circuit keeps track of the counting sequence. Because a completed counting sequence $(11_2, 01_2, 00_2, 10_2)$ represents a complete cycle or 360°, and there are four steps or levels in that cycle, this type of directional sensing is sometimes referred to as

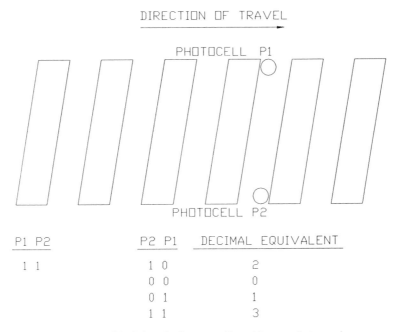

PHOTOCELL P1

PHOTOCELL P2

P1 P2			P2 P1		DECIMAL EQUIVALENT
1 1			1 0		2
			0 0		0
			0 1		1
			1 1		3

Figure 18-7   Obtaining absolute encoding with an optical encoder.

quadrature phasing. There are other schemes that can also be used to create quadrature phasing. Encoder manufacturers have individually developed their own methods for producing such systems. Another quadrature phasing method is explained in Section 18-10.

An added bonus to the system just described is that travel increments of one-fourth the spacing between encoding lines may be measured. Whenever there is a change in the binary count, the photocells are so spaced that a movement of one-fourth the incremental distance will produce one shift in the binary count. Consequently, a resolution four times the line spacing is obtained. Additional resolution may be obtained by adding more properly spaced photocells to the system.

An encoder's resolution is determined mainly by the manufacturer's ability to produce finely ruled opaque lines on a transparent media such as glass. The thinner these lines are, the greater the number of increments per inch or per centimeter that the encoder can respond to. However, one limiting factor in the resolution of encoders is the capability of the sensing system to detect the finely ruled lines that we have been discussing here. The optical encoder must use a light source whose light-beam width is *narrower* than the width of the eclipsing line.

Figure 18-8 shows a schematic of the optical system used in many optical encoders. Assuming a line width of 0.001 in. (0.025 mm), the light source beam is narrowed to less than this value through the use of a light mask and lens system. The light mask is simply a plate having a single hole bored in it to allow the focused light beam to pass through to further restrict the beam's diameter to the desired diameter.

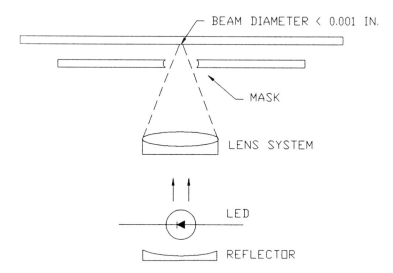

Figure 18-8   Reducing the beam width to increase an encoder's resolution.

18-8.2 Binary Coding Method

The binary coding system is similar to the phasing system just discussed in that line tracks and a photodetection system are required. However, in this case, each line represents one bit of positioning information coded in binary form (see Figure 18-9). Several of these lines, when read in parallel fashion, then represent the exact position of the encoder's slide or shaft. Eight parallel eclipsing lines with eight photocells and light sources can then create an 8-bit digit that represents the encoder shaft's position or location. This particular system would have the capability to produce 2^8 or 256 numbers, which is not very many numbers for, say, a linear encoder having a 15-cm travel. This would result in a measuring capability of only $256 \div 15$ cm or 17.07 divisions/cm. That is, one division would be 1/17.07 cm wide or 0.059 cm. On the other hand, if 13 tracks were used for this same encoder, a measuring capability of $2^{13} \div 15$ cm or 546 divisions/cm would be possible. Each division would only be 1/546 cm or 0.0018 cm wide.

There are several binary encoding schemes that are presently being used. One such scheme is used for rotary-type encoders and is called the *Gray code* system. To understand how this system works, however, it is first necessary to see why a simple binary counting system will *not* work.

Figure 18-10 shows a very simple rotary encoder employing a three-track binary coded wheel with three light sensors and photocells installed as shown in the illustration. The wheel has been divided into eight equal divisions numbering 000_2 through 111_2. Notice what happens as the wheel is rotated in a counterclockwise direction. You would assume that the photocells would be triggered in a smooth binary counting sequence fashion. Unfortunately, what really happens is this: As the transition is made from one binary count to the next (in the case of Figure 18-10, the count is going from 001_2 to 010_2) the three photocells do not "make" the

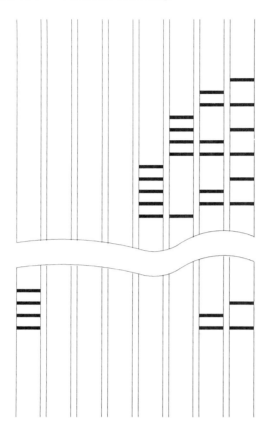

Figure 18-9 Bit pattern on an encoding strip.

transition all at the same time. That is, the three photocells will not switch to their new number values simultaneously. There could very well be intermediate counts such as 000_2 or 011_2 that would be generated just as well during this transitional period. The question then becomes: How do you eliminate these temporary false counts so as not to confuse the encoder's output?

Figure 18-11 shows a redesigned track layout that prevents the generating of spurious counts when moving from one count to the adjacent one. This type of layout is called a *Gray code*. As you move from one count to the next, there are no possibilities of generating another count due to partial switching as was true in the previous case; the only numbers generated during this time are the two numbers that are adjacent to each other.

There are a few methods of coping with the spurious number-generation characteristics of a non-Gray-coded system. One such method uses the fact that since the counting sequence is already known, and since that number can only be one higher or one less than the previous one, any other number that results during the transition period is discarded as being invalid. This is done by comparing the generated number to a counter that adds one count or subtracts one count from the generated count depending on the counting direction. This count is compared to the generated count for authenticity. Any matchup will make that count valid.

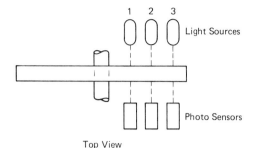

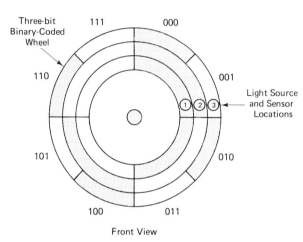

Figure 18-10 Simple 3-bit binary-encoded rotary encoder. (From James R. Carstens, *Automatic Control Systems and Components,* copyright 1990, p. 313. Reprinted by permission of Prentice Hall, Englewood Cliffs, NJ.)

Although this is only a partial explanation of the algorithm that is used for number validation, it demonstrates the general idea.

The major disadvantage of the binary encoding method is that it requires more than one set of line tracks and photodetection devices for it to operate properly. This increases considerably the manufacturing cost of the sliding track. The major advantage of this system is that it is possible to have direct compatibility with existing logic circuitry that is set up for receiving a particular binary counting scheme.

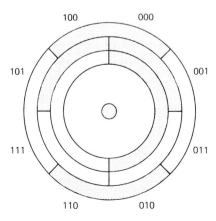

Figure 18-11 3-bit Gray-encoded rotary encoder. (From James R. Carstens, *Automatic Control Systems and Components,* copyright 1990, p. 313. Reprinted by permission of Prentice Hall, Englewood Cliffs, NJ.)

18-9 BRUSH CONTACT ENCODER

The brush contact encoder is an encoder that utilizes a wire or metal brush that makes contact with a moving conductive surface attached to a rotor. The conductive surfaces are arranged in a desired pattern or sequence such that as they pass beneath and make electrical contact with the appropriate brush, a circuit is completed, causing the encoder's output to generate a binary 1 or 0. Brush contact encoders are used only in rotary encoders, not in linear types. Figure 18-12 shows a simplified circuit of a rotary encoder.

It is difficult to construct a brush-type encoder having the resolution capabilities of an optical encoder. It would be very difficult to fit the same number of brushes into the same space inside an encoder housing that is designed to contain a certain number of contactless pickup units. Track construction difficulties would also be compounded by the electrical requirements of insulated parallel paths that would have to be built into the rotating encoder wheel. Additionally, the brush encoder would be more prone to maintenance problems due to the brushes wearing out because of the continuous physical contact made with the encoder wheel. Brushes do not lend themselves very well to direction reversals either. This contributes further to their wear, therefore preventing any sort of absolute positioning ability. Because of these problems, the few rotary brush encoders that do exist are gradually being replaced in industry by other contactless types.

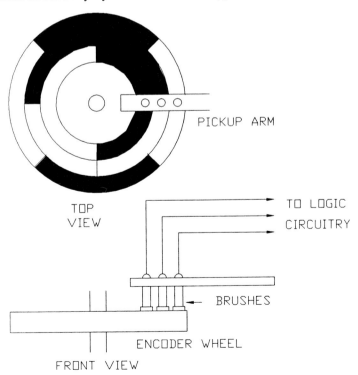

Figure 18-12   Brush contact encoder.

18-10 MAGNETIC ENCODER

Magnetic encoders utilize a sliding or rotating member that contains a magnetic media for the storing of magnetic pulses. These pulses may be interpreted as binary 1's and 0's much like the machine language storage on a computer's floppy or hard disk. A read head suspended immediately over the disk or slide scans this magnetic information while nearby circuitry translates this information into displacement and direction of travel. A variation of this type of encoder construction is illustrated in Figure 18-13. We see a magnetic tape that is coiled inside a spring-loaded drum. As the tape is pulled into and out of this drum, a nearby scanning head containing several reading heads (one reading head for each track of binary information assuming that the encoder is of the digital encoding type) senses that the stored magnetic pulses in the track and sends these data to the signal-conditioning and data-processing circuit board located inside the encoder. The read heads are comprised of magnetic induction coils that generate a voltage pulse as it sweeps through the magnetic field of each stored pulse on the tape. This process is again very similar to the read/write head operation used on floppies and hard disks in computers.

This system is capable of creating a direction-reversal phasing effect due to the fact that the phase of the induced ac voltage in the read head is shifted by 180° when a reversal takes place in the moving magnetic tape; this leaves a convenient signature for tracking reversals. A rotary type of magnetic encoder may also be constructed using the same principles of operation.

Another type of rotary encoder may be constructed again using the principle of sensing the presence of stored magnetic pulses. In this case, however, the magnetic media is a toroidal magnet attached to the rim of a metal disk. This magnet contains several magnetic domains or stored pulses, as discussed in the example just dis-

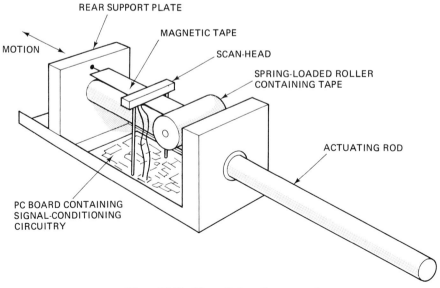

Figure 18-13 Magnetic tape linear encoder.

cussed. Each pulse has the characteristics of a magnet having its own north and south pole. Looking at Figure 18-14, the pickup head is comprised of two magnetoresistive sensing elements. These elements change their electrical resistance whenever encountering a magnetic field. In this particular unit two read heads are displaced from each other by 90°. This produces a sine wave for the output of one pickup while the

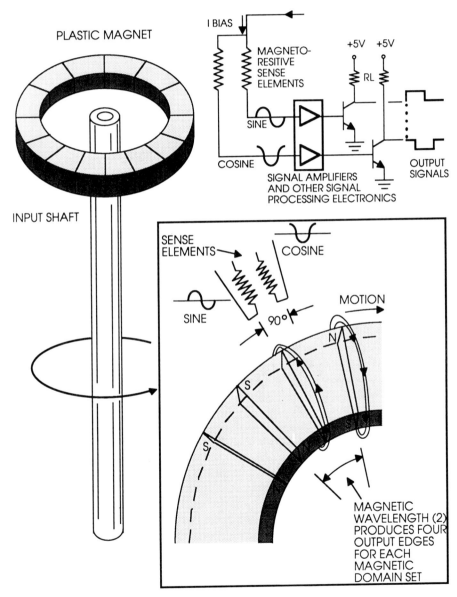

Figure 18-14 High-resolution magnetorestrictive rotary encoder. (Courtesy of Sony Corp. and *Design News*, vol. 43, no. 13. Illustrations reprinted by permission of *Design News* from the article, ''Magnetic encoder boasts high resolution.'')

other produces a cosine wave. When the rotary encoder's disk reverses direction, the pickup heads swap signal characteristics. That is, the sine and cosine functions are reversed in the pickup heads. This swapping of signals produces an immediate indication of motion reversal.

REVIEW QUESTIONS

18-1. What are the advantages of using digital encoding versus using analog encoding?

18-2. Explain the operation of a Schmitt trigger circuit.

18-3. What is a common problem with digital encoding? What is the difference between incremental encoding and absolute encoding?

18-4. Explain how quadrature phasing works in an absolute encoding system.

18-5. How does the Gray code differ from the binary or BCD code?

PROBLEMS

18-6. Calculate the resolution of a linear encoder that has a 50-cm travel span and possesses a 7-bit readout capability.

18-7. A robotics manufacturer wishes to design a robot with a movable arm designed to have a positioning accuracy of 0.005 in. along a 50-in. travel span. Determine the number of binary bits needed from a linear encoder that would allow that type of positioning accuracy.

18-8. A rotary encoder contains an 11-bit output port. Express its resolution in angular seconds per bit.

18-9. Using a rotary encoder with a binary number output, show how a simple circuit could be assembled using a digital comparator so that when the encoder has rotated through a specified angle, an output signal would be created by this circuit.

REFERENCES

CAMPBELL, JOSEPH, *The RS-232 Solution*. Berkeley, CA: Sybex, 1984.

CARSTENS, JAMES R., *Automatic Control Systems and Components*. Englewood Cliffs, NJ: Prentice Hall, 1990.

MALVINO, ALBERT P., and DONALD P. LEACH, *Digital Principles and Applications*. New York: McGraw-Hill, 1981.

MILLER, RICHARD W., *Servomechanisms: Devices and Fundamentals*. Reston, VA: Reston, 1977.

INDEX